BIOTECHNOLOGY FOR WASTE BIOMASS UTILIZATION

BIOTECHNOLOGY FOR WASTE BIOMASS UTILIZATION

Edited by

Prakash K. Sarangi, PhD

Latika Bhatia, PhD

APPLE ACADEMIC PRESS

First edition published 2023

Apple Academic Press Inc.
1265 Goldenrod Circle, NE,
Palm Bay, FL 32905 USA

760 Laurentian Drive, Unit 19,
Burlington, ON L7N 0A4, CANADA

CRC Press
6000 Broken Sound Parkway NW,
Suite 300, Boca Raton, FL 33487-2742 USA

4 Park Square, Milton Park,
Abingdon, Oxon, OX14 4RN UK

Library and Archives Canada Cataloguing in Publication

Title: Biotechnology for waste biomass utilization / edited by Prakash K. Sarangi, PhD, Latika Bhatia, PhD.
Names: Sarangi, Prakash Kumar, editor. | Bhatia, Latika, editor.
Description: First edition. | Includes bibliographical references and index.
Identifiers: Canadiana (print) 2022024023X | Canadiana (ebook) 20220240361 | ISBN 9781774639955 (hardcover) | ISBN 9781774639962 (softcover) | ISBN 9781003277187 (ebook)
Subjects: LCSH: Biomass energy. | LCSH: Biotechnology.
Classification: LCC TP339 .B58 2023 | DDC 662/.88—dc23

Library of Congress Cataloging-in-Publication Data

CIP data on file with US Library of Congress

ISBN: 978-1-77463-995-5 (hbk)
ISBN: 978-1-77463-996-2 (pbk)
ISBN: 978-1-00327-718-7 (ebk)

About the Editors

Prakash Kumar Sarangi, PhD
Scientist in Microbiology, Central Agricultural University in Imphal, India

Prakash Kumar Sarangi, PhD, is a Scientist with specialization in microbiology at the Central Agricultural University in Imphal, India. Dr. Sarangi has a PhD degree in Microbial Biotechnology from the Ravenshaw University, India and MTech degree from Department of Agriculture and Food Engineering, Indian Institute of Technology Kharagpur, India. Dr. Sarangi's research is focused on bioprocess engineering, renewable energy, biochemicals, biomaterials, and sustainable development. His expertise is in bioconversion of biomass into biofuels, biochemicals and nutraceuticals, as well as environmental waste remediation. He has taken leading roles as the principal investigator in different R&D projects on biomass conversion, biofuel from renewable resources, and microbial biodegradation into value-added products. He has more than 12 years of teaching and research experience in biochemical engineering, microbial biotechnology, downstream processing, food microbiology, and molecular biology.

Dr. Sarangi is serving as an editorial board member for many international journals, including *PLOS One, Biotech, Biofuels, Bioproducts, and Biorefining*. He is serving as guest editor for the *Journal of Biomass Conversion and Biorefinery, Chemical Engineering and Technology,* and *Biofuels, Bioproducts and Biorefining*. He has published more than 60 research articles in peer-reviewed journals and has authored more than 25 book chapters. He has presented at many national and international conferences. He is associated with many scientific societies as a Fellow Member (Society for Applied Biotechnology) and Life Member (Biotech Research Society of India; Society for Biotechnologists of India; Association of Microbiologists of India; Indian Science Congress Association; Forum of Scientists, Engineers & Technologists; and International Association of Academicians and Researchers). He is the Editor of three books: *Recent Advancements in Biofuels and Bioenergy Utilization, Fuel Processing and Energy Utilization,* and *Advanced Biotechnology for Sustainable Energy and Products,* published by Springer Nature, CRC Press and I.K. International Publishing House Pvt. Ltd., respectively.

Latika Bhatia, PhD
Assistant Professor, Department of Microbiology and Bioinformatics, Atal Bihari Vajpayee Vishwavidyalaya, Bilaspur, CG, India

Latika Bhatia, PhD, is an Assistant Professor in the Department of Microbiology and Bioinformatics in Atal Bihari Vajpayee Vishwavidyalaya, Bilaspur, C.G., India. She is responsible for academic and administrative responsibilities as a member of various committees, including academic council, board of studies, coordinator of examination, and various others. She has 21 years of teaching experience working at universities, teaching graduates and postgraduates of microbiology and biotechnology. She received her PhD from ITM University, Gwalior, on Biofuel Technology. She is a recipient of a Summer Research Fellowship from the Indian Academy of Sciences, Banglore; Indian National Science Academy, New Delhi; and National Academy of Sciences, Allahabad (IASc-INSA-NASI), from the Indian Institute of Science Education & Research (I.I.S.E.R), Pune, India. She has awards and honors to her credit. She has been invited to deliver a technical talk at the All India Institute of Medical Sciences (AIIMS, New Delhi) conference as well as many other national and international conferences. She has published more than 30 research articles in peer-reviewed journals of international repute, 10 book chapters and three monographs, five books. She has a citation index of more than 400. Her primary research interest is to develop the sustainable process for bioconversion of lignocellulosics into renewable energy and biochemicals. A proactive research lead with an entrepreneurial mindset, she brings hands-on strategic experience to research projects and start-ups to pilot to full commercial scale operations. She has also convened and organized more than ten national and international conferences, seminars, and workshops, funded by various reputed organizations, such as DBT, DST, ICMR, and CGCOST. She has attended and presented papers at more than 30 conferences and workshops of national and international repute in India. She has successfully completed many consultancy and research projects. Dr. Bhatia is reviewer of *PLOS, Sugar Tech* (Springer), *Chemical Engineering & Technology* (Wiley), *Bioprocess & Biosystems Engineering (Springer), Environmental Chemistry Letters* (Springer), and a member of national advisory committees of eminent universities. Dr. Bhatia has been a resource person and a keynote speaker and has chaired sessions at more than 10 national and international conferences including UGC-HRD short-term courses. Dr. Bhatia has developed an online certificate course on immunology in collaboration with the Commonwealth Educational Media Center for Asia (CEMCA), available on Web-portal https://e-atalgyansangum.ac.in/ on UGC's specified four-quadrant approaches. Dr. Bhatia is a recipient of a Best Teacher Award from Atal Bihari Vajpayee University for her academic contributions. She is a convenor of the Bilaspur Chapter of the Indian Science Congress Association, Kolkata, India.

Contents

Contributors

Latika Bhatia
Department of Microbiology & Bioinformatics, Atal Bihari Vajpayee University, Bilaspur, Chhattisgarh, India, E-mail: latikabhatia1@yahoo.co.uk

Debosmita Chakraborty
School of Biotechnology, KIIT University, Bhubaneswar–751024, Odisha, India

Santanu Dasgupta
Reliance Industries Ltd, Reliance Corporate Park, Ghansoli, India

Surajit Debnath
Department of Medical Laboratory Technology, Women's Polytechnic, Govt. of Tripura, E-mail: Surajitdebnath9@gmail.com

Shrestha Debnath
Microbial Engineering and Algal Biotechnology Laboratory, Department of Biotechnology, JIS University, Kolkata,–700109, West Bengal, India

Vinay Dwivedi
Reliance Industries Ltd, Jamnagar, India

Dipankar Ghosh
Microbial Engineering and Algal Biotechnology Laboratory, Department of Biotechnology, JIS University, Kolkata–700109, West Bengal, India, E- mails: dghosh.jisuniversity@gmail.com; d.ghosh@jisuniversity.ac.in

Padan Kumar Jena
Department of Botany, Ravenshaw University, Cuttack–753003, Odisha, India

Harit Jha
Department of Biotechnology, Guru Ghasidas Vishwavidyalaya, Bilaspur, Chhattisgarh, India, E-mail: harit74@yahoo.co.in

Sonia Johri
School of Life Sciences, ITM University, Gwalior, Madhya Pradesh, India

Dowluru S. V. G. K. Kaladhar
Department of Microbiology and Bioinformatics, Atal Bihari Vajpayee University, Bilaspur, C.G., India, E-mail: dkaladhar@gmail.com

Slipa Kanungo
School of Biotechnology, KIIT University, Bhubaneswar–751024, Odisha, India

Khageshwari Karsh
Department of Microbiology & Bioinformatics, Atal Bihari Vajpayee University, Bilaspur, Chhattisgarh, India

Akanksha Kulshreshtha
Department of Biological Sciences and Engineering, Netaji Subhas University of Technology, Delhi–110078, India, E-mail: akankshak@nsut.ac.in

Snehasish Mishra
Bioenergy Lab and BDTC, School of Biotechnology, KIIT University, Bhubaneswar–751024, Odisha, India, E-mail: snehasish.mishra@gmail.com

Neha Namdeo
Department of Biotechnology, Guru Ghasidas Vishwavidyalaya, Bilaspur, Chhattisgarh, India

Mousumi Meghamala Nayak
Procyto Labs Pvt Ltd, KIIT Technology Business Incubator, Bhubaneswa–751024, India

Debasmita Panda
Department of Botany, Banki Autonomous College, Banki, Cuttack, India,
E-mail: debasmita.pinky87@gmail.com

Bhabjit Pattnaik
Department of Botany, Christ College, Cuttack–753008, Odisha, India,
E-mail: bhabjit.int4res@gmail.com

Ritesh Pattnaik
School of Biotechnology, KIIT University, Bhubaneswar–751024, Odisha, India,
E-mail: riteshpattnaik@gmail.com

Raviprasad Podili
Reliance Industries Ltd, Jamnagar, India

Meghna Rajvanshi
Reliance Industries Ltd, Reliance Corporate Park, Ghansoli, India

Hara Prasad Sahoo
Department of Botany, Buxi Jagabandhu Bidyadhar (BJB) College, Bhubaneswar–751014, Odisha, India.

Dilip Kumar Sahu
Department of Microbiology & Bioinformatics, Atal Bihari Vajpayee University, Bilaspur, Chhattisgarh, India

Minakshi Sahu
Centre for Conservation and Utilization of Blue Green Algae, Division of Microbiology, ICAR-Indian Agricultural Research Institute, New Delhi–110012, India

Suman Sahu
Department of Microbiology & Bioinformatics, Atal Bihari Vajpayee University, Bilaspur, Chhattisgarh, India

Debanjan Sanyal
Reliance Industries Ltd, Jamnagar, India, E-mail: Debanjan.Sanyal@ril.com

Prakash Kumar Sarangi
Directorate of Research, Central Agricultural University, Imphal–795004, Manipur, India

Soumya Sasmal
Department of Biological Sciences and Engineering, Netaji Subhas University of Technology, Delhi–110078, India

Ng. Joykumar Singh
Directorate of Research, Central Agricultural University, Imphal–795004 Manipur, India

Puneet K. Singh
Bioenergy Lab and BDTC, School of Biotechnology, KIIT University, Bhubaneswar–751024, Odisha, India

Th. Anand Singh
Directorate of Research, Central Agricultural University, Imphal, Manipur–795004, India,
E-mail: anan4thangjam@gmail.com

Tantravahi Srinivasan
Department of Botany, Indira Gandhi National Tribal University, Amarkantak, M.P., India

Rajesh K. Srivastava
Department of Biotechnology, GIT, GITAM. (Deemed to be University, Visakhapatnam–530045,
A. P. India, E-mail: rajeshksrivastava73@yahoo.co.in

O. N. Tiwari
Centre for Conservation and Utilization of Blue Green Algae, Division of Microbiology,
ICAR-Indian Agricultural Research Institute, New Delhi–110012, India,
E- mail: ontiwari1968@gmail.com

Kondi Vanitha
Department of Pharmaceutics, Vishnu Institute of Pharmaceutical Education and Research, Narsapur,
Medak–502313, Telangana, India

Abbreviations

AAO	aryl alcohol oxidase
ABE	acetone-butanol-ethanol
AD	anaerobic digestion
ADM1	anaerobic digestion model no. 1
AgNPs	silver-based nanoparticles
ALA	α linolenic acid
ARA	arachidonic acid
BE	bromelain enzyme
BGL	β-glucosidase
BOD	biological oxygen demand
Ca	calcium
CBG	compressed biogas
CBH	cellobiohydrolases
CBP	consolidated bioprocessing
CFA	Central Financial Assistance
CGT	cotton ginning stock
CH_4	methane
CM	cow manure
CNG	compressed natural gas
CO_2	carbon dioxide
COD	chemical oxygen demand
CSTR	continuous stirred tank reactor
DGGE	denaturing gradient gel electrophoresis
DHA	docosahexaenoic acid
DM	dry matter
DMCB	dried microalgal cellular biomass
DNES	Department of Non-Conventional Energy Sources
DP	degree of polymerization
EF	electric field
EGL	β-1,4-endoglucanases
EJ	exajoules
EPA	eicosapentaenoic acid
FA	ferulic acid
FAE	ferulic acid esterases

FAME	fatty acid methyl ester
FAO	Food and Agriculture Organization
FBA	flux-balance analysis
FFV	flexible-fuel vehicles
FOS	fructooligosacchrides
GH	glycoside hydrolase
GHG	greenhouse gas
GI	gastrointestinal
GLA	γ linolenic acid
GLOX	glyoxal oxidase
H_2	hydrogen
HBA	hydroxybenzoic acid
HDLs	high-density lipoproteins
HEA	hexane extracted algae
HPH	high pressure homogenization
HPR	hydrogen (H_2) production rate
HRT	hydraulic retention times
HTC	hydrothermal carbonization
HUVECs	human umbilical vein endothelial cell lines
IARI	Indian Agricultural Research Institute
IEA	International Energy Agency
ILEA	ionic liquid extracted algae
ILs	ionic liquids
K	potassium
KJ/mole	kilojoule per mole
KVIC	Khadi and Village Industries Commission
LB	lignocellulosic biomass
LDLs	lower density lipoproteins
LiP	lignin peroxidases
LPG	liquid petroleum gas
MEC	microbial electrolysis cells
MFCs	microbial fuel cells
Mg	magnesium
mmoles/L/D	mole per liter per day
MnP	manganese peroxidases
MNRE	Ministry of New and Renewable Energy
MSW	municipal solid waste
N	nitrogen
NaOH	sodium hydroxide

NBMMP	National Biogas and Manure Management Program
NDOs	non-digestible oligosaccharides
NEDA	Non-Conventional Energy Development Agency
NGS	next generation sequencing
NMMO	N-methyl morpholine-N-oxide
NPBD	National Project on Biogas Development
NREL	National Renewable Energy Laboratory
NRPSs	non-ribosomal peptide synthetases
OFMSW	organic fraction of municipal solid waste
OPEFB	oil palm empty fruit bunch
OPFF	oil palm frond fiber
OPMF	oil palm mesocarp fiber
P	phosphorus
PCR	polymerase chain reaction
PEM	proton exchange membrane
PF	potassium ferrate
PKSs	polyketide synthases
PMS	paper mill sludge
PNSB	purple non-sulfur bacteria
PPW	pineapple processing waste
PRAD	Planning Research and Action Development
PSA	pressure swing adsorption
PSB	purple sulfur bacteria
PUFA	poly unsaturated fatty acid
QTP	Qinghai-Tibetan plateau
RFLP	restriction fragment length polymorphism
RH	rice husk
rRNA	ribosomal RNA
S	sulfur
SCB	sugarcane bagasse
SCFA	short chain fatty acids
SHF	separate hydrolysis and fermentation
SmF	submerged fermentation
SS	solid state fermentation
SS-AD	solid-state anaerobic digestion
SSF	saccharification and fermentation
SWM	solid waste management
Syngas	synthetic gas
TAG	triaceyl glycerol

TAP	tris-acetate phosphate
TCM	traditional Chinese medicine
TDS	total dissolved solids
TEA	tecbhno-economic analysis
TSP	total soluble protein
U.K	United Kingdom
US₵/kWh	United States per kilo watt hour
UV	ultra violet
VA	vanillic acid
VS	volatile solids
W-AD	wet-AD
WH	water hyacinth
WWTP	waste water treatment plants
XOS	xylooligosaccharides
Zα	master plots

Preface

Biomass (food and feed crops and other grassy plants, wood, and algae) is profusely available on Earth in various forms and has the potential to be transformed into biomaterials, food/feed, fuel, and chemicals. These transformations can be manifested by using chemical, physical, thermal, biological, and biotechnological procedures. Solid waste management is a challenging process, especially in developing countries due to the continuous upsurge of population, urbanization, and lack of services and facilities, making solid waste management very difficult to handle. Employing the principles of the biorefinery are the utmost probable modes to valorize such waste biomass, where each of plant biomass component is employed, producing zero or near-zero waste. Various methods and techniques like pyrolysis, gasification, catalytic liquefaction, etc., are in use for solid waste conversion and energy/fuel generation. Furthermore, biochemical techniques like microbial fermentation and anaerobic digestion are of enormous use and have immense potential in the production of biogas and bioethanol/biobutanol, etc. Bioenergy is one of the maximum significant systems in alternative energy systems, which provides potential benefits for sustainable development. Biomass from biowastes such as agriculture crop residues, wood processing residues, forest residues, food waste, industrial waste, and municipal solid waste (MSW) have emerged as potential feedstocks for bioenergy production.

In the twenty-first century, the application of renewable raw resources and effective commercialization of their chemical and biological possibilities have clustered popularity and significance in scientific investigations and industry. The pollution problems associated with agro-industrial wastes have encouraged the usage and bioconversion of these waste into high industrial products. There has been a paradigm shift in the global pattern when nourishment is taken into account.

People are health conscious and prefer the food that would support their overall health. Functional foods are among such foods, and xylooligosaccharides constitute an important part of these foods. Biotechnological advancements are equally important in the exploration of these agro-residues and other lignocellulosic biomasses for the economic and sustainable production of xylooligosaccharides. A broad spectrum of market analysis has shown that diverse ranges of microalgae are the most potential industrial biocatalysts. Microalgae are used as sustainable microbial cell factories to biosynthesis a wide range of promising high value-added products, likely bioactive compounds, biofuels, feed supplements, and many more utilizing cheaper raw materials, i.e., waste biomass.

Microalgae have been used as alternative resources over unsustainable overexploitation of various natural constituents, which gives us various high value-added biomolecules.

We realize that this will be the first book presenting the key features of biotechnology for waste biomass utilization at one point, delivering a valuable source of information not only for students but also for researchers involved in bioenergy and environmental programs chiefly and an essential source of data for academicians, researchers, economists, policymakers, and policy analysts. We also comprehend that this book explicitly is a first distinctively intended scientific and technical literature on waste biomass management in the context of India and covers the advancements towards biomass conversion for sustainable management of solid waste, novel applications and products of water hyacinth through biotechnological interventions, biotechnological modes of xylooligosaccharides production from waste biomass in an economic and ecological way, multifaceted utilization of microalgal biomass towards industrial applications, methods of converting waste biomass to energy, production of commercial products by vermiculture and vermicomposting thus making it an exclusive source of wealth of knowledge.

We are extremely happy to present this book to readers, who we are confident will find it a very valuable and exclusive source of information. The editors of this book have huge experience and global expertise and are eminent researchers in the field of waste biomass management not only in India but internationally. We also thank the authors of the various chapters for their contributions.

—Editors

CHAPTER 1

Waste Biomass to Bioenergy: A Compressive Review

MEGHNA RAJVANSHI,[1] RAVIPRASAD PODILI,[2] VINAY DWIVEDI,[2] DEBANJAN SANYAL,[2] and SANTANU DASGUPTA[1]

[1]*Reliance Industries Ltd, Reliance Corporate Park, Ghansoli, India*

[2]*Reliance Industries Ltd, Jamnagar, India,*
E-mail: Debanjan.Sanyal@ril.com (Debanjan Sanyal)

ABSTRACT

The increase in global energy demand, limited availability of fossil fuel, and adverse effects on climate due to greenhouse gas emissions from fossil fuel burning create tremendous pressure in developing alternative sustainable energy sources. Energy from biomass is gaining momentum because it is a cleaner and sustainable form of energy and has the potential to address the dual problem of energy security and waste management. Biomass from biowastes such as agricultural crop residues, wood processing residues, forest residues, food waste, industrial waste, and municipal solid waste (MSW) have emerged as potential feedstocks for bioenergy production. The chapter mainly focuses on biochemical conversion processes for bioenergy production as these processes are efficient, clean, operate in moderate conditions, and are environmentally friendly. However, structuring of biomass supply chain and substantial technological advancements in biochemical conversion technologies are needed to make biomass to bioenergy a commercially viable option. Concerted efforts from industry and academia and government support in terms of funding, subsidies, and supportive policies will ultimately unveil the true potential of bioenergy from waste biomass.

Biotechnology for Waste Biomass Utilization. Prakash Kumar Sarangi & Latika Bhatia (Eds.)

1.1 INTRODUCTION

Energy is vital for economic development and growth. Fossil fuels such as coal, oil, and natural gas are the primary energy resources worldwide. However, their excessive use poses a significant threat to climate and living beings. The majority (~70%) of global greenhouse gas (GHG) emission is due to the heavy use of fossil fuels for electricity, heat, and transportation. It is estimated that CO_2 emission of more than 420 gigatons will lead to overshooting of the global warming threshold of 1.5 ^{0}C, which will result in irreversible detrimental climate changes (United-Nations, 2019). Therefore, it is imperative to arrive at net zero CO_2 emission by 2050 to prevent the detrimental rise in global temperature (IEA, 2019). Also, it is projected by the U.S. Energy Information Administration that by 2050, energy requirements will be nearly doubled (Kahan, 2019). Therefore, to fulfill the increasing energy requirement and to address the possible harmful effects associated with fossil fuels, it is crucial to develop alternative and cleaner energy sources. In the past several decades, hydroelectric power, solar, wind, nuclear, geothermal, and biomass energy are the alternative energy sources of fossil fuels that evolved as cleaner energy options. Energy from biomass has been gaining importance due to its potential to be a sustainable alternative energy source.

Global primary energy consumption in the year 2019 was 583.9 exajoules (EJ), with China alone consuming 141.7 EJ, accounting for 24% of the global energy consumption. China was followed by the U.S., European Union, and India consuming 94.7, 68.8, and 34.1 EJ, respectively. Currently, 84% of the total energy needs are fulfilled by petroleum (33%), coal (27%), and natural gas (24%) and renewable energy constitutes only ~16% of the current energy needs (Koyama, 2020). However, it is projected that by 2050, renewable energy will surpass petroleum and other liquid energy sources to become the most used energy form (Kahan, 2019). In renewable energies, the present global contribution of bioenergy is around 56 EJ, which fulfills approximately 10% of the global energy needs (WBA, 2016). But concerted efforts towards developing green energy from biomass can increase the share of bioenergy substantially. Biomass as an energy source offers multiple advantages. Energy from biomass is considered carbon-neutral because it takes CO_2 from the air during the photosynthesis process, and this CO_2 is returned to the atmosphere when biomass is utilized in energy generation. Thus, bioenergy can help in reducing net greenhouse gas (GHG) emissions and negative environmental changes. Biomass as an energy source also has the potential to lessen the dependency on fossil fuel and thus can play a significant role in improving energy security and saving foreign exchange due to less import of crude oil. In addition, valorization of residual and waste biomass for energy also helps in better resource utilization, waste minimization, tackling waste disposal problems, and can generate opportunities and employment for the upliftment of rural economies (Kumar and Pandey, 2019).

Traditionally, biomass is used in the form of firewood to provide energy through direct combustion. However, it can also be utilized for the production of liquid or gaseous biofuels for transportation. Depending on the feedstock used, biofuels from biomass would be classified as the first, second, and third generations. First-generation biofuels mostly encompass ethanol production from food grains, whereas second-generation biofuels utilize energy crops and waste biomass mainly comprising lignocellulosic residues from agriculture and forestry. Third-generation biofuels utilize sewage sludge, municipal solid wastes (MSW), and algae as feedstocks for biofuel production (Nanda et al., 2018). Second and third-generation biofuels are attractive options, as they do not compete with food. Multiple technologies are developed for biomass transformation into a usable form of energy. Key technologies include transesterification, thermochemical and biochemical conversion processes (Lee et al., 2019). The transesterification process converts extracted oils into biodiesel through a chemical reaction. Whereas the thermochemical process reforms biomass into syngas, bio-oil, and biochar under high-temperature conditions. The main thermochemical processes are gasification (syngas), pyrolysis (syngas, bio-oil, and biochar), liquification (bio-oil), and combustion (electricity). Microorganisms convert biomass into gaseous fuels or liquid through a biochemical conversion process. Alcoholic fermentation, anaerobic digestion for biogas production, and biological hydrogen gas production are some of the examples of biochemical conversion processes (Lee et al., 2019). This chapter mainly focuses on a review of waste biomass conversion into various energy forms. In specific, it focuses on various types of waste biomass feedstocks used for bioenergy production, different biochemical conversion processes, and the key challenges associated with the commercialization of waste biomass to bioenergy.

1.2 SOURCES OF WASTE BIOMASS AS FEEDSTOCK FOR ENERGY

Waste biomass for energy originates from crop waste, forest residues, wood processing residues, food processing waste, sorted municipal solid waste, algae biomass, etc. (US-DOE, 2020). They are categorized as primary, secondary, and tertiary residues, based on the extent of processing involved in generating biomass residues. Primary residues are biomass residues that are left after harvesting. For example, crop harvest residues (corn stover, wheat straw, rice straw, oat straw, etc.) and forest residues. Food and wood processing residues come under the secondary residues category. Some examples of secondary residues are wood chips, palm kernel press cake, sugarcane bagasse, etc. Tertiary residues are non-food organic wastes (Brosowski et al., 2016). The availability of biomass residues is quite variable. Table 1.1 summarizes the annual availability of some of the waste biomass feedstocks.

Bioethanol, biohydrogen, and biomethane production have been demonstrated with several types of feedstocks. For instance, lignocellulosic biomass, sawdust, corn stover, sugarcane bagasse, palm kernel press cake, and de-oiled microalgae were used for alcoholic fermentation (Cai et al., 2016; Cerveró et al., 2010; Krishnan et al., 2010; Qureshi et al., 2018). Several food wastes, like potato peel and banana, grape and sugar beet pomace, and cafeteria and household food wastes, have also been used for the production of bioethanol (Arapoglou et al., 2010; Matsakas et al., 2014; Oberoi et al., 2011; Rodríguez et al., 2010). Another prevalent method of utilizing food and organic municipal wastes is through anaerobic absorption for biogas production (Paritosh et al., 2017).

1.3 BIOTECHNOLOGICAL APPROACHES FOR BIOFUEL PRODUCTS

In the past few decades, biotechnological approaches have gained momentum for converting biomass into bioenergy, primarily because biochemical processes are considered safer, cheaper, environment-friendly, and less energy-intensive, compared to thermochemical processes (Zhang et al., 2007; Thakur et al., 2013). Moreover, studies aiming to improve the efficiency of conventional thermochemical conversions did not give satisfactory results. These factors developed a greater interest in improving biochemical conversion processes by modifying the existing enzymes for higher enzymatic efficiency and complete utilization of biomass. Significant advancements have been made in the field of biotechnology to improve biomass to biofuel conversions, particularly in the case of processing of lignocellulosic biomass (Mishra et al., 2017; Mathibe et al., 2020; Riyadi et al., 2020). Fülöp and Ecker stated that modern computational, biochemical, and biotechnological methods would be helpful in improving enzyme function, thus making it more efficient for biomass conversion (Fülöp and Ecker, 2020). Apart from

TABLE 1.1 Annual Availability of Biomass Residues (Million Tons/Annum)

Biomass Residue	Global	U.S.	Europe	India	China	Reference
Rice straw	731	37.2	3.9 m	139.2.6	266.55	FAOSTAT (2014); Karimi et al., (2006)
Wheat straw	529	83.3	103.9	133.30	173.18	FAOSTAT (2014); Govumoni et al., (2013)
Corn cob	1114.7	366.6	60.7	26	257.7	Kopp, (2019)
Sugarcane Bagasse	540	181	5.7	80	67.4	Nemerow and Agardy (1998) Bezerra and Ragauskas (2016)
Palm kernel pressed cake	1115	-	2.6	-	-	Ng et al., (2002); Omoti, (2009)
Food waste	1300	40	88	67	17–18	FAO, (2013); Worldwide Food Waste, (2020)
Organic municipal solid waste	2009.9	267.8	88	90	215	Yan et al., (2020)

enzyme modification, another approach to improving biomass conversion is through the use of Recombinant organisms (GMOs). These organisms are found to be more efficient than their wild counterpart (Mohapatra et al., 2019a; Wang et al., 2016a; Cai et al., 2016). Bioethanol, biobutanol, biogas, biohydrogen, and bioelectricity are the key energy products obtained through microbial conversion processes and are discussed below.

1.3.1 BIOETHANOL

Bioethanol is a renewable alternative liquid fuel to gasoline (Mohapatra et al., 2019; Patni et al., 2013; Roy and Dutta, 2019). Bioethanol is formed through a fermentation procedure, using simple sugars, starch, and lignocellulosic biomass as feedstock. Lignocellulosic biomass mostly contains lignin, hemicellulose, and cellulose (Shi et al., 2019; Ma et al., 2019). Complex biomass feedstock is pretreated and hydrolyzed through acid or enzymatic treatment to break down the complex polysaccharides into sugars, which are subsequently used in bioethanol production through a microbial fermentation process. The ideal pretreatment process must be cost-effective, should enhance the release of fermentable sugars, minimize carbohydrate degradation and inhibitor formation during the pretreatment process (Chen et al., 2017; Kim et al., 2019; Carrasco et al., 2010). Huang et al. (2019) reported that genetic modification of plant cell walls is another potential strategy for enhancing enzymatic hydrolysis of lignocellulosic biomass. In addition, a comprehensive understanding of the mechanism of enzyme function (ligninases and cellulases) and rational modification of enzymes will be helpful for the efficient production of bioethanol from biowaste (Fülöp and Ecker, 2020; Zang et al., 2018; Veradi et al., 2020; Shah et al., 2019). There are multiple ways through which fermentation and hydrolysis may be carried out for ethanol production. This include; simultaneous saccharification and fermentation (SSF), separate hydrolysis and fermentation (SHF), consolidated bioprocessing (CBP), and simultaneous saccharification and co-fermentation (Jahnavi et al., 2017; Özçimen and İnan, 2015; Xiros et al., 2013; Conde-Mejía et al., 2012; Taherzadeh and Karimi, 2007). A lot of studies are focused on utilizing lignocellulosic feedstocks for bioethanol production. Different lignocellulosic feedstocks and their processing conditions are summarized in Table 1.2. Apart from lignocellulosic crop waste, food waste is also evaluated for bioethanol production. For instance, pumpkin peel waste has been identified as a novel potential feedstock for bioethanol production. With

TABLE 1.2 Ethanol Production from Waste Biomass As Feedstock

Biomass source	Organism	Ethanol (g/L)	Reference
Pumpkin peel wastes (SHF)	*Saccharomyces cerevisiae*	50.69	Moncef et al., (2020)
Carbohydrate accumulated green microalga by PPW (SHF)	*S. cerevisiae*	0.258	Chandra et al., (2020)
Liquid hydrolysate of *Chlorella vulgaris ESP-31* (SHF)	*S. cerevisiae*	0.025–0.127	Yu et al., (2020)
Liquid hydrolysate *Chlorella* sp *GD* (SHF)	*S. cerevisiae*	0.008–0.011	Yu et al., (2020)
Rice Straw (SHF)	*S. cerevisiae* NGY10	7.18	Pandey et al., (2019)
Rice Straw (SSF)	*S. cerevisiae* NGY10	30.22	Pandey et al., (2019)
Carbohydrate accumulated *Chlorella vulgaris* by PPW (SHF)	*S. cerevisiae* and *A. niger*	0.33	Agwa et al., (2017)
Soybean residue and Cooked rice (SSF)	*S. cerevisiae* TISTR 5339	40.7	Salakkam et al., (2017)
Eucalyptus globulus wood	*S. cerevisiae* IR2T9-a	42	Yáñez-S et al., (2013)
Rice straw (Batch SSCF)	*S. cerevisiae, Candida tropicalis, S. stipites*	28.6	Suriyachai et al., (2013)
Corn stover (SHF)	Recombinant *S. cerevisiae* Y35	45.5	Jin et al., (2013)
Bagasse (SSF process)	*Zymomonas mobilis* immobilized with calcium alginate	5.44	Wirawan et al., (2012)
Bagasse (SSF process)	*Z. mobilis* immobilized with polyvinyl alcohol	5.53	Wirawan et al., (2012)
Bagasse (SHF process)	*Z. mobilis* immobilized with calcium alginate	5.52	Wirawan et al., (2012)
Bagasse (SHF process)	*Z. mobilis* immobilized with polyvinyl alcohol	6.24	Wirawan et al., (2012)
Corn stover (CBP)	*Clostridium phytofermentans*	7.0	Jin et al., (2012)
Cellulosic material (CBP)	Recombinant *Kluyveromyces marxianus* K1	4.24	Yanase et al., (2010)

SHF: separate hydrolysis and fermentation; SSF: simultaneous saccharification and fermentation; SSCF: simultaneous saccharification and co-fermentation; CBP: consolidate bioprocess.

pumpkin peels, the maximum bioethanol concentration reached was 50.69 g/L, which was higher than that obtained from potato peel wastes, sugarcane leaf waste, and sorghum leaves. Bioethanol concentration was 5.30 g/L, 30.49 g/L, and 17.15 g/L for potato peel waste, sugarcane leaf waste, and sorghum leaves, respectively (Moncef et al., 2020; Chohan et al., 2019; Moodley and Gueguim-Kana, 2019). Li et al. (2007) evaluated biodegradable MSW fractions consisting of carrot and potato peels, grass, newspaper, and scrap paper for the production of ethanol (Li et al., 2007). In another study, the combined effect of acid and glucoamylase treatment was evaluated for the conversion of kitchen waste constituting rice, meats, and vegetables. The process resulted in 94.46 g/L of total sugars, out of which 93.25 g/L was glucose for possible use in bioconversion to bioethanol and other valuable products (Hafid et al., 2015). Likewise, hydrolysis and fermentation of kitchen wastes yielded 32.2 g/L ethanol (Uncu and Cekmecelioglu, 2011). Salakkam et al. (2017) evaluated the co-fermentation of soybean residue and cooked rice. Soybean residues also acted as a low-cost nitrogen source for *Saccharomyces cerevisiae* TISTR 5339. Salakkam et al. (2017) reported that ethanol concentration reached 40.7 g/L. Similarly, pie waste, being a cheap sugar source, can produce fermentable sugar post-treatment with a mixture of enzymes (alpha amylase 45%, gamma amylase 45%, and pectinase 10%). A 20% pie waste solids yielded 32.9% (w/w) ethanol (Magyar et al., 2017). Australian cotton ginning stock (CGT) is an emerging feedstock in the Australian biofuel industry. However, to improve the release of glucose from CGT for fermentation, acid and enzyme hydrolysis is needed (McIntosh et al., 2014). Coffee waste is also a good feedstock and produced 6.12 g/L ethanol post-acid hydrolysis and fermentation process. This study demonstrated the feasibility of utilizing wet coffee waste for ethanol production (Woldesenbet et al., 2016). Sugar cane distillery waste has been evaluated for ethanol production. Here, *Saccharomyces cerevisiae* NGY10 in SSCF and CBP processes transformed sugar cane distillery waste into bioethanol. Similarly, acid-treated rice straw was also efficiently used for bioethanol production, with 86.43% and 92.81% fermentation effectiveness throughout SSF and SHF process, respectively (Pandey et al., 2019). Another study shows that H_2O_2 pretreatment followed by enzymatic hydrolysis enabled utilization of horticultural waste by *Saccharomyces cerevisiae* for ethanol production (Geng et al., 2012). These studies indicated that biowaste coming from various sources may be exploited for bioethanol production, as an effective feedstock provided biomass is pretreated suitably to make fermentable sugars available for the fermentation process.

1.3.2 BIOBUTANOL

Like bioethanol, biobutanol is also a renewable alternative to gasoline. Higher energy content, lower volatility, and lesser hygroscopicity make biobutanol superior compared to bioethanol (Dürre, 2007). Solventogenic strains like *Clostridium acetobutylicum,* under anaerobic conditions, produce biobutanol through acetone-butanol-ethanol (ABE) fermentation (Dürre, 2007). Traditional substrates for ABE fermentation are molasses, cereals, potatoes, corn starch, etc. (Ezeji and Blashek 2010; Ezeji et al., 2007; Syed et al., 2008). However, due to competition with food and the high cost of these feedstocks, research interest has diverted towards the utilization of various biomass residues as feedstocks (Zheng et al., 2013). In general, lignocellulosic biomass is the most promising feedstock but they could not be utilized by most solventogenic strains like *Clostridia* because of inexpression lignocellulose humiliating enzymes (Xin et al., 2014). However, the use of a combination of enzymes in enzymatic hydrolysis had a synergistic effect on the process. For example, enzymatic degradation of lignin and hemicellulose in lignocellulosic biomass improved the overall economics of the biobutanol production (Klein-Marcuschamer et al., 2012; Ibrahim et al., 2017; Jang et al., 2012). Recently, food waste, particularly with higher water content has also attracted attention for biobutanol production as feedstock (Abo et al., 2019; Stoeberl et al., 2011). Huang et al. (2015) compared food waste and glucose as carbon sources for ABE production. ABE concentration was 18.9 g/L and 14.2 g/L when food waste and glucose were used as the substrate, respectively (Huang et al., 2015). In another study, potato peel, an industrial waste from a snack factory was suggested for biorefineries aimed at butanol production. Six *Clostridium* strains (*C. beijerinckii* CECT 508, *C. acetobutylicum* DSM 1732, DSM 1733 and DSM 1738, *C. saccharobutylicum* DSM 13864 and *C. saccharoperbutylacetonicum* DSM 2152) were used to evaluate the conversion of potato hydrolysate into butanol. Out of these, *C. saccharobutylicum* DSM 13864 produced 2.1 g/L, 7.6 g/L, and 0.6 g/L acetone, butanol, and ethanol, respectively in 96 h with a yield of 0.186 gB/gS, whereas *C. saccharoperbutylacetonicum* DSM 2152 produced 1.8 g/L, 8.1 g/L, and 1.0 g/L acetone, butanol, and ethanol, respectively in 120 h with a yield of 0.203 gB/gS (Hijosa-Valsero et al., 2018). Recently, Wu et al. (2020) have developed a novel, efficient, and less energy-intensive integrated process from rice straw for butanol production, which was pretreated with cellulase enzyme from *Trichoderma viride* to release monomeric sugar molecules for fermentation. The resultant hydrolysate was subjected to mono and co-culture fermentation with *Clostridium beijerinckii* f-6, *C. beijerinckii*, and *Saccharomyces cerevisiae.* Co-culture resulted in 115.4% higher production than the monoculture system (Wu et al., 2020). Papermill sludge (PMS) has been used as a feedstock for biobutanol production, where fermentation using *C. acetobutylicum* resulted in 9.7 g/L butanol with a yield and productivity of 0.13 g/g and 0.10 g/L/h, respectively (Guan et al., 2016). The advantage of these inedible and low-cost feedstocks is that they help in improving the economics of butanol fermentation (Abo et al., 2019). However, the ineffectiveness of lignocellulose degrading enzymes and the negative effect of high solvent concentration, especially butanol, on enzyme activity affect efficient utilization of lignocellulosic residues by *Clostridia* during fermentation and thus affect process economics (Xin et al., 2014; Quershi and Eller, 2018; Baral and Shah 2016; Mariano et al., 2012). Several recent studies have reported the advances in the butanol production process from different biomass feedstocks. Strain engineering through synthetic biology played a key role in the improvement of butanol production and reduction in by-product formation. Synthetic biology overcomes the bottlenecks of the process, like toxicity caused by high acidity, high solvent conditions, and due to by-product formation (Wang et al., 2017a; Lu et al., 2017; Lütke-Eversloh and Bahl, 2011). Moreover, the focus has shifted on organisms like *E. coli* as a production host for high butanol production due to difficulties associated with genetic manipulation of *Clostridia* (Kumar and Gayen 2011). Ohtake et al. (2017) reported high butanol concentration (18.3 g/L) obtained through the application of metabolic engineering (Ohtake et al., 2017). Recent studies revealed that, engineered *C. tyrobutyricum* Δack-adhE2 produced 16.1 g/L and 15.8 g/L butanol from cassava bagasse and cotton stalk hydrolysates, respectively (Huang et al., 2019; Li et al., 2019). Likewise, engineered *C. tyrobutyricum* Δcat1::adhE2 using CRISPR Cas system showed enhanced production of biobutanol from PMS and corn steep liquor. These feedstocks are a good alternative to expensive nitrogen sources and also supply lactic acid to be used in fermentation for butanol production (Cao et al., 2020). Thus, GMO strains help in the efficient utilization of substrate as well as in the enhancement of biobutanol production.

1.3.3 BIOMETHANE

Biomethane is another promising biofuel produced by anaerobic digestion (AD) of biomass. AD is also a biochemical conversion process, which occurs at slight conditions like pressure using microorganisms and lower temperatures. The mechanism involves bacterial degradation of biomass in the absence of oxygen. This is one of the most widely used technologies in wastewater treatment plants (WWTP) because of its low process cost and at the same time, it generates revenue by producing energy (Nghiem et al. 2014; Ferreira et al. 2018; Ferreira et al., 2020; Semblante et al. 2014). Some of the biowaste used in this process are manure, organic waste, grasses, and processed residual algal biomass (Ward et al.; 2014). The primary produce from the anaerobic ingestion is biogas which is a blend of CO_2 and methane. Biogas qualifies to be a second-generation biofuel as the feedstock does not compete with the feed and food (Cherubini, 2010). In general, anaerobic digestion is a three-step process, which involves hydrolysis, fermentation, and methanogenesis. Hydrolysis yields simple sugars from the breakdown of carbohydrates. Then in fermentation, these sugars are used by bacteria to produce alcohols, acetic acid, volatile fatty acid, and gas. In the final methanogenesis step, these products are metabolized in methane and CO_2. Thus, the anaerobic digestion process involves the interaction of multiple microorganisms with the substrate (Gonzalez-Fernandez et al., 2015). Though the organic fraction of municipal solid waste (OFMSW) and sewage sludge is rich in carbohydrates, proteins, lipids as well as other substances, the complex nature of OFMSW can hinder methanogenesis and in turn biogas production (Panigrahi and Dubey, 2019). Developing alternative technologies is important for biogas production with high methane content (Panigrahi and Dubey, 2019; Ferreira et al. 2018; Tyagi et al. 2018). The use of proteases, lipases, hydrolytic enzymes in the anaerobic co-digestion process is gaining interest among other alternatives to produce better-quality biogas (Ferreira et al., 2020). Enzymatic treatment of effluents with high fat content is an economically feasible option (Soares et al., 2019). Thus, the application of hybrid technology is an important approach to produce methane from the co-digestion of deposits with fat content and other compounds (Ferreira et al., 2020). The enzymatic treatment allows the good performance of the bacteriological population in anaerobic digestion, as it improves the interaction between microbes and the biomass, thus helping in enhancing the productivity of biomethane (Mendes et al. 2010; Soares et al. 2019; Valladao et al. 2011). Valladao et al. (2009) treated effluent with lipases from *Penicillium* sp. and observed a 50% increase in biogas generation, compared to unprocessed effluents (Valladao et al., 2009). *Aspergillus niger* produced 78% and 75% methane in solid-state fermentation from OFMSW and wheat bran as substrates, respectively (Mlaik et al., 2019). Though biological pretreatments are an environment-friendly option, it requires long incubations and highly specific enzymes which leads to increased production cost. Therefore, to obtain a high methane yield, further research on hybrid technologies (enzymatic pretreatment before co-digestion) is essential to achieve an economically viable solution at an industrial scale (Agabo-García et al., 2019; Ferreira et al., 2020).

1.3.4 BIOHYDROGEN

Hydrogen is a clean renewable fuel. Biological hydrogen production is environment-friendly, harmless, and safe. Dark or photo fermentation methods are promising ways of producing biohydrogen from biomass using microbes (Argun and Dao, 2017). In dark fermentation, the carbohydrate-rich feedstock is converted into biohydrogen by anaerobic bacteria in the unavailability of light, while in photo-fermentation, photosynthetic bacteria utilize the energy of light for biohydrogen production from organic feedstocks (Bharathiraja et al., 2016; Hallenbeck et al., 2002). *Clostridia* sp. and *Enterobacter* break down the carbohydrates and other intermediates like alcohols and volatile fatty acids to produce biohydrogen through dark fermentation (Levin et al., 2004). Industrial wastewater from the chemical industry, sugar industry, palm oil effluent, beverage industry, and distillery industry effluent are used as feedstocks for biohydrogen production, which are subjected to pretreatment to enhance the production (Boodhun et al., 2017; Kumar et al., 2017). Substrate composition, pretreatment of the substrate, optimization of C/N ratio, mode of reactor operation, and bacterial consortium are some of the factors which affect biohydrogen yield (Kumar et al., 2017; Venkata Mohan et al., 2008; Mota et al., 2018; Hernández-Mendoza et al., 2014). The optimization of these factors helps in improved biohydrogen production. Industrial wastes generate various inhibitory compounds, which hampers the production of biohydrogen. Lay et al. (2012), evaluated the effect of various inhibitory compounds generated from textile and pharmaceutical wastes on hydrogen production potential (Lay et al., 2012). Various pretreatment studies have been reported to improve hydrogen yield in dark fermentation (Wang et al., 2003). Pretreatment of substrate suppressed the activity of methanogens but promoted the activity of hydrogen-producing bacteria, thus improving overall hydrogen yield during dark fermentation (Venkata Mohan et al., 2008). Recently a study revealed an innovative strategy of pretreatment with potassium ferrate (PF), a strong oxidant, from dark fermentation of food waste to improve the production of hydrogen. The results showed that PF treatment resulted in 173.5 mL/g hydrogen production when the dosage of PF was 0.4 g/g of total suspended solids. The obtained yield was about three times compared to the untreated control (Yang et al., 2020). Another study showed that fermentation of sago industrial effluent by pure cultures of anaerobic bacteria; *Clostridium sartagoforme* and *Enterobacter cloacae* produced 156.7 ml hydrogen/g of glucose (Nizzy et al., 2020).

In the photo-fermentation process of biohydrogen production, small organic molecules are converted into carbon dioxide and hydrogen in the availability of light by photosynthetic microorganisms, like purple non-sulfur bacteria (PNSB) and purple sulfur bacteria (PSB). *Rhodopseudomonas capsulata, Rhodopseudomonas sphaeroides, Rhodopseudomonas palustris, Rhodospirillum rubrum*, etc. are common photosynthetic bacteria used in the production process (Tang et al., 2008). Tables 1.3 and 1.4 consolidate some of the recent studies targeting biohydrogen production through dark and photo fermentation, respectively. For further details, readers can refer recent review summarising the latest advancements in H_2 production from starchy wastes (Das and Basak, 2020). It is envisaged that overcoming technical and commercial issues like biohydrogen storage, distribution, and high infrastructure costs will be a leap forward in large-scale hydrogen production (Singh and Rathore, 2017).

1.3.5 BIOELECTRICITY (MICROBIAL FUEL CELL TECHNOLOGY)

Microbial fuel cells (MFCs) is a bio-electrochemical process in which electricity is produced through the movement of electrons derived from biochemical reactions mediated by bacteria. MFC has anode and cathode chambers, which are detached by a proton exchange membrane (PEM). The oxidation of organic matter by electrogenic bacteria generates electrons, which are transferred to the anode of MFC while protons are released into the solution. Protons travel to the cathode through PEM, while electrons reach the cathode chamber through an external circuit (Rahimnejad et al., 2015). Ramaprasad et al. (2018) reported a novel bacterium belonging to the genus, *Rhodococcus* (JC435T) and phylum Actinobacteria, which produced higher electricity compared to other reported strains, *R. ruber* and *R. aetherivorans*. The power output was 94.5±2.6 mW for *Rhodococcus* JC435T, which was relatively higher than other tested strains. MFCs inoculated with *R. ruber* KCTC 9806T and *R. aetherivorans* JCM 14343T resulted in power output of 11.9±0.9 and 8.5±0.4 mW, respectively (Ramaprasad et al. 2018). In another study, *Alkanivorax xenomutans* were used to degrade hydrocarbons, both under toxic and anoxic conditions. Anoxic conditions favored high power

TABLE 1.3 Biohydrogen Production Through Dark Fermentation Using Different Biomass Residues As Feedstocks

Waste activated sludge	**Hydrogen consuming microbes**	**19.2 ml H2/g VSS**	**Yang et al., (2020)**
Sago wastewater	*Enterobacter aerogenes*	7.42 mmol H_2 mol/1 glucose	Ulhiza et al., (2018)
Wheat	Sludge	1.6 mol H_2 mol/1 TS	Kirli and Karapinar, (2018)
Wheat	Sludge	1.96 mol H_2 mol/glucose	Gokfiliz and Karapinar, (2017)
Waste peach pulp	Anaerobic Sludge	123.27 mLH_2/gTOC	Argun and Dao, (2017)
Potato	*C. butyricum* NRRL-B-1024v and *E. aerogenes* NRRL-B-115	7.21 mmol H_2 g/COD	Hitit et al., (2017)
Cassava	*Bacillus cereus* and *Brevumdimonas naejangsanensis*	1.72 mol H_2 mol/glucose	Wang et al., (2017b)
Corn	*Clostridium acetobutylicum* DSM 792	1.59 L H_2 L/medium	Zagrodnik and Łaniecki, (2017)
Starch reagent	*Enterobacter sp.*	0.56 mol H_2 mol/hexose	Maintinguer et al., (2017)
Corn wastewater	*Bacillus cereus* and *Brevumdimonas naejangsanensis*	1.88 mol H_2 mol/glucose	Wang et al., (2016b)
Cassava	*Enterobacter aerogenes* ATCC13408	124.3 mL H_2 g/starch	Lin et al., (2016)
Wheat	*Biohydrogenbacterium* R3	2.34 mol H_2 mol/glucose	Han et al., (2015)
Wheat	*Caldicellulosiruptor saccharolyticus* DSM 8903	2.5 mmol H_2 L/medium	Panagiotopoulos et al., (2013)

VSS: volatile suspended solids; TS: total solids; TOC: total organic carbon; COD: chemical oxygen demand.

TABLE 1.4 Photo Fermentative Production of Hydrogen from Various Dark Fermentative Effluent

Substrate	Organism	Hydrogen Concentration	References
SMW	*Rhodobacter sp.* KKU-PS1	1.92 mol H_2 mol/substrate	Assawamongkholsiri et al., (2019)
Bread dark effluent	*Rhodopseudomonas palustris* 42OL	3.1 mol H_2 mol/glucose	Adessi et al., (2018)
Corn	*Rhodobacter sphaeroides* O U 001	4.79 mol H_2 mol/substrate	Zagrodnik et al., (2017)
Potato	*R. palustris* GCA009	7.21 mmol H_2 g/ COD	Hitit et al., (2017)
DEPOME	*R. palustris*	1.7 ml H2/ml POME	Mishra et al., (2016)
Wheat	*Rhodobacter sphaeroides* NRLL-B1727	1.2 L H_2 g/ TVFA	Sagnak and Kargi, (2011)
Wheat	*Rhodobacter sphaeroides* RV	1.23 mol H_2 mol/ glucose	Kapdan et al., (2009)
Wheat	*Rhodobacter sphaeroides* and *R. palustris*	1.25 mol H_2 mol/ glucose	Argun et al., (2009)
Wheat	*Rhodobacter sphaeroides*	693 mL H_2 g/ TVFA	Argun et al., (2008)
Sweet potato	*Rhodobacter sphaeroides* M-19	4.5 mol H_2 mol/ glucose	Yokoi et al., (2002)

SMW: sugar manufacturing wastewater; COD: chemical oxygen demand; TVFA: total volatile fatty acids; POME: palm oil mill effluent; DEPOME: dark fermentative effluent palm oil mill effluent.

output by the electrogenic bacterium (Mahidhara et al., 2019). Dark fermentation followed by MFC to produce electricity from food processing water was first shown by Oh and Logan (2005). High hydrogen yield was obtained in the electrohydrogenic process, where several carbon sources viz. butyric, acetic, lactic, valeric acids, propionic, cellulose, and glucose were used as substrates (Cheng and Logan 2007). Microbial electrolysis cell (MEC) is another approach to capture H_2 using reverse electrodialysis (Baek et al., 2021). MEC has a pair of electrodes, which were inserted in an anaerobic digestion system for the degradation of organic residues in wastewater using the electrode reaction of MEC (Zhang et al., 2013). In another study, an integrated system (fermentation, MFC, MEC) produced 14.3 mmol H_2/g cellulose which was 41% more compared to fermentation alone (Wang et al., 2011). Further developments in these technologies are needed to improve process efficiency and to lower the capital cost of MFC, which are the main impediments in the large-scale implementation and commercialization of bioelectricity.

1.4 CHALLENGES IN INDUSTRIAL SCALE CONVERSION OF WASTE BIOMASS TO BIOENERGY

Though a clear understanding of biomass conversion is the first major step in the realization of waste biomass to bioenergy production, making this process economically viable and competitive to fossil fuel is very critical. The major bottlenecks in upscaling of waste biomass are the underdeveloped waste biomass supply chain and technological hurdles in biomass conversion technologies. In addition, relatively inexpensive and well-developed fossil energy technologies make this situation even more grave.

The main challenges in the biomass supply chain are seasonal availability, variable quantities of residual biomass, availability on scattered locations, the high moisture content in biomass making it susceptible for degradation, low biomass density leading to increased storage and transportation cost and thus demanding to follow biomass densification process. These reasons render biomass logistics for the year-round operation of biorefinery quite cost-intensive and complex (Souza et al., 2015). The complex structure of lignocellulosic biomass often results in low biomass conversion efficiencies and scarcity of matured technologies aggravates the difficulties further (Balan, 2014). Additionally, insufficient funds and investments to run biomass projects, non-standard and fluctuating residual biomass pricing, high risk due to uncertainties in raw material availability, and low returns on investments in combination with limitations in downstream conversion technologies limit commercialization of biomass to bioenergy (Malladi and Sowlati, 2018).

Utilizing municipal solid waste (MSW) as a feedstock for generating energy appears to be a very lucrative option both economically and ecologically. However, MSW management is another largely unorganized and mismanaged sector. Global MSW generation accounts for nearly two billion tons per annum and it is projected to reach 9.5 billion tons by 2050 (WBA, 2017). Approximately, 40–45% of MSW is an organic biodegradable material that can be used for energy generation. However, open burning and disposing in landfills are the most common practices of waste management around the globe causing serious concerns about space, climate change, and health. Approximately 50% of the MSW end up in landfills (WBA, 2017). MSW is also highly heterogeneous in nature, consisting of compostable, recyclable, inert, and toxic substances (Kalyani and Pandey, 2014). Ignorance at the community level, improper or limited preparation of waste separation at source are the key operational challenges in the conversion of MSW to energy. This was one of the reasons for the failure of a large biomethane project in India. As for the effectiveness of the anaerobic digestion process, well-segregated organic waste is desirable (Kalyani and Pandey, 2014). Moreover, waste-to-energy plants are not enough and existing plants are underutilized or nonoperational to harness the energy from organic waste (Padella et al., 2019). Apart from that, poor logistic planning, limited financial resources, and lack of firm implementation of policy guidelines for waste management are the key stumbling blocks in MSW to bioenergy generation (Lakshmikanthan, 2019; Moharir et al., 2019).

1.5 PATH AHEAD

Dedicated and focused efforts are needed for a cleaner and sustainable energy. The paradigm shift towards renewable energy use is a necessary step in this direction. Increased bioenergy contribution in renewable energy portfolio will not only address the global warming problem but also provides a fruitful solution for the waste management problem. Following are some of the areas which need focused attention to make biomass to bioenergy an economically viable option.

- Coordinated partnerships and commitment among government, investors, researchers, and corporates to address scientific, technological, and scaleup challenges in biomass conversion technologies.

- Building awareness and sensitizing people towards the world's welfare leading to making the right choices to tackle the issue of climate change and energy security.
- Government focus on providing subsidies on waste to energy plants, the introduction of zero waste policy, setting clear renewable energy targets, and strengthening the drive of sustainable bioenergy through a strong policy framework.
- Address key bottlenecks in currently fragmented residual biomass and waste supply chains to ensure sustainability in the supply of biomass in an economically feasible way. These include the development of robust business models through which biomass supply chain can be well-coordinated, mechanization of the agriculture sector, especially in developing and less developed counties to improve process efficiency, improving awareness among the farming community, and providing incentives so that instead of burning and discarding the crop residues, they channelize it to bioenergy production.
- In the case of MSW management, the worldwide introduction of strict policies to ban organic waste dumping in landfills, heavy landfill taxes to promote waste minimization, recycling, and reusing are some of the steps in ensuring proper waste segregation and its routing to waste-to-energy generation. Most importantly, it is necessary to bridge the gaps between policy-making and its implementation.

Thus, thoughtful steps towards bringing innovation in biomass conversion technologies, a structured policy-driven framework to manage residual biomass and waste supply chains, and active community support will ultimately make waste biomass to bioenergy a commercially viable option and can successfully contribute towards energy security, environmental sustainability, and waste management.

KEYWORDS

- **bioenergy**
- **biohydrogen**
- **biomass**
- **biomethane**
- **municipal solid wastes**

REFERENCES

Abo, B. O., Gao, M., Wang, Y., Wu, C., Wang, Q., & Ma, H., (2019). Production of butanol from biomass: Recent advances and future prospects. *Environ. Sci. Pollut. Res. Int., 26*(20), 20164–20182.

Adessi, A., Venturi, M., Candeliere, F., Galli, V., Granchi, L., & De Philippis, R., (2018). Bread wastes to energy: Sequential lactic and photo-fermentation for hydrogen production. *Int. J. Hydrogen Energy, 43*(20), 9569–9576.

Agabo-García, C., Pérez, M., Rodríguez-Morgado, B., Parado, J., & Solera, R., (2019). Biomethane production improvement by enzymatic pretreatments and enhancers of sewage sludge anaerobic digestion. *Fuel, 255*, 115713.

Agwa, O. K., Nwosu, I. G., & Abu, G. O., (2017). Bioethanol production from *Chlorella vulgaris* biomass cultivated with plantain (Musaparadisiaca) peels extract. *Advances in Bioscience and Biotechnology, 8*, 478–490.

Arapoglou, D., Varzakas, T., Vlyssides, A., & Israilides, C., (2010). Ethanol production from potato peel waste (PPW). *Waste Management, 30*(10), 1898–1902.

Argun, H., & Dao, S., (2017). Bio-hydrogen production from waste peach pulp by dark fermentation: Effect of inoculum addition. *Int. J. Hydrog. Energy, 42*(25), 69–74.

Argun, H., Kargi, F., & Kapdan, I. K., (2008). Light fermentation of dark fermentation effluent for bio-hydrogen production by different *Rhodobacter* species at different initial volatile fatty acid (VFA) concentrations. *Int. J. Hydrogen Energy, 33*(24), 7405–7412.

Argun, H., Kargi, F., & Kapdan, I. K., (2009). Hydrogen production by combined dark and light fermentation of ground wheat solution. *Int. J. Hydrogen Energy, 34*(10), 4305–4311.

Assawamongkholsiri, T., Reungsang, A., & Sittijunda, S., (2019). Photo-hydrogen and lipid production from lactate, acetate, butyrate, and sugar manufacturing wastewater with an alternative nitrogen source by *Rhodobacter* sp. KKU-PS1. *Peer J., 7*.

Baek, G., Shi, L., Rossi, R., & Logan, B. E., (2021). The effect of high applied voltages on bioanodes of microbial electrolysis cells in the presence of chlorides. *Chemical Engineering Journal, 405*(1), 126742.

Balan, V., (2014). Current challenges in commercially producing biofuels from lignocellulosic biomass. *ISRN Biotechnology*, 1–33.

Baral, N. R., & Shah, A., (2016). Techno-economic analysis of cellulosic butanol production from corn stover through acetone-butanol-ethanol fermentation. *Energy Fuel, 30*(7), 5779–5790.

Bezerra, T. L., & Ragauskas, A. J., (2016). A review of sugarcane bagasse for second-generation bioethanol and biopower production*: Biofuels Bioprod. Biorefin., 10*, 634–647.

Bharathiraja, B., Sudharsanaa, T., Bharghavi, A., Jayamuthunagai, J., & Praveenkumar, R., (2016). Biohydrogen and Biogas – an overview on feedstocks and enhancement process. *Fuel, 185*, 810–828.

Boodhun, B. S. F., Mudhoo, A., Kumar, G., Kim, S. H., & Lin, C. Y., (2017). Research perspectives on constraints, prospects and opportunities in biohydrogen production. *Int. J. Hydrog. Energy, 42*(45), 27471–27481.

Brosowski, A., Thrän, D., Mantau, U., Mahro, B., Erdmann, G., Adler, P., Stinner, W., et al., (2016). A review of biomass potential and current utilization – status quo for 93 biogenic wastes and residues in Germany. *Biomass and Bioenergy, 95*, 257–272.

Cai, Y., Zhang, K., Kim, H., Hou, G., Zhang, X., Yang, H., Feng, H., et al., J., (2016). Enhancing digestibility and ethanol yield of Populus wood via expression of an engineered monolignol 4-O-methyltransferase. *Nature Communications, 7*, 1–14.

Cao, X., Chen, Z., Liang, L., Guo, L., Jiang, Z., Tang, F., Yun, Y., & Wang, Y., (2020). Co-valorization of paper mill sludge and corn steep liquor for enhanced n-butanol production with *Clostridium tyrobutyricum* Δcat1: Adh E2. *Bioresour Technol., 296*, 122347.

Carrasco, C., Baudel, H. M., Sendelius, J., Modig, T., Roslander, C., Galbe, M., Hahn-Hägerda, B., et al., (2010). SO_2-catalyzed steam pretreatment and fermentation of enzymatically hydrolyzed sugarcane bagasse. *Enzyme and Microbial Technology, 46*, 64–73.

Cerveró, J. M., Skovgaard, P. A., Felby, C., Sørensen, H. R., & Jørgensen, H., (2010). Enzymatic hydrolysis and fermentation of palm kernel press cake for production of bioethanol. *Enzyme and Microbial Technology, 46*(3), 177–184.

Chandra, N., Shukla, P., & Mallick, N., (2020). Role of cultural variables in augmenting carbohydrate accumulation in the green microalga *Scenedesmus acuminatus* for bioethanol production. *Biocatalysis and Agricultural Biotechnology, 26*, 101632.

Chen, H., Liu, J., Chang, X., Chen, D., Xue, Y., Liu, P., Lin, H., & Han, S., (2017). A review on the pretreatment of lignocellulose for high-value chemicals. *Fuel Process. Technol., 160*, 196–206.

Cheng, S., & Logan, B. E., (2007). Sustainable and efficient biohydrogen production via electrohydrogenesis. *Proc Natl Acad Sci., 104*, 18871–18873.

Cherubini, F., (2010). The biorefinery concept: Using biomass instead of oil for producing energy and chemicals. *Energy Convers Manag., 51*, 1412–1421.

Chohan, N. A., Aruwajoye, G. S., Sewsynker-Sukai, Y., & Gueguim, K. E. B., (2019). Valorization of potato peel wastes for bioethanol production using simultaneous saccharification and fermentation: Process optimization and kinetic assessment. *Renew. Energy, 146*, 1031–1040.

Conde-Mejía, C., Jimenez-Gutierrez, A., & El-Halwagi, M., (2012). A comparison of pretreatment methods for bioethanol production from lignocellulosic materials. *Process Saf. Environ. Prot., 90*, 189–202.

Das, S. R., & Basak, N., (2020). Molecular biohydrogen production by dark and photo fermentation from wastes containing starch: Recent advancement and future perspective. *Bioprocess Biosyst Eng.*, 1–25.

Dürre, P., (2007). Biobutanol: An attractive biofuel. *Biotechnol. J., 2*, 1525–1534.

Ezeji, T. C., & Blaschek, H. P., (2010). Butanol production from lignocellulosic biomass. In: Blaschek, H. P., (ed.), *Biofuels from Agricultural Wastes and Byproducts* (1st edn., pp. 19–37) Wiley, New York.

Ezeji, T. C., Qureshi, N., & Blaschek, H. P., (2007). Production of acetone butanol (AB) from liquefied corn starch, a commercial substrate, using *Clostridium beijerinckii* coupled with product recovery by gas stripping. *J. Ind. Microbiol. Biotechnol., 34*(12), 771–777.

FAOSTAT, (2014). *Mushrooms and Truffles*. Rome: Food and Agriculture Organization of the United Nations.

Ferreira, J. D. S., Oliveira, D. D., Maldonado, R. R., Kamimura, E. S., & Jr, A. F., (2020). Enzymatic pretreatment and anaerobic co-digestion as a new technology to high-methane production. *Appl Microbiol Biotechnol., 104*(10), 4235–4246.

Ferreira, J. S., Volschan, I. Jr., & Cammarota, M. C., (2018). Enhanced biogas production in pilot digesters treating a mixture of sewage sludge, glycerol and food waste. *Energy Fuel, 32*, 6839–6846.

Food and Agriculture Organization (FAO) of the United Nations Rome (2013). The State of Food and Agriculture, 1–114.

Fülöp, L., & Ecker, J., (2020). An overview of biomass conversion: Exploring new opportunities. *Peer J., 8*(1), 1–21.

Geng, A., Xin, F., & Ip, J., (2012). Ethanol production from horticultural waste treated by a modified organosolv method. *Bioresource Technology, 104*, 715–721.

Gokfiliz, P., & Karapinar, I., (2017). The effect of support particle type on thermophilic hydrogen production by immobilized batch dark fermentation. *Int. J. Hydrogen Energy, 42*(4), 2553–2561.

Gonzalez-Fernandez, C., Sialve, B., & Molinuevo-Salces, B., (2015). Anaerobic digestion of microalgal biomass: Challenges, opportunities and research needs. *Bioresource Technology, 198*, 896–906.

Govumoni, S. P., Koti, S., Kothagouni, S. Y., Venkateshwar, S., & Linga, V. R., (2013). Evaluation of pretreatment methods for enzymatic saccharification of wheat straw for bioethanol production. *Carbohydr. Polym., 91*(2), 646–650.

Guan, W., Shi, S., Tu, M., & Lee, Y. Y., (2016). Acetone-butanol-ethanol production from kraft paper mill sludge by simultaneous saccharification and fermentation. *Bioresour. Technol., 200*, 713–721.

Hafid, H. S., Rahman, N. A., Shah, M. D., & Baharudin, A. S., (2015). Enhanced fermentable sugar production from kitchen waste using various pretreatments. *J. Environ. Manage., 156*, 290–298.

Hallenbeck, P. C., & Benemann, J. R., (2002). Biological hydrogen production; Fundamentals and limiting processes. *Int. J. Hydrogen Energy, 27*, 1185–1193.

Han, W., Liu, D. N., Li, Y. F., Zhao, H. T., & Ren, N. Q., (2015). Utilization of wheat for biohydrogen production by a combination of solid-state fermentation and batch fermentation. *Int. J. Hydrogen Energy, 40*(17), 5849–5855.

Hernández-Mendoza, C. E., Moreno-Andrade, I., & Buitrón, G., (2014). Comparison of hydrogen producing bacterial communities adapted in continuous and discontinuous reactors. *Int. J. Hydrog. Energy, 39*, 14234–14239.

Hijosa-Valsero, M., Paniagua-García, A. I., & Díez-Antolínez, R., (2018). Industrial potato peel as a feedstock for biobutanol production. *N. Biotechnol., 46*, 54–60.

Hitit, Z. Y., Lazaro, C. Z., & Hallenbeck, P. C., (2017). Increased hydrogen yield and COD removal from starch/glucose based medium by sequential dark and photo-fermentation using *Clostridium butyricum* and *Rhodopseudomonas palustris*. *Int. J. Hydrogen Energy, 42*(30), 18832–18843.

Huang, J., Du, Y., Bao, T., Lin, M., Wang, J., & Yang, S. T., (2019). Production of n-butanol from cassava bagasse hydrolysate by engineered *Clostridium tyrobutyricum* overexpressing adhE2: Kinetics and cost analysis. *Bioresour Technol., 292*, 121969.

Huang, Y., Wei, X. Y., Zhou, S. G., Liu, M. Y., Tu, Y. Y., & Li, A., (2015). Steam explosion distinctively enhances biomass enzymatic saccharification of cotton stalks by largely reducing cellulose polymerization degree in *G. barbadense* and *G. hirsutum*. *Bioresour. Technol., 181*, 224–230.

Ibrahim, M. F., Ramli, N., Bahrin, E. K., & Abd-Aziz, S., (2017). Cellulosic biobutanol by Clostridia: Challenges and improvements. *Renew. Sust. Energ. Rev., 79*, 1241–1254.

IEA, (2019). World Energy Outlook. https://www.iea.org/reports/world-energy-outlook-2019.

Jahnavi, G., Prashanthi, G. S., Sravanthi, K., & Rao, L. V., (2017). Status of availability of lignocellulosic feed stocks in India: Biotechnological strategies involved in the production of Bioethanol. *Renewable and Sustainable Energy Reviews, 73*, 798–820.

Jang, Y. S., Malaviya, A., Cho, C., & Lee, S. Y., (2012). Butanol production from renewable biomass by clostridia. *Bioresour. Technol., 123*, 653–663.

Jin, M., Gunawan, C., Balan, V., & Dale, B. E., (2012). Consolidated bioprocessing (CBP) of AFEX™-pretreated corn stover for ethanol production using *Clostridium phytofermentans* at a high solids loading. *Biotechnol. Bioeng., 109*(8), 1929–1936.
Jin, M., Sarks, C., Gunawan, C., Bice, B. D., Simonett, S. P., & Narasimhan, R. A., (2013). Phenotypic selection of a wild *Saccharomyces cerevisiae* strain for simultaneous saccharification and co-fermentation of AFEX™ pretreated corn stover. *Biotechnol. Biofuels, 6*, 108.
Kahan, A., (2019). *Today in Energy*. U.S. Energy Information Administration.
Kalyani, K. A., & Pandey, K. K., (2014). Waste to energy status in India: A short review. *Renewable and Sustainable Energy Reviews, 31*, 113–120.
Kapdan, I. K., Kargi, F., Oztekin, R., & Argun, H., (2009). Bio-hydrogen production from acid hydrolyzed wheat starch by photo-fermentation using different *Rhodobacter* sp. *Int J. Hydrogen Energy, 34*(5), 2201–2207.
Karimi, K., Kheradmandinia, S., & Taherzadeh, M. J., (2006). Conversion of rice straw to sugars by dilute-acid hydrolysis. *Biomass and Bioenergy, 30*(3), 247–253.
Kim, J. Y., Lee, H. W., Lee, S. M., Jae, J., & Park, Y. K., (2019). Overview of the recent advances in lignocellulose liquefaction for producing biofuels, bio-based materials and chemicals. *Bioresour. Technol., 279*, 373–384.
Kirli, B., & Karapinar, I., (2018). The effect of HRT on biohydrogen production from acid hydrolyzed waste wheat in a continuously operated packed bed reactor. *Int. J. Hydrogen Energy, 43*(23), 10678–10685.
Klein-Marcuschamer, D., Oleskowicz-Popiel, P., Simmons, B. A., & Blanch, H. W., (2012). The challenge of enzyme cost in the production of lignocellulosic biofuels. *Biotechnol. Bioeng., 109*(4), 1083–1087.
Kopp, C. P., (2019). *The World's 6 Biggest Corn Producers.* https://www.investopedia.com/articles/markets-economy/090316/6-countries-produce-most-corn.asp (accessed on 12 December 2021).
Koyama, K., (2020). *Global Energy Situation Indicated by BP Statistics.* Institute of Energy Economics.
Krishnan, C., Sousa, L. D. C., Jin, M., Chang, L., Dale, B. E., & Balan, V., (2010). Alkali-based AFEX pretreatment for the conversion of sugarcane bagasse and cane leaf residues to ethanol. *Biotechnology and Bioengineering, 107*(3), 441–450.
Kumar, G., Sivagurunathan, P., Pugazhendhi, A., Thi, N. B. D., Zhen, G., Chandrasekhar, K., & Kadier, A., (2017). A comprehensive overview on light independent fermentative hydrogen production from wastewater feedstock and possible integrative options. *Energy Convers. Manag., 141*, 390–402.
Kumar, M., & Gayen, K., (2011). Developments in biobutanol production: New insights. *Appl. Energy, 88*(6), 1999–2012.
Kumar, S., & Pandey, A., (2019). Current developments in biotechnology and bioengineering and waste treatment processes for energy generation: An introduction. In: Kumar, S., Kumar, R., & Pandey, A., (eds.), *Current Developments in Biotechnology and Bioengineering* (pp. 1–9) Elsevier.
Lakshmikanthan, P., (2019). Value creation with waste to energy: Economic considerations. *Current Developments in Biotechnology and Bioengineering*, 307–318.
Lay, C. H., Kuo, S. Z., Sen, B., Chen, C. C., Chnag, J. S., & Lin, C. Y., (2012). Fermentative biohydrogen production from starch containing textile wastewater. *Int. J. Hydrog. Energy, 37*(2), 2050–2057.

Lee, S. Y., Sankaran, R., Chew, K. W., Tan, C. H., Krishnamoorthy, R., Chu, D. T., & Show, P. L., (2019). Waste to bioenergy: A review on the recent conversion technologies. *BMC Energy, 1*(1), 4.

Levin, D. B., Pitt, L., & Love, M., (2004). Biohydrogen production prospects and limitations to practical application. *Int. J. Hydrog. Energy, 29*, 173–185.

Li, A., Antizar-Ladislao, B., & Khraisheh, M., (2007). Bioconversion of municipal solid waste to glucose for bio-ethanol production. *Bioprocess Biosyst. Eng., 30*(3), 189–196.

Li, J., Du, Y., Bao, T., Dong, J., Lin, M., Shim, H., & Yang, S. T., (2019). N-butanol production from lignocellulosic biomass hydrolysates without detoxification by *Clostridium tyrobutyricum* Δack-adhE2 in a fibrous-bed bioreactor. *Bioresour Technol., 289*, 121749.

Lin, R., Cheng, J., Ding, L., Song, W., Liu, M., Zhou, J., & Cen, K., (2016). Enhanced dark hydrogen fermentation by addition of ferric oxide nanoparticles using *Enterobacter aerogenes*. *Biores Technol., 207*, 213–219.

Lu, C., Yu, L., Varghese, S., Yu, M., & Yang, S. T., (2017). Enhanced robustness in acetone-butanol-ethanol fermentation with engineered *Clostridium beijerinckii* overexpressing adhE2 and ctfAB. *Bioresour. Technol., 243*, 1000–1008.

Lütke-Eversloh, T., & Bahl, H., (2011). Metabolic engineering of *Clostridium acetobutylicum*: Recent advances to improve butanol production. *Curr. Opin. Biotechnol., 22*(5), 634–647.

Ma, J., Shi, S., Jia, X., Xia, F., Ma, H., Gao, J., & Xu, J., (2019). Advances in catalytic conversion of lignocellulose to chemicals and liquid fuels. *J. Energy Chem., 36*, 74–86.

Magyar, M., Da Costa, S. L., Jayanthi, S., & Balan, V., (2017). Pie waste–a component of food waste and a renewable substrate for producing ethanol. *Waste Manag., 62*, 247–254.

Mahidhara, G., Burrow, H., Sasikala, C., & Ramana, C. V., (2019). Biological hydrogen production: Molecular and electrolytic perspectives. *World J. Microbiol. Biotechnol., 35*(8), 116.

Maintinguer, S. I., Lazaro, C. Z., Pachiega, R., Varesche, M. B. A., Sequinel, R., & De Oliveira, J. E., (2017). Hydrogen bioproduction with *Enterobacter sp.* isolated from brewery wastewater. *Int J. Hydrogen Energy, 42*(1), 152–160.

Malladi, K. T., & Sowlati, T., (2018). Biomass logistics: A review of important features, optimization modeling and the new trends. *Renewable and Sustainable Energy Reviews, 94*, 587–599.

Mariano, A. P., Dias, M. O. S., Junqueira, T. L., Cunha, M. P., Bonomi, A., & Filho, R. M., (2012). Butanol production in a first-generation Brazilian sugarcane biorefinery: Technical aspects and economics of greenfield projects. *Bioresour. Technol., 135*, 316–323.

Mathibe, B. N., Malgas, S., Radosavljevic, L., Kumar, V., Shukla, P., & Pletschke, B. I., (2020). Lignocellulosic pretreatment-mediated phenolic by-products generation and their effect on the inhibition of an endo-1,4-β-xylanase from *Thermomyces lanuginosus* VAPS-24. 3 *Biotech., 10*(8), 349.

Matsakas, L., Kekos, D., Loizidou, M., & Christakopoulos, P., (2014). Utilization of household food waste for the production of ethanol at high dry material content. *Biotechnology for Biofuels, 7*(1), 4.

McIntosh, S., Vancov, T., Palmer, J., & Morris, S., (2014). Ethanol production from cotton gin trash using optimized dilute acid pretreatment and whole slurry fermentation processes. *Bioresour Technol., 173*, 42–51.

Mendes, A. A., Ernandes, B. P., Furigo, Jr. A., & Heizir, F. C., (2010). Anaerobic biodegradability of dairy wastewater pretreated with porcine pancreas lipase. *Braz. Arch Biol. Technol., 53*(6), 1279–1284.

Mishra, P., Thakur, S., Singh, L., Wahid, Z. A., & Sakinah, M., (2016). Enhanced hydrogen production from palm oil mill effluent using two stage sequential dark and photo fermentation *Int. J. Hydrogen Energy, 41*, 18431–18440.

Mishra, V., Jana, A. K., & Jana, M. M., (2017). Gupta A. Fungal pretreatment of sweet sorghum bagasse with supplements: Improvement in lignin degradation, selectivity and enzymatic saccharification. *3 Biotech, 7*(2), 110.

Mlaik, N., Khoufi, S., Hamza, M., Ali, M. M., & Sayadi, S., (2019). Enzymatic pre-hydrolysis of organic fraction of municipal solid waste to enhance anaerobic digestion. *Biomass Bioenergy, 127*, 105–286.

Mohapatra, S., Mishra, S. S., Bhalla, P., & Thatoi, H., (2019a). Engineering grass biomass for sustainable and enhanced bioethanol production. *Planta, 250*(2), 395–412.

Mohapatra, S., Ray, R. C., & Ramachandran, S., (2019b). Bioethanol from biorenewable feedstocks: Technology, economics, and challenges. In: Ray, R. C., & Ramachandran, S. B. T., (eds.), *BP from FC* (pp. 3–27). Academic Press.

Moharir, M., Pourkargar, D. B., Almansoori, A., & Daoutidis, (2019). Graph representation and distributed control of diffusion-convection-reaction system networks. *Chemical Engineering Science, 204*, 128–139.

Moncef, C., Daoued, K. B., Riguane, K., Rouissi, T., & Ferrari, G., (2020). Production of bioethanol from pumpkin peel wastes: Comparison between response surface methodology (RSM) and artificial neural networks (ANN). *Industrial Crops & Products, 155*, 112822.

Moodley, P., & Gueguim-Kana, E. B., (2019). Bioethanol production from sugarcane leaf waste: Effect of various optimized pretreatments and fermentation conditions on process kinetics. *Biotechnol. Rep., 22*, e00329.

Mota, V. T., Ferraz, Jr. A. D. N., Trably, E., & Zaiat, M., (2018). Biohydrogen production at pH below 3.0: Is it possible? *Water Res., 128*, 350–361.

Nanda, S., Rana, R., Sarangi, P. K., Dalai, A. K., & Kozinski, J. A., (2018). A broad introduction to first-, second-, and third-generation biofuels. In: Sarangi, P. K., Nanda, S., & Mohanty, P., (eds.), *Recent Advancements in Biofuels and Bioenergy Utilization* (pp. 1–25). Springer Singapore.

Nemerow, N. L., & Agardy, F. J., (1998). *Strategies of Industrial and Hazardous Waste Management*. International Thomson publishing company, USA.

Ng, W. K., Lim, H. A., Lim, S. L., & Ibrahim, C. O., (2002). Nutritive value of palm kernel meal pretreated with enzyme or fermented with *Trichoderma koningii* (Oudemans) as a dietary ingredient for red hybrid tilapia (*Oreochromis* sp.). *Aquaculture Research, 33*, 1199–1207.

Nghiem, L. D., Manassa, P., Dawson, M., & Fitzgerald, S. K., (2014). Oxidation reduction potential as a parameter to regulate micro-oxygen injection into anaerobic digester for reducing hydrogen sulphide concentration in biogas. *Bioresource Technology, 173*, 443–447.

Nizzy, A. M., Kannan, S., & Anand, S. B., (2020). Identification of hydrogen gas producing anaerobic bacteria isolated from sago industrial effluent. *Curr. Microbiol., 77*(9), 2544–2553.

Oberoi, H. S., Vadlani, P. V., Saida, L., Bansal, S., & Hughes, J. D., (2011). Ethanol production from banana peels using statistically optimized simultaneous saccharification and fermentation process. *Waste Management, 31*(7), 1576–1584.

Oh, S., & Logan, B. E., (2005). Hydrogen and electricity production from a food processing wastewater using fermentation and microbial fuel cell technologies. *Water Res., 39*, 4673–4682.

Ohtake, T., Pontrelli, S., Laviña, W. A., Liao, J. C., Putri, S. P., & Fukusaki, E., (2017). Metabolomics-driven approach to solving a CoA imbalance for improved 1-butanol production in *Escherichia coli*. *Metab. Eng., 41*, 135–143.

Omoti, U., (2009). *Oil Palm Sector Analysis in Nigeria* (Vol. 64, No. 18, p. 275). Main Report, submitted to the United Nations Industrial Development Organization (UNIDO). Abuja.

Özçimen, D., & İnan, B., (2015). An overview of bioethanol production from algae. *Biofuels-Status and Perspective*, 141–162. Intech Open.

Padella, M., Connell, A. O., & Prussi, M., (2019). What is still limiting the deployment of cellulosic ethanol? analysis of the current status of the sector. *Appl. Sci., 9*, 1–13.

Panagiotopoulos, I., Bakker, R., De Vrije, T., Claassen, P., & Koukios, E., (2013). Integration of first and second generation biofuels: Fermentative hydrogen production from wheat grain and straw. *Biores. Technol., 128*, 345–350.

Pandey, A. K., Kumar, M., Kumari, S., Kumari, P., Yusuf, F., Jakeer, S., Naz, S., et al., (2019). Evaluation of divergent yeast genera for fermentation-associated stresses and identification of a robust sugarcane distillery waste isolate Saccharomyces cerevisiae NGY10 for lignocellulosic ethanol production in SHF and SSF. *Biotechnol Biofuels, 12*(40), 1–23.

Panigrahi, S., & Dubey, B. K., (2019). A critical review on operating parameters and strategies to improve the biogas yield from anaerobic digestion of organic fraction of municipal solid waste. *Renew Energy, 143*, 779–797.

Paritosh, K., Kushwaha, S. K., Yadav, M., Pareek, N., Chawade, A., & Vivekanand, V., (2017). Food waste to energy: An overview of sustainable approaches for food waste management and nutrient recycling. *BioMed Research International*, 2370927.

Patni, N., Pillai, S. G., & Dwivedi, A. H., (2013). Wheat as a promising substitute of corn for bioethanol production. *Procedia Eng., 51*, 355–362.

Qureshi, N., & Eller, F., (2018). Recovery of butanol from *Clostridium beijerinckii* P260 fermentation broth by supercritical CO_2 extraction. *J. Chem. Technol. Biotechnol., 93*(4), 1206–1212.

Qureshi, N., Saha, B. C., Hector, R. E., Dien, B., Hughes, S., Liu, S., Iten, L., et al., (2018). Production of butanol (a biofuel) from agricultural residues: Part II – Use of corn stover and switchgrass hydrolysates. *Biomass and Bioenergy, 34*(4), 566–571.

Rahimnejad, M., Adhami, A., Darvari, S., Zirepour, A., & Oh, S., (2015). Microbial fuel cell as new technology for bioelectricity generation: A review. *Alexandria Engineering Journal, 54*, 745–756.

Ramaprasad, T., Kumar, R. J., Naresh, U., Prakesh, M., Kothandan, D., & Naidu, K. C. B., (2018). Effect of pH value on structural and magnetic properties of $CuFe_2O_4$ nanoparticles synthesized by low temperature hydrothermal technique. *Materials Research Express, 5*(9), 095025.

Riyadi, F. A., Tahir, A. A., Yusof, N., Sabri, N. S. A., Noor, M. J. M. M., Akhir, F. N. M. D., Othman, N., et al., (2020). Enzymatic and genetic characterization of lignin depolymerization by Streptomyces sp. S6 isolated from a tropical environment. *Sci. Rep., 10*(7813), 1–9.

Rodríguez, L. A., Toro, M. E., Vazquez, F., Correa-Daneri, M. L., Gouiric, S. C., & Vallejo, M. D., (2010). Bioethanol production from grape and sugar beet pomaces by solid-state fermentation. *International Journal of Hydrogen Energy, 35*(11), 5914–5917.

Roy, P., & Dutta, A., (2019). Life cycle assessment (LCA) of bioethanol produced from different food crops: economic and environmental impacts. In: Ray, R. C., & Ramachandran, S. B. T., (eds.), *BP from FC* (pp. 385–399). Academic Press.

Sagnak, R., & Kargi, F., (2011) Photo-fermentative hydrogen gas production from dark fermentation effluent of acid hydrolyzed wheat starch with periodic feeding. *Int. J. Hydrogen Energy, 36*(7), 4348–4353.

Salakkam, A., Kingpho, Y., Najunhom, S., Aiamsonthi, K., Kaewlao, S., & Reungsang, S., (2017). Bioconversion of soybean residue for use as alternative nutrient source for ethanol fermentation *Biochem. Eng. J., 125*, 65–72.

Semblante, G. U., Hai, F. I., Ngo, H. H., Guo, W., You, S. J., Price, W. E., & Nghiem, L. D., (2014). Sludge cycling between aerobic, anoxic and anaerobic regimes to reduce sludge production during wastewater treatment: Performance, mechanisms, and implications. *Bioresource Technology, 155*, 395–409.

Shah, K., Vyas, R., & Patel, G. B., (2019). Bioethanol Production from pulp of fruits. *Bioscience Biotechnology Research Communications, 12*(2), 464–471.

Shi, N., Liu, D., Huang, Q., Guo, Z., Jiang, R., Wang, F., Chen, Q., et al., (2019). Product-oriented decomposition of lignocellulose catalyzed by novel polyoxometalates-ionic liquid mixture. *Bioresour Technol., 283*, 174–183.

Singh, A., & Rathore, D., (2017). *Biohydrogen Production: Sustainability of Current Technology and Future Perspective, XVIII*. 320.

Soares, J. L., Magali, C. C., Melissa, L. E. G., & Volschan, I. Jr., (2019). Reduction of scum accumulation through the addition of low-cost enzymatic extract in the feeding of high-rate anaerobic reactor. *Water Sci Technol., 80,* 67–74.

Souza, S. P., Gopal, A. R., & Seabra, J. E. A., (2015). Life cycle assessment of biofuels from an integrated Brazilian algae-sugarcane biorefinery. *Energy, 81*(1), 373–381.

Stoeberl, M., Werkmeister, R., Faulstich, M., & Russ, W., (2011). Biobutanol from food wastes-fermentative production, use as biofuel the influence on the emissions. *Procedia Food Sci., 1*, 1867–1874.

Suriyachai, N., Weerasaia, K., Laosiripojana, N., Champreda, V., & Unrean, P., (2013). Optimized simultaneous saccharification and co-fermentation of rice straw for ethanol production by Saccharomyces cerevisiae and *Scheffersomyces stipites* co-culture using design of experiments. *Bioresour. Technol., 142*, 171–180.

Syed, Q., Nadeem, M., & Nelofer, R., (2008). Enhanced butanol production by mutant strains of Clostridium acetobutylicum in molasses medium. *Turk. J. Biochem., 33*(1), 25–30.

Taherzadeh, M. J., & Karimi, K., (2007). Enzyme-based hydrolysis processes for ethanol from lignocellulosic materials: A review. *Bio. Resources, 2*(3), 472–499.

Tang, G., Tang, Q., Huang, J., Liu, G., & Sun, Z., (2008). Effects of substrate species on fermentative hydrogen production. *Huanjing Kexue, 29*(8), 2345–2349.

Thakur, S., Shrivastava, B., Ingale, S., Kuhad, R. C., & Gupte, A., (2013). Degradation and selective ligninolysis of wheat straw and banana stem for an efficient bioethanol production using fungal and chemical pretreatment. *3 Biotech, 3*, 365–372.

Tyagi, V. K., Fdez-Güelfo, L. A., Zhou, Y., Álvarez-Gallego, C. J., Garcia, L. I. R., & Jern, N. W., (2018). Anaerobic co-digestion of organic fraction of municipal solid waste (OFMSW): Progress and challenges. *Renew. Sust. Energ. Rev., 93*, 380–399.

Ulhiza, T. A., Puad, N. I. M., & Azmi, A. S., (2018). Optimization of culture conditions for biohydrogen production from sago wastewater by Enterobacter aerogenes using response surface methodology. *In.t J. Hydrogen Energy, 43*(49), 22148–22158.

Uncu, O. N., & Cekmecelioglu, D., (2011). Cost-effective approach to ethanol production and optimization by response surface methodology. *Waste Manag., 31*(4), 636–643.

United-Nations, (2019). *The Future is Now Science for Achieving Sustainable Development*. United Nations.

US-DOE, (2020). Biomass resources. In: *Bioenergy* (Vol. 2020). Office of Energy Efficiency and Renewable Energy. Washington. DC-20585.

Valladao, A. B. G., Cammarota, M. C., Torres, A. G., & Freire, D. M. G., (2011). Profiles of fatty acids and triacylglycerols and their influence on the anaerobic biodegradability of effluents from poultry slaughterhouse. *Bioresour. Technol., 102*, 7043–7050.

Valladao, A. B. G., Sartore, P. E., Freire, D. M. G., & Cammarota, M. C., (2009). Evaluation of different pre-hydrolysis times and enzyme pool concentrations on the biodegradability of poultry slaughterhouse wastewater with a high fat content. *Water Sci Technol., 60*(1), 243–249.

Venkata, M. S., Lalit, B. V., & Sarma, P. N., (2008). Effect of various pretreatment methods on anaerobic mixed microflora to enhance biohydrogen production utilizing dairy wastewater as substrate. *Bioresour. Technol., 99*, 59–67.

Veradi, A., Lopresto, C. G., Blasi, A., Chakraborty, S., & Calabrò, V., (2020). Bioconversion lignocellulosic biomass to bioethanol and biobutanol. In: Yousuf, A., Pirozzi, D., & Sannino, F., (eds.), *Lignocellulosic Biomass to Liquid Biofuels* (pp. 67–125). Cambridge: Academic Press.

Wang, A., Sun, D., Cao, G., Wang, H., Ren, N., Wu, W. M., & Logan, B. E., (2011). Integrated hydrogen production process from cellulose by combining dark fermentation, microbial fuel cells, and a microbial electrolysis cell. *Bioresour. Technol., 102*, 4137–4143.

Wang, C. C., Chang, C. W., Chu, C. P., Lee, D. J., Chang, B. V., Liao, C. S., & Tay, J. H., (2003). Using filtrate of waste biosolids to effectively produce bio-hydrogen by anaerobic fermentation. *Water Research, 37*, 2789–2793.

Wang, S., Dong, S., & Wang, Y., (2017a). Enhancement of solvent production by overexpressing key genes of the acetone-butanol-ethanol fermentation pathway in *Clostridium saccharoperbutylacetonicum* N1-4. *Bioresour. Technol., 245*, 426–433.

Wang, S., Ma, Z., Zhang, T., Bao, M., & Su, H., (2017b). Optimization and modeling of biohydrogen production by mixed bacterial cultures from raw cassava starch. *Front Chem. Sci. Eng., 11*(1), 100–106.

Wang, S., Zhang, T., & Su, H., (2016a). Enhanced hydrogen production from corn starch wastewater as nitrogen source by mixed cultures. *Renew Energy, 96*, 1135–1141.

Wang, Y., Fan, C., & Hu, H., (2016b). Genetic modification of plant cell walls to enhance biomass yield and biofuel production in bioenergy crops. *Biotechnol. Adv., 34*(5), 997–1017.

Ward, A. J., Lewis, D. M., & Green, F. B., (2014). Anaerobic digestion of algae biomass: A review. *Algal Research, 5*, 204–214.

Wirawan, F., Cheng, C. L., Kao, W. C., Lee, D. J., & Chang, J. S., (2012). Cellulosic ethanol production performance with SSF and SHF processes using immobilized *Zymomonas mobilis*. *Appl. Energy, 100*, 19–26.

Woldesenbet, A. G., Woldeyes, B., & Chandravanshi, B. S., (2016). Bio-ethanol production from wet coffee processing waste in Ethiopia. *Springer Plus, 5*(1), 1903.

World Bioenergy Association (WBA), (2016). *Global Biomass Potential Towards 2035* (Vol. 2020). World Bioenergy Association.

World Bioenergy Association (WBA), (2017). Global Bioenergy Statistics. http://www.indiaenvironmentportal.org.in/content/446437/global-bioenergy-statistics-2017/ (accessed on 31 December 2021).

Worldwide Food Waste, (2020). https://www.unenvironment.org/thinkeatsave/get-informed/worldwide-food-waste (accessed on 12 December 2021).

Wu, J., Dong, L., Liu, B., Xing, D., Zhou, C., Wang, Q., Wu, X., et al., (2020). A novel integrated process to convert cellulose and hemicellulose in rice straw to biobutanol. *Environ. Res., 186*, 1–6.

Xin, F. X., Wu, Y. R., & He, J. Z., (2014). Simultaneous fermentation of glucose and xylose to butanol by *Clostridium* sp. strain BOH_3. *Appl. Environ. Microbiol., 80*(15), 4771–4778.

Xiros, C., Topakas, E., & Christakopoulos, P., (2013). Hydrolysis and fermentation for cellulosic ethanol production. *Wires Energy Environ., 2*, 633–654.

Yan, M., Treu, L., Campanaro, S., Tian, H., Zhu, X., Khoshnevisan, B., Tsapekos, P., et al., (2020). Effect of ammonia on anaerobic digestion of municipal solid waste: Inhibitory performance, bioaugmentation and microbiome functional reconstruction. *Chemical Engineering Journal, 401*, 126159.

Yanase, S., Hasunuma, T., Yamada, R., Tanaka, T., Ogino, C., Fukuda, H., & Kondo, A., (2010). Direct ethanol production from cellulosic materials at high temperature using the thermotolerant yeast *Kluyveromyces marxianus* displaying cellulolytic enzymes. *Appl. Microbiol. Biotechnol., 88*(1), 381–388.

Yáñez-S, M., Rojas, J., Castro, J., Ragauskas, A., Baeza, J., & Freer, J., (2013). Fuel ethanol production from *Eucalyptus globulus* wood by autocatalized organosolv pretreatment ethanol–water and SSF. *J. Chem. Technol Biotechnol., 88*(1), 39–48.

Yang, J., Liu, X., Xu, Q., Wang, W., Wang, D., Yang, G., Fu, Q., et al., (2020). Enhanced dark fermentative hydrogen production from waste activated sludge by combining potassium ferrate with alkaline pretreatment. *Sci. Total Environ., 707*, 1–8.

Yokoi, H., Maki, R., Hirose, J., & Hayashi, S., (2002). Microbial production of hydrogen from starch-manufacturing wastes. *Biomass Bioenerg., 22*(5), 389–395.

Yu, K. L., Chen, W. H., Sheen, H. K., Chang, J. S., Lin, C. S., Ong, H. C., Show, P. L., et al., (2020). Production of microalgal biochar and reducing sugar using wet torrefaction with microwave-assisted heating and acid hydrolysis pretreatment. *Renew Energy, 156*, 349–360.

Zagrodnik, R., & Łaniecki, M., (2017). Hydrogen production from starch by co-culture of *Clostridium acetobutylicum* and *Rhodobacter sphaeroides* in one step hybrid dark-and photofermentation in repeated fed-batch reactor. *Biores. Technol., 224*, 298–306.

Zagrodnik, R., & Łaniecki, M., (2017). The effect of pH on cooperation between dark-and photo-fermentative bacteria in a co-culture process for hydrogen production from starch. *Int. J. Hydrogen Energy, 42*(5), 2878–2888.

Zang, L., Qiao, X., Hu, L., Yang, C., Liu, Q., Wei, C., Qiu, J., et al., (2018). Preparation and evaluation of coal fly ash/chitosan composites as magnetic supports for highly efficient cellulase immobilization and cellulose bioconversion. *Polymers, 10*(5), 523.

Zhang, J., Zhang, Y., Quan, X., Chen, S., & Afzal, S., (2013). Enhanced anaerobic digestion of organic contaminants containing diverse microbial population by combined microbial electrolysis cell (MEC) and anaerobic reactor under Fe (III) reducing conditions. *Bioresour Technol., 136*, 273–280.

Zhang, X., Xu, C., & Wang, H., (2007). Pretreatment of bamboo residues with *Coriolus versicolor* for enzymatic hydrolysis. *J. Biosci. Bioeng., 104*, 149–151.

Zheng, J., Tashiro, Y., Yoshida, T., Gao, M., Wang, Q., & Sonomoto, K., (2013). Continuous butanol fermentation from xylose with high cell density by cell recycling system. *Bioresour Technol., 129*, 360–365.

CHAPTER 2

Biotechnological Modes of Xylooligosaccharides Production from Waste Biomass: An Economic and Ecological Approach

LATIKA BHATIA,[1] KHAGESHWARI KARSH,[1] SUMAN SAHU,[1] DILIP KUMAR SAHU,[1] and SONIA JOHRI[2]

[1]*Department of Microbiology & Bioinformatics, Atal Bihari Vajpayee University, Bilaspur, Chhattisgarh, India, E-mail: latikabhatia1@yahoo.co.uk (Latika Bhatia)*

[2]*School of Life Sciences, ITM University, Gwalior, Madhya Pradesh, India*

ABSTRACT

In the twenty-first century, the application of renewable raw resources and effective commercialization of their chemical and biological possibilities have clustered popularity and significance in scientific investigations and industry. The pollution problems associated with agro-industrial wastes have encouraged the usage and bioconversion of these wastes into high industrial products. There has been a paradigm shift in the global pattern when nourishment is taken into account. People are health-conscious and prefer the food that would support their overall health. Functional foods are among such foods and xylooligosaccharides constitute an important part of these foods. Derived from xylan, xylooligosaccharides offer many benefits and are considered important for many industrial engrossments. Xylan, being the second most available polysaccharide on earth, is present in almost all lignocellulosic biomasses including agricultural waste, forest, industrial

Biotechnology for Waste Biomass Utilization. Prakash Kumar Sarangi & Latika Bhatia (Eds.)

waste, etc. Employing the cost-effective methods and crude enzymes, xylo-oligosaccharides can be effectively accessed from xylan. Moreover, these enzymes can be produced from xylan, providing an opportunity to handle these residues economically and ecologically. Being the source of potential enzymes, microorganisms are explored for novel enzymes and metabolites. Biotechnological advancements are equally important in the exploration of these agro-residues and other lignocellulosic biomasses for the economic and sustainable production of xyloologosaccharides.

2.1 INTRODUCTION

Agricultural waste is the term given to the residues generated from the growing and post-processing of agro-products like fruits, vegetables, meat, poultry, dairy products, and crops. The nature of agricultural actions governs the form of waste generated, which could be either solid, liquid, or slurries. The agricultural sector and agro-based industries generate approximately 620 million tons of waste in India (Singh and Prabha, 2017). About seventy percent of this agro-waste is consumed as fodder and as a fuel for the domestic sector, industrial sector, etc. (MNRE in association with Indian Institute of Science, Bangalore, 2014). The remaining 30% (about 120–150 million tons) of agro-residues can be potentially accessed to generate numerous value-added products like biofuels, animal feeds, chemicals, enzymes, etc. as this percentage of agro-waste is an excess per year. There are enormous possibilities for the value addition and exploitation of this agro-waste or residues for food applications for instance manufacture of xylooligosaccharides (XOS), xylanase, and xylose (Thakuria and Sheth, 2018).

Groundnut cake, rice bran, rice straw, wheat bran, sugarcane bagasse, cotton leaf scraps, fruits and vegetable wastes, etc. are generated in substantial amounts (about 625 million tons per year) and are considered as an accessible agro-waste in India (Rosmine et al., 2019). Among these, sugarcane bagasse (SCB) is considered an excellent source of cheap carbon for the commercial production of vital enzymes, xylooligosaccharides, food supplements, and others (Bhatia et al., 2019). Xylooligossacharides production is feasible from SCB, as SCB is a non-wood fiber plant material that contains a huge amount of xylan (28–30%). Xylan acts as a starting material for xylooligosaccharide production. It is the need of the hour to manage this agro-waste efficiently and economically. This will not only curtail their detrimental effect on the environment but will also ensure their effective usage for the commercial creation of value-added goods of industrial importance. Probable enzymes can be produced out of these agro-wastes, as they are an important source of carbon in fermentation. This is a significant cost-effective approach for managing these residues. An enormous amount of cellulose and hemicellulose are found in agricultural residues that can be exploited as active and cheap sources for xylanase production also (Hauli et al., 2013).

The biotechnology industry is not only dynamic but has proven itself to be capable of fulfilling the various global needs of the 21st century. Microbial biotechnology firms constitute an important segment of the biotechnology industry. Pharmaceuticals, agriculture, chemicals, computer, medical device, environmental industries, etc. are the end-users of the biotechnology industry. The employment of technology has a vital role in manifesting the needs of these industries (Cantley, 2004).

Human healthcare has witnessed the profound and significant impact of microbial biotechnology. Avenues of microbial biotechnology have expanded themselves to various applications of animal health, industries, and the environment. Specialized industry subsectors have evolved as an outcome of the amalgamation of microbial biotechnology, high-tech, and traditional industries. These specialized industry subsectors rely upon novel fusion technologies (Hall and Bagchi-Sen, 2007). Microbial biotechnology plays a crucial role both for developed and developing nations. In the former, it forms the basis for making policies, whereas, in the latter, it supports them to triumph the game in the revolutionary of technology (Hamedi and Azimi, 2013).

2.2 CONCEPT OF BIOREFINERY AND XYLOOLIGOSACCHARIDES

Biomass biorefinery is an optimistic and holistic approach to unlock the immense potential of biomass, as the biomass is well-thought-out to be a basis of not only various value-added products, such as chemicals, pharmaceuticals, ingredients of food and feed, etc. but has also been researched as a vital source of various forms of fuel. Sustainable development of human civilization becomes feasible with biomass biorefinery (Dotsenko et al., 2017).

Prevailing time demands the development of sustainable methodologies/schemes towards the alteration of LB into biofuels, bioenergy, and a wide spectrum of value-added products in the form of bioplastics, biochemicals, advanced biofuels, etc. This is one of the major bottlenecks for researchers around the globe who are exploring solutions to combat climate alteration attenuation. The formation of these value-added products from LB engaging the platform of lignocellulose biorefinery not only curtails the pollution but supports sustainability too. Sustainability is an important issue of the society, environment, and economics that directly affects people, the planet, and profit. Lignocellulose biorefinery is said to be equivalent to petroleum refinery. The former is capable of sufficing the future needs of renewable fuels, chemicals, and materials. Sustainable development ensures the manufacture of these value-added products along with societal development (Bhatia et al., 2019).

Industrial biotech has an unimpeachable part in the complete accomplishment of biorefineries. According to National Renewable Energy Laboratory (NREL), Colorado, USA, and International Energy Agency (IEA), Bioenergy- France, biorefinery refers to a sustainable procedure that converts biomass into a wide spectrum of merchantable energy and products. The word biorefinery incorporates a system of amenities that assimilate various pieces of knowledge (procedures and equipment) either to disperse the components of lignocellulosic biomass or its usage as integral for bio-constructed products (chemicals, materials, energy, fiber) (Chandel et al., 2018a). Recent years have witnessed immense growth of biorefineries, that holds the potential to valorize each constituent of lignocellulosic biomass (LB) for socio-economic and environmental development in a sustainable way. Biorefining can potentially replace conservative petrorefinery, being a platform for renewable generation of various majority and specific biochemicals, plus biofuels (FitzPatrick et al., 2010). Renewable generation of any product needs a renewable source as a starting material, and lignocellulosic biomass qualifies this criterion. Some of the important bases of this lignocellulose include agro-residues, dry energy grasses (annual and perennial), sugarcane bagasse (SB), and straw and municipal solid waste (MSW) (Silveira et al., 2018). These biomasses are abundantly accessible renewable feedstock on this planet. LB not only supports the sustainable production of biofuels or biochemicals but can also be considered as a crucial foundation of oligosaccharides production (Bhatia et al., 2019).

Xylooligosaccharides (XOS) is a chain of 2 to 10 xylose and is an oligomer. They cannot be digested when employed as food ingredients. Hydrolysis of xylan leads to the formation of xylooligosaccharides and β-1,4-glycosidic linkage joins the xylose units to constitute linear chains of xylooligosaccharides. Overall, the associated least of 2–20 molecules of xylose is considered as XOS (Fang et al., 2018). XOS is considered an excellent prebiotic. They can potentially promote the development of probiotic organisms. XOS supports gastrointestinal (GI) health and biological accessibility of calcium, thereby exhibiting various physiological effects. It has anti-cancerous properties

(Gupta et al., 2017). Commercial XOS production from agro-residues provides countless possibilities to the industries like nutraceutical and pharmaceutical, as the base material is cost-effective and copiously accessible. According to the Stowell classification, the xylooligosaccharides (XOS) are characterized under evolving prebiotics (Stowell, 2006). The worldwide market for functional foods was US$300 billion in the year 2017 and is presumed to hike to US$440 billion in 2022 (www.statista.com, accessed in November 2018). XOS plays a prominent character in the expansion of the inclusive oligosaccharides market. The purity of XOS plays a critical role in deciding its market price that varies from US$22 to 50/kg (Taniguchi, 2004).

XOS derived from agro-residues finds various applications in many sectors like food, cosmetic, pharmaceutical, and nutraceutical. For instance, xylo-oligomers (obtained from the hemicellulosic substrate) having a degree of polymerization (DP) from 2 to 12 are considered the outstanding basis of a soluble dietary fiber exhibiting organoleptic properties with no adverse consequence on human health (Bhatia et al., 2019).

This chapter critically reviews the health effects of xylooligosaccharides, its feedstocks, methods for the xylooligosaccharides production from lignocellulosic biomass, its commercial production, and the existing marketplace.

2.3 HEALTH BENEFITS OF XYLOOLIGOSACCHARIDES

In the current scenario, where the health of the majority of the population around the globe is compromised due to many reasons, the attention has now shifted to consuming functional foods. These foods are categorized as special food as they promote health and prevent diseases. Prebiotics are one of the chief categories of functional foods. Few factors govern when an ingredient is to be recognized as prebiotic. Prebiotic can tolerate gastric acidity. It gets fermented by intestinal microbiota. Prebiotics has the property to discriminately excite the development and/or physiological action of those intestinal microbiotas that support health and well-being (Manap, 2012). Though the microbiota of individuals differs extensively, there are a few organisms like bifidobacteria and lactobacilli that are common in all individuals. Their prominence or deficiency are deciding factors of health, as their dominance specifies good health. The prebiotic activity supports the growth of these vital organisms, and hence, is considered a basic principle (Saman et al., 2016; Bode, 2009).

Oligosaccharides are the small chain of monomers that are midways of monosaccharides and polysaccharides. They constitute a vital group of carbohydrates present in all living organisms. A few examples of oligosaccharides are fructooligosaccharide, galactooligosaccharides, lactosucrose, isomalto-oligosaccharides, glucooligosaccharides, xylooligosaccharide, mannan oligosaccharides, etc. These are considered functional foods or functional oligosaccharides (Cui et al., 2013). Chief sources of these functional oligosaccharides are milk, honey, sugarcane juice, soybean, lentils, mustard, fruits, vegetables, and bamboo shoots, however, their precise concentration is yet to be explored. Pseudomonas aurantiaca is a source of lactosucrose. Oligosaccharides play important roles as bulking agents, sweeteners, and humectants that are employed in food products during their preparation (Kothari et al., 2014). These oligosaccharides are excellent prebiotics. They are able to resist digestive enzyme and supports the probiotic population like bifidobacterium in mice and the human gut. These findings exhibit that these oligosaccharides are potential prebiotics and can be utilized for the preparation of prebiotic or symbiotic food products. Xylan is the second most profound biopolymer in the plant kingdom. Xylan hydrolysis leads to the formation of XOS. (Jnawali et al., 2017).

XOS is a popular prebiobiotic that confers many health benefits to the host. Growth of Bifidobacterium spp. (beneficial gut microbiota) is extensively supported in presence of XOS along with the stimulation of its activity (Finegold et al., 2014). Besides this benefit, there are many plus points in the credit of XOS, as it is anti-cancerous, anti-microbial, antioxidant, anti-allergic, anti-infection, anti-inflammatory, etc. It lowers cholesterol and is immunomodulatory too (Aachary and Prapulla, 2011). These properties gave acceptance to XOS as a nutraceutical and feed additive. XOS is able to withstand elevated temperatures and low pH and is considered as an appropriate source of carbon of bifidobacteria, ultimately improving digestion and ingestion of nutrients. XOS have also been reported for their potential to synthesize active compounds that can efficiently combat viral and cancer-leading compounds. The structurally modified of when oligosaccharides are structurally modified, they exhibit various properties like antiviral activities, antitumor, anticoagulant, and tissue/cartilage regenerating properties (Nabarlatz et al., 2007).

Non-digestible oligosaccharides (NDOs) are oligosaccharides that have a low degree of polymerization (DP 2–20 monomers). They are potentially beneficial as dietary fibers. These NDOs also possess features like sweetening ability, water binding capacity, and fat replacement value. They confront to digestion in the upper sections of the GI tract and deserve appreciation. The

NDOs are reported to be employed in processed food and are considered capable functional components in nutraceutical products (Rossi et al., 2011). These nutraceutical products are getting attention not only because of their wide utilities, but the source of their production has shifted to agriculture or fruit wastes. This makes the production of these nutraceutical products economic and environment friendly too. It becomes feasible for agro-based and food manufacturing to generate value-added products out of agriculture or fruit residues (Gupta et al., 2017).

2.4 PHARMACEUTICAL APPLICATIONS OF XYLOOLIGOSACCHARIDES

XOS are considered to be well known for their antiviral and antitumor properties, which makes them suitable for pharmaceutical applications. In-vitro studies have revealed that survival of leukemia cell lines resultant from acute lymphoblastic leukemia is adversely affected in presence of fractions of xylose, XOS, and water-soluble lignin. This is due to their cytotoxic effects on these cell lines. Research on xylooligosaccharides has revealed that these oligosaccharides exhibit a variety of activities such as immunomodulatory, anti-cancerous, anti-microbial, growth regulatory, etc. XOS have many outstanding biological activities also. XOS have antioxidant, anti-allergic, anti-inflammatory, anti-hyperlipidemic activity and cosmetics, and a variety of other properties. XOS are used for curing GI disorders, in which they are employed in the formation of micro or nanoparticles and hydrogels for drug delivery and treatments. Few pieces of research reveal that XOS stimulate mineral absorption in the intestines, as they possess non-cariogenic properties and save insulin secretion. Some researchers have also explored that these XOS potentially stimulate bacterial growth and fermentation, and being mildly laxative, these oligosaccharides can alter the bowel norm. Bifidobacterium population is extremely benefited by XOS, as they flourish in its presence. This has formed the grounds of formulation of a nutritional product, in which oil is blended with eicosapentaenoic acid or docosahexaenoic acid and XOS (a source of indigestible carbohydrate). Microbes present in the human colon metabolizes this nutritional product to fatty acids of short-chain. People suffering from ulcerative colitis are extremely benefited by these nutritional products. Formation of short-chain fatty acids (SCFA) like acetate, propionate and carbon di-oxide, hydrogen, butyrate, and lactate takes place due to the fermentation of xylooligosaccharides in the colon along with the profuse growth of beneficial microbes. Intake of XOS reduces constipation in pregnant women with no antagonistic effects. When infants are fed with nutritional formula comprising XOS, their gut barrier maturation gets improved (Taeko et al., 2016). The formation of secondary bile acids and physiologically active fatty acids get suppressed in course of digestion when dietary supplements contain the XOS. Sulfated xylogalactans derived from Nothogenia fastigiate (red seaweed) and a xylomannan derived from algae are antiviral against herpes simplex virus type 1. Antiviral activity is also exhibited by a polysaccharide mixture obtained from a marine alga. This polysaccharide constitutes xylose, glucose, glucuronic acid, and mannose. This polysaccharide, when combined with a protein functions as a biological response modifier, significantly influencing the immune system (Kazumitsu et al., 2016).

2.5 NOVEL APPLICATION OF XYLOOLIGOSACCHARIDES

Xylans and high molecular weight XOS are the sources of ethers and esters in addition to their usage in food and pharmaceutical. These ethers and esters are employed as thermoplastic compounds for biodegradable plastics, water-soluble films, coatings, capsules, and tablets, and the preparation of chitosan–xylan hydrogels as well (Glasser et al., 2016). XOS finds its utility in agricultural applications also. XOS are excellent growth promoters, acting as a ripening agent, and thereby, enhancing the yield. Food and Agriculture Organization (FAO) has stated the employment of XOS as a feed to domestic animals and fish (Yu et al., 2015). XOS can be efficiently employed as a non-natural sweetener in food applications and as an alternative for antibiotics in the production of animals (Samanta et al., 2015).

When used as an antioxidant and gelling agent in food products, XOS supports in treating colon cancer, arteriosclerosis, and reducing overall cholesterol and low-density lipids in patients having type 2 diabetes mellitus. Other than these multidimensional applications, XOS are advantageous for diabetes too (Belorkar and Gupta, 2016).

When compared with the monosaccharides, the assimilation of oligosaccharides by microorganisms is challenging as an organism needs extra hydrolytic enzymes and transportation systems. Incomplete assimilation of oligosaccharides in course of microbial transformation may lead to a reduction of product yield and enhancement in water pollution. Consumption of oligosaccharides becomes feasible, if the microbe secretes or displays glycosidase on its surface, or has a portal of the devoted transport system to assimilate oligosaccharides for intracellular operation. Microorganisms that have the potential to directly assimilate biomass-derived oligosaccharides for producing biofuels or biochemicals would be much more beneficial (Dotsenko et al., 2017).

2.6 LIGNOCELLULOSIC BIOMASSES, THEIR AVAILABILITY, AND COMPOSITION

Lignocellulosic biomass (LB) is the most copious organic matter found on this planet. It exists in various forms that are valorized into a variety of products like biomaterial, fuel, chemicals, food/feed, etc. Largely, the LB comprises agro-residues (sugarcane bagasse, sugarcane straw, corn stover, rice straw, wheat straw, among others), forestry residues, fruit and vegetable residues, etc. Cellulose, hemicellulose, and lignin are the important constituents of LB. The conformation of these three principal constituents rests on the biomass type, cultivation, and climatic situations. Hemicellulose shares nearly 20%–30% of lignocellulosic plant biomass. Hemicellulose has the privilege of being the second most copious natural polysaccharide. It is a heteropolysaccharide that chiefly consists of xylose as key support associated with glucose, galactose, arabinose, and sugar acids (Chandel et al., 2018a). Table 2.1 shows the cell wall arrangement of numerous lignocellulosic biomasses explored for oligosaccharide production.

TABLE 2.1 Cell Wall Arrangement of Numerous Lignocellulosic Materials (Bhatia et al., 2019)

Lignocellulosic substrate	Lignin (%)	Hemicellulose (%)	Cellulose (%)
Wheat	8.25	28.24	43.68
Corn stover	13.64	26.34	43.31
Switch grass	17.35	26.10	33.48
Rice Straw	25	25	38
Sugarcane bagasse	21.10	27	45.5
Sugarcane leaves	18	25	45
Bamboo	28.1	24.6	46.7
Soya stalk	25.4	17.3	37.6
Sorghum husk	14.6	41.2	40.2
Sugarcane straw	25.8	30.8	40.8
Wheat straw	20.0	33.0	33.0
Rice Husk	20.0	25.0	25.0
Coconut Husk	28.48	16.15	39.31
Peels of *Litchi chinensis*	25.0	28.0	40.0

In annual plants, xylan shares thirty percent of cell wall material, in hardwood and softwood this accounts for upto 15–30% and 7–10%, respectively. Xylan acts as the main support in all these LBs (Quinones et al., 2015). Important sugars found in hemicellulose are acetylated or methylated D-xylose, L-arabinose, D-mannose, D-glucose, D-galactose, L-rhamnose, L-fructose, and D-glucuronic acid (Figure 2.1). The most viable method to utilize hemicellulose is through the elementary biorefinery principles, assimilating the exploitation of every constituent of biomass, producing zero or near-zero left-over. Production of xylooligosaccharides from xylans of hemicellulose is one of the best cost-effective and eco-friendly approaches of utilizing this component of LB (Chandel et al., 2018b). Figure 2.2 depicts the various forms of agro-wastes employed for xylan production. Various forms of XOs can be generated from xylan, such as xylobiose (X2), xylotriose (X3), xylotetraose (X4), and xylopentaose (X5). Corncobs, cotton stalks, tobacco stalks, sunflower stalks, wheat straw, sugarcane bagasse, oil palm empty fruit bunch (OPEFB), oil palm mesocarp fiber (OPMF), and oil palm frond fiber (OPFF) have been reported as an important starting material for XOS production (Ahmad et al., 2018).

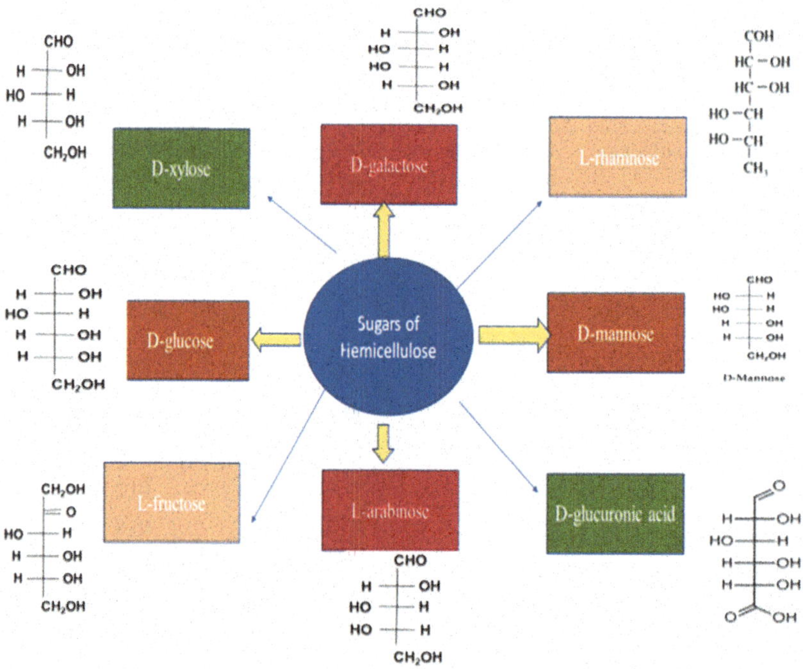

FIGURE 2.1 Sugars of hemicellulose.

FIGURE 2.2 Substrates employed for xylan production 1–Peels of *Ananas cosmosus*, 2– Coconut husk, 3 – Peels of *Citrus sinesis* var mausambi, 4 – Peels of *Litchi chinensis*.

Aspergillus, Trichoderma, Penicillium, Bacillus, and *Streptomyces* are also important producers of XOS. XOS is prevalent in plant sources like Bengalgram husk, wheat bran and straw, spentwood, barley hulls, brewery spent grains, almond shells, bamboo, and corn cob (Belorkar and Gupta, 2016). However, valorization of agro-waste into a beneficial product curtails the problems associated with the management, treatment, and disposal of agro residues (Khat-udomkiri et al., 2018). Currently, various lignocellulosic rich biomasses/agro residues have been effectively explored by many researchers for XOS extraction. Chief among them are corn cob, corn stalks, sugarcane bagasse, cotton stalks, wheat straw, sunflower stalks, tobacco stalks, pigeon pea stalks, green coconut husks, corn fiber, barley husk, and rice hulls, etc. Various methods like chemical, enzymatic, autohydrolysis, and a combination of chemical and enzymatic approaches are employed to achieve this goal (Chakdar et al., 2016).

Husk constitutes the major portion of the coconut fruit, which accounts for 50% of the fruit. Important constituents of coconut husk are cellulose (39.31%), hemicellulose (16.15%), and lignin (28.48%) along with some water-soluble substances (26%). A substantial quantity of hemicellulose is found in a coconut husk, which makes it a potential source of xylan (Figure 2.3) and therefore, xylooligosaccharides (Finegold et al., 2014). Researchers have reported that by applying physicochemical treatment (20% of NaOH along with steam treatment for 60 min) on coconut husk, 93% xylan could be extracted, that upon enzymatic hydrolysis yields xylooligosaccharide (XOS) (Jnawali et al., 2017).

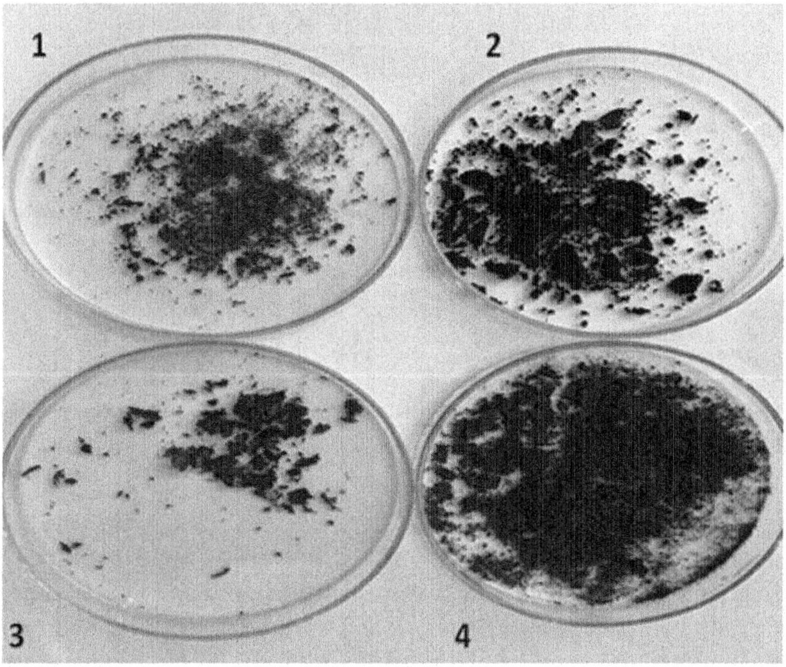

FIGURE 2.3 Xylan produced from 1–Peels of *Ananas cosmosus*, 2– Coconut husk, 3–Peels of *Citrus sinesis* var mausambi, 4 – Peels of *Litchi chinensis*.

Waste generated after processing orange fruit is considered a valuable source of xylooligosaccharides because other than soluble sugars and starch, fibers of this agro-waste are rich in cellulose, hemicellulose, lignin, and pectins. Fats, proteins, and ashes are also present in this waste. The overall cellulose share in the wastes of orange fruit varies from 12.7 to 13.6% and hemicellulose varies from 5.3 to 6.1% (Rivas et al., 2008). Employment of this agro-residue for XOS production is a cost-effective approach. Moreover, the fat concentration of dried orange fruit wastes is comparatively lower, which further enhances the prebiotic components. Hemicellulose and pectin prevalent in orange peels could be commercially employed to yield prebiotics like XOS, fructooligosacchrides (FOS), and pectin (Gupta et al., 2017).

Rice husk (RH) is also considered an excellent and sustainable repertoire of xylooligosaccharide (XOS). It is generated during the rice hulling process and is considered to be a prominent agro-waste. Firing is the usual practice of clearing this waste, and hence, one of the major reasons for air pollution. RH is made up of 25% cellulose, 25% hemicellulose, and 20% lignin (Chandel et al., 2018b). Xylose (17.35%) and glucose (4.42%) are the prominent reducing sugars obtained post-acid hydrolysis of rice husk. The extreme XOS production of 17.35 ± 0.31 mg XOS per mL xylan was obtained. The researchers revealed that the xylan can be efficiently taken out from rice husk employing an operative base couple with the steam application and the enzymatic hydrolysis that aids to maximize the XOS yield, which can be further employed in functional foods and dietary supplements (Khat-udomkiri et al., 2018). Likewise, sugarcane bagasse (SCB) and straw are rich sources of hemicellulose (24%), and often researched biomass in labs, pilot plants, demos, and large-scale processes.

2.7 PROCESS OF XYLOOLIGOSACCHARIDE PRODUCTION

As mentioned earlier, lignocellulosic biomass is the source of hemicellulose. Hemicellulose is the source of xylans, and xylans are the source of XOS. Recovery of XOS from LB can be mediated through enzymatic, chemo-enzymatic, and fractional xylan hydrolysis from innumerable sources such as barley hulls, rice hulls, corn cobs, peanut pods, sugarcane bagasse, wheat straw, cotton stalks, orange peels, mango peels, etc. (Gupta et al., 2016). This technology is equally relevant to obtain XOS from agro-residues of fruit and vegetable, nuts and oilseeds industries, etc., thereby supporting the country's economic progress by making this product accessible not only in the national market but for export at the global level. Being a biopolymer, hemicellulose is priorly transformed into xylans, which are easily accessed by enzymes to produce XOS. Thermo-chemical methods, or enzymatic attack, harness sugar oligomers from hemicellulose (Gupta et al., 2015).

Pretreatment is an unavoidable step for hemicellulose bioconversion into sugar oligomers and other products. Pretreatment manifests the depolymerization of hemicellulose (Waghmare et al., 2018). Steam explosion allows the depolymerization of hemicellulose into its components as such, or in the existence of an acid catalyst and dilutes acid hydrolysis. Commercially, XOS is produced from xylan-rich lignocellulose agro-waste, in which the raw material undergoes physiochemical pretreatment tailed by enzymatic hydrolysis. Hydrothermolysis is one of the most employed physiochemical pretreatment procedures that operate under mild conditions. The product of this process is enriched with xylan having XOS of DP (Degree of polymerization) characteristically 20. Chemical pretreatment for xylan extraction followed by enzymatic hydrolysis is the feasible approach to derive XOS from LB (Qing et al., 2013).

Autohydrolysis is another approach to valorizing the agro-waste for commercial production of XOS. It is a chemical process, mediated in presence of an acid or an alkali. In this process, LB is heated with water in precise equipment under a controlled state. The process becomes cost-effective when autohydrolysis occurs in the presence of an acid (Qing et al., 2013). Lignin, monosaccharides, and furfural (an aldehyde of furan) are undesirable products that contaminate the XOS in both the autohydrolysis and acid-mediated hydrolysis processes. These products are responsible for many health issues like respiratory irritation, lung congestion, hyperplasia, kidney, and olfactory epithelial damage, edema, and inflammation. Purification of XOS is a prerequisite, if XOS is used in food formulations (Carvalho et al., 2013). Refinement of these extracts is a tough and cost-intensive affair and may involve multistage processing and fractionation. Nevertheless, limited techniques such as vacuum evaporation, solvent extraction, solvent precipitation, chromatographic separation, ion exchange, and membrane separation have been exploited by various researchers and have been reviewed (Moure et al., 2006).

Toxicity and purification issues can be resolved and overcome, by adopting enzymatic procedures. These procedures are mostly preferred because it neither produces lethal compounds nor involves distinct equipment for XOS generation (Akpinar et al., 2007); and have the possibility to be exploited at a pilot scale. There are some prominent factors that govern the XOS yield through enzymatic extraction methods. They are viz. source of xylan, enzyme activity as well as incubation conditions such enzyme concentration, pH, reaction time, and temperature. Enzymes with endoxylanase activity are preferred over the use of exoxylanase activity to curtail the production of monosaccharide xylose and later intensify the XOS production (Samanta et al., 2014).

Enzymatic hydrolysis is a cost-effective approach to obtain XOS. Moreover, it is mediated in a short span. It is an often preferred method, executed after chemical pretreatment for xylan extraction. The alkaline extraction is a significant exercise to isolate xylan. The separation of xylan from other lignocellulosic biomass augments the efficacy of the enzyme during the production of XOS (Carvalho et al., 2013; Samanta et al., 2015). Moreover, alkaline pretreatment can advance the digestibility of substrate, the chief anticipated characteristic in XOS production via enzymatic hydrolysis (McIntosh and Vancov, 2011). It has been reported that xylan can be recovered from dried orange fruit wastes powder by sodium hydroxide pretreatment. This method supports the separation of hemicellulose content from other cellulosic waste stuffs like lignin, pectins, fat, and protein materials (Samanta et al., 2012a, b).

2.8 XYLANASE, ITS IMPORTANCE, AND PRODUCTION

Adoption of enzymatic degradation of xylan to produce XOS is economic and environment friendly when compared with physio-chemical processing methods (Dotsenko et al., 2017). Enzymes selectively attack the substrate, producing the desired product only. Xylanase is a class of enzymes that manifest the degradation of the linear polysaccharide β-1, 4-Xylan into xylose. Xylanase is the extracellular product of bacteria, yeast, and filamentous fungi. Trichoderma, Aspergillus, Fusarium, and Pichia are a few fungal genera that are considered to be the potential producers of xylanases. Xylanases of microbial origin have significant utilization in xylan breakdown. Substrate xylan is a biopolymer made up of D-xylose monomers associated through β1–4 glycozyl bond, which prevail profusely in lignocellulosic biomass. Xylanases can be classified as endo and exoxylanases. Exoxylanases (β D-xylopyranosidase) are sometimes known as extracellular xylanase. There are three ways to classify Xylanases: based on the molecular weight and isoelectric point (pI), the structure of the crystal and kinetic properties, or the specificity of the substrate and product profile (Mane et al., 2018). Xylanases are inducible enzymes. In presence of xylan, the microorganism produces this enzyme in extracellular surroundings. Xylan is broken down and gets metabolized as a source of carbon. Xylan cannot enter the cell. Oligosaccharide is produced in this process (Mane et al., 2018).

Xylans are abundant as well as structurally heterogeneous, which demands the diversified enzymes for their degradation. Distinctive xylan–degrading enzymes are endo-β-xylanases (EC 3.2.1.8), which act on the key chain of xylans and β-xylosidases (EC 3.2.1.37), which breakdown xylooligo-saccharides into D-xylose (Karunakaran et al., 2014). Both these enzymes are formed by various bacteria and fungi that mediated the complete hydrolysis of xylans during biomass conversion. Xylanase supports potential applications in biotechnology such as the production of hydrolysates from agro-industrial residues, nutrient enhancement of lignocellulosic feedstuff, clarification of juices and wines, and biobleaching of kraft pulp in the paper industry (Karunakaran et al., 2014).

Xylanase (E.C. 3.2.1.8) allows the formation of XOS with the desired degree of polymerization (mostly 2–4) by hydrolyzing XOS of high DP. Endohydrolysis of xylosidic bonds in xylan, takes place in presence of xylanases, generating XOS of varied DPs. Crude xylanase obtained from *Pichia stipitis* could potentially produce XOS with DP of 2 to 4 as the key constituents in xylan hydrolysates. Therefore, high activity xylanase formed by *P. stipitis* can intensify efficiency of hydrolysis and forms XOS with desired DP distributions.

Hydrolysis of xylan is considered an important step in carbon recycling in nature and has been an area of intensive research for exploring this material for the renewable formation of value-added products. Xylan breakdown is a complicated process accomplished by the involvement of a variety of enzymes. Xylanase (1,4--D-xylan xylanohydrolase; EC 3.2.1.8), which cleaves the internal bonds on the -1,4-xylose backbone, plays a crucial role. Familiar xylanases are clustered into glycoside hydrolase (GH) families 10 and 11 (CAZy [Carbohydrate-Active enZYmes] database), but recently limited xylanases have been recognized to glycoside hydrolase families 5, 7, 8, and 43. The enzymes XynA obtained from *Erwinia chrysanthemi* and XynC obtained from Bacillus subtilis cause the hydrolysis of glucuronoxylan to branched xylooligosaccharides. These two xylanases belong to family GH5 and have been biochemically characterized in length. XynA obtained from *Erwinia chrysanthemi* manifests the exclusive cleavage of methyl-glucuronic acid-branched oligosaccharides, whereas the cleavage of straight-chain xylooligosaccharides or arabinoxylans is unfeasible by this enzyme. This type of xylanases must exhibit a prominent part in supplementing the act of GH10 and GH11 enzymes during depolymerization of glucuronoxylans in lignocellulosic fibers (Gallardo et al., 2010).

Maximal hydrolysis of variable feedstock can be ensured by the involvement of a complete xylanolytic system. Xylanases, endo-xylosidases, and many debranching enzymes are included in such an enzyme system. There are a few research works that have explored the xylooligosaccharides production from the agro residues by integrating xylanase treatment (Kanimozhi and Nagalakshmi, 2014). Xylan biodegradation employing xylanases [Endo-1,4-b-D- xylanohydrolase, EC 3.2.1.8] produces XOS, that can be employed as constituents of functional food, cosmetics, biofuel, pharmaceuticals, or agricultural products without causing environmental pollution. The last few years have witnessed the potent applications of, xylanases in various industrial zones such as pulp bleaching, baking and brewing, animal feeding, waste-treating, and bioenergy conversion (Hauli et al., 2013). Numerous bacterial and fungal species are described to be potential producers of xylanases when cultivated in the existence of xylan.

There is a wide variety of microorganisms that can thrive well at an elevated temperature and are physiologically active. The enzymes produced from these organisms are thermostable and can operate specifically at elevated temperatures. Chief among these are enzymes like chitinases, amylase, protease, lipases, xylanases, etc. that are industrially significant. These thermostable enzymes maintain their structural integrity above 55°C. These enzymes are superior to their mesophilic counterparts in many aspects like (1) specific activity and stability of these enzymes is higher that allows extended times of hydrolysis and saccharification at their low concentration; (2) cost involved in the cooling process is curtailed, making the process cost-effective; (4) cost involved in mass transport are reduced due to decreased fluid viscosity; (5) biorefinery process can be flexibly configured; (6) risks associated with microbial contamination are significantly condensed; and finally (7) it is easy to store these enzymes at room temperature without activity inactivation. These plus points are significant because almost one-half of the anticipated process costs in conversions of biomass is projected to be related to the production of enzymes, and all these merits credited to thermostable enzymes will result in an upgradation to the total economy of the procedure (Yeoman et al., 2010).

2.9 PRODUCTION OF XYLANASE FROM AGROWASTE

As described earlier, the major components of LB are cellulose, hemicellulose, and lignin. Hemicellulose is the second most plentiful polysaccharide of earth and xylan is its chief constituent, hence xylan can be equally regarded as an important carbon source that can be utilized by diversified microorganisms to suffice their physiological metabolism. One of the most important physiological metabolites of many microorganisms is an enzyme known as xylanase. As discussed earlier, xylanase is crucial for the formation of XOS from xylan. The use of pure xylan for xylanase production is a cost-intensive affair. Hence, the production of xylanase from an agro-waste is a feasible option to further curtail the cost involved in XOS production, making XOS production further sustainable, cost-effective and ecofriendly. Large-scale production of xylanase is a must for it to be utilized for commercial applications. To meet this prerequisite, cheap lignocellulosic residuals are the best option. Sugarcane bagasse, wheat bran, rice bran, and corn cob are some of the natural xylan sources that are profusely accessible in numerous countries are the possible starting substances employed as the source of carbon for the production of xylanase (Figure 2.4). Reusing and recycling these organic agro-residues, would convert them into resources, and can be considered as a better substitute auxiliary as the source of carbon for industrial production of xylanase. The utilization of agro-residues supports in curtailing the production cost in addition to providing massive amounts of growth-promoting nutrients. Moreover, they are the inducer for the production of xylanase (Ho, 2015).

FIGURE 2.4 Agro-waste employed for Xylanase production: Wheat bran and rice bran.

Bacteria, fungi, and yeast have the potential to produce xylanase from xylan either by submerged fermentation (SmF) or solid state fermentation (SsF). There are few reports available on the optimization of xylanase production by *B. subtilis*, employing agro-residues and other carbon sources. It is reported that the production of xylanase by Bacillus subtilis was improved by substituting carbon sources with sustainable and economic LB or agro-residues (Nwodo-Chinedu et al., 2005). Bacillus sp. strain BP-7 is a potential xylan-degrading strain obtained from agricultural soils, that exhibits various enzymatic systems for degrading xylan, including GH11 xylanase (Gallardo et al., 2010). Production of xylanase via solid state fermentation (SSF) is a better method as compared to its production via submerged fermentation. Capital and operating costs are low in solid state fermentation (SSF). Moreover, SSF is a less energy-intensive process that needs less water and has high enzyme yield and activity. Though the incubation time is more in SSF, it is a cost-effective approach to producing xylanase compared to submerged fermentation. Filamentous fungi are the known potential producers of a broad spectrum of plant cell wall degrading enzymes such as xylanases. They liberate substantial levels of the enzymes into the culture medium. Potential producers of xylanases under SSF are *Paecilomyces themophila*, *Thermomyces lanuginosus*, *Trichoderma orientalis*, *Bacillus circulans*, and *Steptomyces remeus*. *Pitchia stipitis* also produces high activity xylanase (Gong et al., 2014). Figure 2.5 depicts the cultural characteristics of organisms employed for xylanase production. Purification of xylanase is crucial for the assessment of its catalytic activities and its sensitivity to regulatory molecules that raise or lower activity. Crude xylanase accessed from the solid-state culture can be purified and characterized biochemically (Figure 2.6). As the prebiotic properties of XOS depend upon XOS degree of polymerization (DP), the quantification of DPs of XOS broken by crude and unadulterated xylanases is important (Ding et al., 2018).

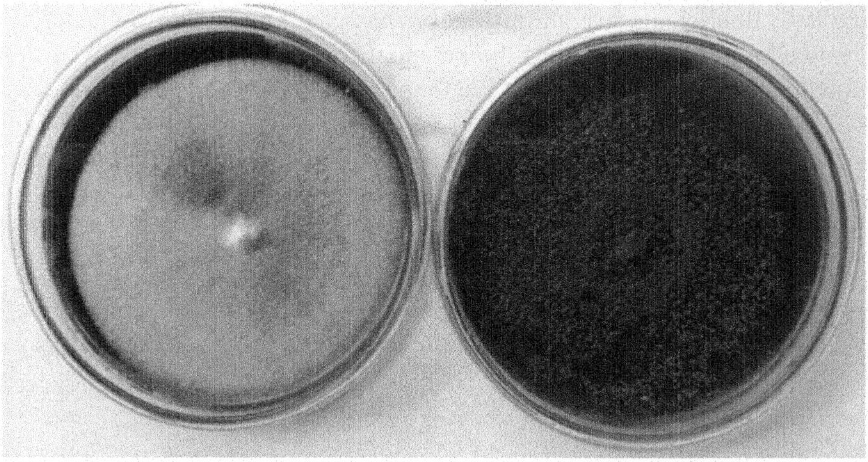

FIGURE 2.5 Cultural characteristics of organisms employed for xylanase production.

Various bacteria, archaea, and fungi have gathered substantial consideration as potential sources for thermostable enzymes. The extensiveness of thermophilic microorganisms with enzymatic characteristics amenable to lignocellulose destruction has also been reviewed (Yeoman et al., 2010).

FIGURE 2.6 Crude xylanase.

2.10 CONCLUSION

The availability of lignocellulosic biomasses is abundant and their disposal is an issue for the global environment. Being the renewable source of a variety of sugars there has been excessive attention of many industries to biovalorize this waste economically and ecologically. Lignocellulose biorefineries are gradually flourishing, thriving on these biomasses as a renewable raw material. Hemicellulose biorefinery has the potential for complete expansion of the global bioeconomy as it is the source for producing xylooloigosaccharides as well as xylanases, thereby supporting the holistic and sustainable progress of society. The employment of cost-effective agricultural and food processing by-products is extremely preferred to advance the commercial viability of bioprocess technology.

KEYWORDS

- **agro-residues**
- **biorefinery**
- **oligosaccharides**
- **xylanases**
- **xylooligosaccharides**

REFERENCES

Aachary, A. A., & Prapulla, S. G., (2011). Xylooligosaccharides (XOS) as an emerging prebiotic: Microbial synthesis, utilization, structural characterization, bioactive properties, and applications. *Compr. Rev. Food Sci. Food Saf., 10*(1), 2–16.

Ahmad, N., Zakaria, M. R., Zulkhairi, M., Yusoff, M., Fujimoto, S., Inoue, H., Ariffin, H., & Shirai, Y., (2018). Subcritical water-carbon dioxide pretreatment of oil palm mesocarp fiber for xylooligosaccharide and glucose production. *Molecules, 23*, 1310. doi: 10.3390/molecules23061310.

Akpinar, O., Ak, O., Kavas, A., Bakir, U., & Yilmaz, L., (2007). Enzymatic production of xylooligosaccharides from cotton stalks. *J. Agr. Food. Chem., 55*, 5544–5551.

Belorkar, S. A., & Gupta, A. K., (2016). Oligosaccharides: A boon from nature's desk. *AMB Expr., 6*, 82. doi: 10.1186/s13568-016-0253-5.

Bhatia, L., Sharma, A., Bachetti, R. K., & Chandel, A. K., (2019). Lignocellulose derived functional oligosaccharides: Production, properties and health benefits: A Review. *Process Biochemistry & Biotechnology.* https://doi.org/10.1080/10826068.2019.1608446.

Bode, L., (2009). Human milk oligosaccharides: Prebiotics and beyond: A review. *Nutrition Reviews, 67*(2), 183–191.

Cantley, M., (2004). "How should public policy respond to the challenges of modern biotechnology?" *Current Opinions in Biotechnology, 15*, 258–263.

Carvalho, A. F. A., Neto, P. O., Da Silva, D. F., & Pastore, G. M., (2013). Xylo-oligosaccharides from lignocellulosic materials: Chemical structure, health benefits and production by chemical and enzymatic hydrolysis. *Food Res. Int., 51*, 75–85.

Chakdar, H., Kumar, M., Pandiyan, K., Singh, A., Nanjappan, K., Kashyap, P. L., & Srivastava, A. K., (2016). Bacterial xylanases: Biology to biotechnology. *3 Biotech, 6*, 150.

Chandel, A. K., Antunes, F. A. F., Hilares, R. T., Cota, J., Ellila, S., Silveira, M. H. L., & Da Silva, S. S., (2018b). Bioconversion of hemicellulose into ethanol and value-added products: Commercialization, trends, and future opportunities. In: *Advances in Sugarcane Biorefinery* (pp. 97–134). Elsevier: Amsterdam, Netherlands.

Chandel, A. K., Garlapati, V. K., Singh, A. K., Antunesa, F. A. F., & Da Silvaa, S. S., (2018a). The path forward for lignocellulose biorefineries: Bottlenecks, solutions, and perspective on commercialization. *Bioresource Technology, 264*, 370–381. https://doi.org/10.1016/j.biortech.2018.06.004.

Cui, S. W., Wu, Y., Ding, H., Delcour, J. A., & Poutanen, K., (2013). The range of dietary fiber ingredients and a comparison of their technical functionality. In*: fiber-rich and Wholegrain Foods: Improving Quality* (pp. 96–119). Elsevier, Amsterdam.

Ding, C., Li, M., & Yuqi, H., (2018). High-activity production of xylanase by *Pichia stipitis*: Purification, characterization, kinetic evaluation and *xylooligosaccharides* production. International *Journal of Biological Macromolecules, 117*, 72–77.

Dotsenko, G., Meyer, A. S., Canibe, N., Thygesen, Nielsen, M. K., & Lange, L., (2017). Enzymatic production of wheat and ryegrass derived xylooligosaccharides and evaluation of their *in vitro* effect on pig gut microbiota. *Biomass Conversion and Biorefinery.* https://doi.org/10.1007/s13399-017-0298-y 2017.

Fang, H., Kandhola, G., Rajan, K., Djioleu, A., Carrier, D. J., Hood, K. R., & Hood, E. E., (2018). Effects of oligosaccharides isolated from pinewood hot water pre-hydrolyzates on recombinant cellulases. *Front. Bioeng. Biotechnol., 6*, 55. doi: 10.3389/fbioe.2018.00055.

Finegold, S. M., Li, Z., Summanen, P. H., Downes, J., Thames, G., Corbett, K., Dowd, S., et al., (2014). Xylooligosaccharide increases bifidobacteria but not lactobacilli in human gut microbiota. *Food. Funct., 5*, 436–445.

FitzPatrick, M., Champagne, P., Cunningham, M. F., & Whitney, R. A., (2010). A biorefinery processing perspective: Treatment of lignocellulosic materials for the production of value-added products. *Bioresour. Technol., 101*(23), 8915–8922.

Food and Agriculture Organization of the United Nations (FAO), (2007). *Technical Meeting Report on Prebiotics. 42*(3), 156–159. DOI: 10.1097/MCG.0b013e31817f184e.

Gallardo, O., Valls, C., Valenzuela, S. V., Roncero, M. B., Vidal, T., Díaz, P., & Pastor, F. I. J., (2010). Characterization of a family GH5 xylanase with activity on neutral oligosaccharides and evaluation as a pulp bleaching aid. *Applied and Environmental Microbiology, 76*(18), 6290–6294. doi: 10.1128/AEM.00871-10.

Glasser, et al., (2016). *Thermoplastic Pentosan- Rich Polysaccharides from Biomass.* Unit. States Patent, 5430142.

Gong, Z., Wang, Q., Shen, H., Wang, L., Xie, H., & Zhao, Z. K., (2014). Conversion of biomass-derived oligosaccharides into lipids. *Biotechnology for Biofuels, 7*, 13.

Gupta, P. K., Agrawal, P., & Hegde, P., (2015). Extraction of xyloloigosaccharides by using *Aspergillus niger* from orange wastes. *International Journal of Pharm. Tech. Research, 7*(3), 488–496.

Gupta, P. K., Agrawal, P., & Hegde, P., (2017). Value addition of orange fruit wastes in the enzymatic production of xylooligosaccharides. *African Journal of Biotechnology, 16*(24), 1324–1330. doi: 10.5897/AJB2017.15927.

Gupta, P. K., Agrawal, P., Hegde, P., Shankarnarayan, N., Vidyashree, S., Singh, S. A., Ahuja S., (2016). Xylooligosaccharide – a valuable material from waste to taste: A review. *Journal of Environmental Research and Development, 10*(3) 555–563.

Hall, L. A., & Bagchi-Sen, S., (2007). An analysis of firm-level innovation strategies in the US biotechnology industry. *Technovation, 27*(1), 4–14.

Hamedi, J., & Azimi, A., (2013). Identification and prioritization of interorganizational success factors in microbial biotechnology firms. *Journal of Virology & Microbiology*. doi: 10.5171/2013.190528.

Hauli, I., Sarkar, B., Mukherjee, T., Chattopadhyay, A., & Mukhopadhyay, S. K., (2013). Alkaline extraction of xylan from agricultural waste, for the cost-effective production of xylooligosaccharides, using thermoalkaline xylanase of thermophilic *Anoxybacillus* sp. Ip-C. *Int. J. Pure App. Biosci., 1*(6), 126–131.

Ho, H. L., (2015). Xylanase production by bacillus subtilis using carbon source of inexpensive agricultural wastes in two different approaches of submerged fermentation (SmF) and solid-state fermentation (SsF). *J. Food Process Technol., 6*, 437. doi: 10.4172/2157-7110.1000437.

Jnawali, P., Kumar, V., Tanwar, B., Hirdyani, H., & Gupta, P., (2017). Enzymatic production of xylooligosaccharides from brown coconut husk treated with sodium hydroxide. *Waste Biomass Valor.* doi: 10.1007/s12649-017-9963-4.

Kanimozhi, K., & Nagalakshmi, P. K., (2014). Xylanase production from *Aspergillus Niger* by solid state fermentation using agricultural waste as substrate. *Int. J. Curr. Microbiol. App. Sci., 3*(3), 437–446.

Karunakaran, S., Saravanan, A., Dhanasekaran, S., Senbagam, D., & Senthilkumar, B., (2014). Xylanase production from *Aspergillus Niger. Int. J. Chem. Tech. Res., 6*(9), 4206–4211.

Kazumitsu, S., et al., (2006). Production of food and drink. Japanese Patent JP 9248153, 1997. Capturado em 20 Set. On line. Disponivel na internet: http://www.freepatentsonline.com/9248153.html (accessed on 12 December 2021).

Khatudomkiri, N., Sivamaruthi, B. S., Sirilun, S., Lailerd, N., Peerajan, S., & Chaiyasut, C., (2018). Optimization of alkaline pretreatment and enzymatic hydrolysis for the extraction of xylooligosaccharide from rice husk. *AMB Expr., 8*, 115 https://doi.org/10.1186/s135680 1806459.

Kothari, D., Patel, S., & Goyal, A., (2014). Therapeutic spectrum of nondigestible oligosaccharides: Overview of current state and prospect. *J. Food Sci., 79*, 1491–1498.

Manap, Y. A., (2012). Prebiotic activity of polysaccharides extracted from *Gigantochloa levis* (Buluhbeting) shoots. *Molecules, 17*(2), 1635–1651.

Mane, R. S., Patil, G. S., & Patil, A. M., (2018). Isolation, production, characterization and purification of alakaline xylanase from paper and pulp watse water of sangli kupwad midc(mh). *International Journal of Advanced Multidisciplinary Scientific Research, 1*(3), https://doi.org/10.31426/ijamsr.2018,1.3.139.

McIntosh, S., & Vancov, T., (2011). Optimization of dilute alkaline pretreatment for enzymatic scarification of wheat straw. *Biomass Bioenergy, 35*(7), 3094–3103.

Moure, A., Gullon, P., Domíngue, H., & Parajo, J. C., (2006). Advances in the manufacture, purification and applications of xylo-oligosaccharides as food additives and nutraceuticals. *Process Biochem., 41*, 1913–1923.

Nabarlatz, D., Montane, D., Kardosova, A., Bekesova, S., Hrıbalova, V., & Ebringerova, A., (2007). Almond shell xylooligosaccharides exhibiting immunostimulatory activity. *Carbohydr. Res., 342*, 1122. doi: 10.1016/j.carres.2007, 02,017.

Nwodo-Chinedu, S., Okochi, V. I., Smith, H. A., & Omidiji, O., (2005). Isolation of cellulolytic microfungi involved in wood-waste decomposition: Prospect for enzymatic hydrolysis of cellulosic wastes. *Int. J. Biomed. Health Sci., 1*(2), 41–51.

Qing, Q., Li, H., Kumar, R., & Wyman, C. E., (2013). Xylooligosaccharides production, quantification, and characterization in context of lignocellulosic biomass pretreatment. In: Wyman, C. E., (ed.), *Aqueous Pretreatment of Plant Biomass for Biological and Chemical Conversion to Fuels and Chemicals*. Wiley, Chichester.

Quinones, T. S., Retter, A., Hobbs, P. J., Budde, J., Heiermann, M., Pl€ochl, M., & Ravella, S. R., (2015). Production of xylooligosaccharides from renewable agricultural lignocellulose biomass. *Biofuels, 6*, 147–155. doi: 10.1080/ 17597269.2015.1065589.

Rivas, B., Torrado, A., Torre, P., Converti, A., & Domínguez, J. M., (2008). Submerged citric acid fermentation on orange peel autohydrolysate. *J. Agric. Food Chem., 56*, 2380–2387.

Rosmine, E., Changan, N., Sainjan, E., Silvester, R., & Varghese, S. A., (2019). Utilization of agrowaste xylan for the production of industrially important enzyme xylanase from aquatic *Streptomyces* sp. and potential role of xylanase in deinking of newsprint. *Int. J. Curr. Microbiol. App. Sci., 8*(1), 2061–2076.

Rossi, M., Amaretti, A., & Raimondi, S., (2011). Folate production by probiotic bacteria. *Nutrients, 3*, 118–134.

Saman, P., Chaiongkarn, A., Moonmangmee, S., Sukcharoen, J., Kuancha, C., & Fungsin, B., (2016). Evaluation of prebiotic property in edible mushrooms. *Biological and Chemical Research, 3*, 75–85.

Samanta, A. K., Jayapal, N., Jayaram, C., Roy, S., Kolte, A. P., Senani, S., & Sridhar, M., (2015). Xylooligosaccharides as prebiotics from agricultural by products: Production and applications. *Bioactive Carbohydrates and Dietary fiber, 5*, 62–71.

Samanta, A. K., Jayapal, N., Kolte, A. P., Senani, S., Sridhar, M., Suresh, K. P., & Sampath, K. T., (2012a). Enzymatic production of xylooligosaccharides from alkali solubilized xylan of natural grass (*Sehima nervosum*). *Bioresour. Technol., 112*, 199–205.

Samanta, A. K., Jayapal, N., Kolte, A. P., Senani, S., Sridhar, M., Dhali, A., Suresh, K. P., et al., (2014). Process for enzymatic production of xylooligosaccharides from the xylan of corn cobs. *J. Food. Process. Preserv., 39*, 729–736.

Samanta, A. K., Senani, S., Kolte, A. P., Sridhar, M., Sampath, K. T., Jayapal, N., & Devi, A., (2012b). Production and *in vitro* evaluation of xylooligosaccharides generated from corn cobs. *Food Bioproducts Process, 90*, 466–474.

Silveira, M. H. L. B., Vanelli, B. A., & Chandel, A. K., (2018). Second generation ethanol production: Potential biomass feedstock, biomass deconstruction, and chemical platforms for process valorization. *Advances in Sugarcane Biorefinery*. doi: https://doi.org/10.1016/B978-0-12-804534-3.00006-9 ©.

Singh, D. P., & Prabha, R., (2017). Bioconversion of agricultural wastes into high value biocompost: A route to livelihood generation for farmers. *Advances in Recycling & Waste Management., 2*(3).

Stowell, J., (2006). Calorie control and weight management. In: Mitchell, H., (ed.), *Sweeteners and Sugar Alternatives in Food Technology*. Blackwell Publishing Ltd.

Taeko, I., et al., (2016). Food and drink effective in anti-obesity. Japanese Patent JP 10290681, 1998.http://www.freepatentsonline.com/10290681.html (accessed on 12 December 2021).

Taniguchi, H., (2004). Carbohydrate research and industry in Japan and the Japanese society of applied glycoscience. *Starch., 56*, 1. doi: 10.1002/star.200300258.

Thakuria, A., Sheth, M., (2018). Alkaline extraction of xylan from agro waste and determining xylooligosaccharide (xos) using enzymatic hydrolysis. *Paripex – Indian Journal of Research, 7*(11).

Waghmare, P. R., Khandare, R. V., Jeon, B. H., & Govindwar, S. P., (2018). Enzymatic hydrolysis of biologically pretreated sorghum husk for bioethanol production. *Biofuel Res. J., 19*, 846–853.

Yeoman, C. J., Han, Y., Dodd, D., Schroeder, C. M., & Cann, I. K. O., (2010). Thermostable enzymes as biocatalysts in the biofuel industry. *Adv Appl Microbiol., 70*, 1–55. doi: 10.1016/S0065-2164(10)70001-0.

Yu, X., Yin, J., Li, L., Luan, C., Zhang, J., Zhao, C., & Li, S., (2015). Prebiotic potential of xylooligosaccharides derived from corn cobs and their *in vitro* antioxidant activity when combined with *Lactobacillus*. *J. Microbiol. Biotechnol., 25*(7), 1084–1092. https://doi.org/10.4014/ jmb.1501.01022.

CHAPTER 3

Microbial Bioconversion of Agro-Waste Biomass into Useful Phenolic Compounds

BHABJIT PATTNAIK,[1] PRAKASH KUMAR SARANGI,[2]
PADAN KUMAR JENA,[3] and HARA PRASAD SAHOO[4]

[1]*Department of Botany, Christ College, Cuttack-753008, Odisha, India, E-mail: bhabjit.int4res@gmail.com*

[2]*Directorate of Research, Central Agricultural University, Imphal, Manipur, India*

[3]*Department of Botany, Ravenshaw University, Cuttack-753003, Odisha, India*

[4]*Department of Botany, Buxi Jagabandhu Bidyadhar (BJB) College, Bhubaneswar-751014, Odisha, India*

[5]*Department of Microbiology and Bioinformatics, Atal Bihari Vajpayee University, Bilaspur, India*

ABSTRACT

There is a massive annual accumulation of agro-industrial wastes that represent most of the vital energy resources. The major constituents of such wastes are cellulose and hemicellulose (75–80%) whereas lignin accounts for only 14%. Agricultural residues generated from every crop are of great concern following the problems of environmental pollution, recycling, utilization, and rural sanitation. At the same time, the accumulation results in the loss of a huge amount of potentially important materials. This has led to the concern of extracting useful residues to counterbalance the cost of treating and disposing of the wastes. Agro-industrial waste is greatly nutritious in nature

Biotechnology for Waste Biomass Utilization. Prakash Kumar Sarangi & Latika Bhatia (Eds.)

and expedites microbial growth. Agricultural residues can thus be employed for the production of several value-added products, such as useful phenolic compounds. Agro-industrial wastes, such as rice bran, corn cob, and sugar cane bagasse, have been extensively scrutinized via diverse fermentation approaches for the production of phenolics. Most of these waste products are presently employed as animal feed, but the focus for generating rich value compounds from these trashes is being emphasized by several industrial as well as academic researchers. Biological degradation, for both economic as well as ecological aims, has turned out to be a progressively prevalent alternative for the treatment of industrial, agricultural, organic as well as toxic residues into several value-added products. Agricultural remains can be recycled both naturally as well as artificially by microorganisms. Aerobic microorganisms like bacteria, fungi, as well as some anaerobic organisms have exhibited potency to degrade major components of these residues. Several species of microorganisms are capable to degrade agro-waste residues into diverse value-added phenolic compounds. In recent times, there has been a growing interest to develop flavor production using biotechnological processes employing microorganisms. There exists a chemical resemblance between ferulic acid and vanillin following which the bioconversion of ferulic acid to vanillin was a study of interest. But for the production of "natural vanillin," it is obligatory to utilize "natural ferulic acid" which can be extracted from raw materials by GRAS (Generally Regarded As Safe) enzymes.

3.1 INTRODUCTION

The most widely available and most used energy source is the agro-waste biomass in India. A huge amount of paddy straw, cane trash, and other farm wastes are simply misused by burning in the field. Due to constraints, the farmers usually burn huge quantities of biomass that contribute to global warming. Besides, a huge power generation capacity of about 50,000 MW, this agro-wastes biomass can be employed for the manufacture of a wide array of useful phenolic compounds which are of immense economic usage for mankind. The agro-wastes are chemically lignocelluloses that correspond to a rich, economical, and renewable energy source occurring in a variety of wastes belonging to agricultural, forestry, and agro-industrial sectors. Resources like sugarcane bagasse, sawdust, poplar trees, paper waste, switch grass, as well as stalks, straws, leaves, husks, peels, and shells derived from cereal crops like corn, rice, barley, wheat, and sorghum correspond to various forms of agro wastes (Mussatto and Teixeira, 2010). Cellulose, hemicelluloses, and lignin form the major constituents of lignocelluloses, occurring in an enormously intricate arrangement that is quite tough to disintegrate. The composition of various lignocellulosic feedstock is shown in Table 3.1.

The huge annual accumulation of lignocellulosic biomass poses environmental impacts at the same time discarder of such resources creates a dearth of potentially useful substances. Ferulic acid, a constituent of lignin occurs in the cell walls of several plants at a reasonably high concentration (Mussatto and Teixeira, 2010). Different useful compounds from lignin, cellulose, hemicelluloses are shown in Table 3.2. Other phenolic compounds like Vanillin, Vanillic acid, coumaric acid, sinapic acid, salicyclic acid, etc., constitute the other components of the lignin constituent in agro-waste biomass.

Only quite recently, biomass power production has attracted serious attention throughout the world. Given green catalysis, bioconversion of renewable forms of lignocellulosic materials into useful goods is highly necessary (Menon and Rao, 2012). Farmers need to shift their attention from primary agricultural activities to secondary processing technology wherein they can enhance the value of their produce and utilize all possible by-products.

3.2 BIOCONVERSION OF AGRO-WASTES

Bioconversion, otherwise known as *biotransformation* or *microbial transformation,* is the transformation of organic resources like plant or animal waste utilizing biological methods involving living organisms or some microorganisms, detritivores, or enzymes into energy sources and/or usable products. It broadly refers to the processes which involve the conversion of organic compounds into structurally-related products by microorganisms. In other words, bioconversion deals with microbial (enzymatic) conversion of a substrate into product(s) with a limited number (one or a few) of enzymatic reactions. Microorganisms possess the inherent capability to enzymatically transform a wide array of organic compounds. These microbes during the bioconversion provide enzymes that act upon and convert the organic compounds into other compounds or modify them. Although there are hundreds of bio-conversions known, only a few selected ones are useful for the synthesis of commercially important products.

Bioconversion can be elucidated as the precise alteration of a particular organic compound into a typical product(s) possessing structural similarity, employing biological catalysts as well as microorganisms such as bacteria and

TABLE 3.1 (Composition of Representative Lignocellulosic Feedstock)

Agro-industrial wastes	Chemical composition (% w/w)						References
	Cellulose	Hemicellulose	Lignin	Ash (%)	Total solids (%)	Moisture (%)	
Rice straw	39.2	23.5	36.1	12.4	98.62	6.58	El-Tayeb et al., (2012)
Sunflower stalks	42.1	29.7	13.4	11.17	_	_	Motte et al., (2013)
Sugarcane bagasse	30.2	56.7	13.4	1.9	9.66	4.8	El-Tayeb et al., (2012) and Nigam et al., (2009)
Corn stalks	61.2	19.3	6.9	10.8	97.78	6.40	El-Tayeb et al., (2012)
Oat straw	39.4	27.1	17.5	8	_	_	Martin et al., (2012)
Sawdust	45.1	28.1	24.2	1.2	98.54	1.12	El-Tayeb et al., (2012) and Martin et al., (2012)
Sugar beet waste	26.3	18.5	2.5	4.8	87.5	12.4	El-Tayeb et al., (2012)
Soya stalks	34.5	24.8	19.8	10.39	_	11.84	Motte et al., (2013)
Cotton stalks	58.5	14.4	21.5	9.98	_	7.45	Nigam et al., (2009)
Wheat straw	32.9	24.0	8.9	6.7	95.6	7	Nigam et al., (2009) and Martin et al., (2012)

TABLE 3.2 Useful Compounds Potentially Consequential from Lignin, Cellulose, and Hemicelluloses

Cellulose	Ethanol, Glucuronic acid, 3-hydroxy-propanoic acid, Glutamic acid	–
	Lactic acid	Acetaldehyde 2,3-pentanedione, pyruvic acid, acrylic acid.
	Levulinic acid	1,4 butanediol, succinic acid, lactones.
	Succinic acid	1,4-butanediol, tetrahydrofurane, 2-pyrrolidones.
	Itaconic acid	3-methyl pyrrolidone 2, 3-methyl THF, Itaconic diamide, methyl-1,4-butane diamide.
Hemicellulose	Ferulic acid	Vanillic acid, protacatechuic acid, vanillin.
	Ethanol, Xylitol, Butanol, Lactic acid, Chitosan, Xylo oligosaccharides, Furfural	–
Lignin	Phenols	Cresols, coniferols, syringol, eugenols.
	Syngas	Ethanol, methanol/dimethy ether.
	Hydrocarbon	Higher alkylates, cyclohexane.
	Marcomolecules	Activated carbon, carbon fibers, substituted lignins, polyelectrolites, resins, and neutraceuticals/drugs.
	Oxidized products	Vanillic acid, vanillin, DMSO, quinones, aliphatic and aromatic acids.

fungi. Biological catalyst refers to either an enzyme or an entire, incapacitated microorganism containing a single enzyme or numerous enzymes synthesized in it. These involve simple, chemically defined reactions catalyzed by enzymes present in the microbial cells.

Numerous microbial strains and enzymes possess the potential of specific biotransformation required for the bioconversion of a multitude of various materials into anticipated products, predominantly main pharmaceutical components that are optically active. But biotransformation characteristically signifies the second-generation method choice in the creation of a small pharmaceutical molecule following the missing broad strain and enzyme choice as well as timeline constraints in the development cycle of the pharmaceuticals. Novel biocatalysts are necessary primarily for the bioconversion process, mainly oxidoreductases and lyases.

Bioconversion of agro-wastes can lead to the generation of a variety of phenolics because lignocellulosic biomasses are a storehouse of a wide array of such compounds. Phenolic substances are largely dispersed phytochemicals found in plant tissues, including fruits as well as vegetables (De la Rosa et al., 2019). These compounds perform a vital part in the growth control and display a wide array of functions including protective, antioxidant, structural, attractant, and signaling functions, in plants (Babenco et al., 2019). On consumption in the diet, these compounds confer health-protective effects, although they are not nutrients simply because these compounds are a storehouse of numerous bioactive properties (De la Rosa et al., 2019). Much of the attention is devoted to the phenolic compounds following their effect on diverse organoleptic parameters, like color or taste. Nevertheless, much focus is predominantly concerted on these components due to the health-beneficial properties that they are estimated to impart. Characteristically, such compounds are effective antioxidants, while some of them are also considered anti-carcinogenic or anti-microbial (Herrero et al., 2013). Among the various phenolic compounds occurring in plant tissues, vanillin, vanillic acid, and ferulic acid are the important ones.

Ferulic acid is the most abundant organic acid (hydroxyl cinnamic acid) consequential from phytochemical phenolic compounds. The acid is mainly found in plants belonging to the family Poaceae. Ferulic acid is quite popular following its anti-oxidant and anti-inflammatory properties (Graf, 1992). Consequently, it exhibits enormous prospects in industrial and medicinal fields. In addition, it also displays numerous therapeutic effects against several ailments like diabetes, neurodegenerative and cardiovascular diseases, and cancer (Srinivasan et al., 2007).

Vanillic acid is a significant hydroxybenzoic acid (HBA) and has a role as a plant metabolite. This acid is a phenolic compound occurring in various plant extracts and forms of vanilla. This compound is one of the chief components of the 'natural vanilla' flavor. Vanillic acid is employed in a variety of products (Bock and Anderson, 1955) while its derivatives display antibacterial activity (Rai and Maurya, 1966). Vanillic acid can serve as a snake venom inhibitor (Dhananjaya et.al., 2009) as well as display antisickling and anthelmintic activities (Itoh et al., 2010).

Vanillin is a phenolic compound, an aldehyde occurring in vanilla beans, naturally. It is employed extensively as an aromatic additive for incense, candles, perfumes, and air fresheners, as a flavoring additive for cooking and beverages (Kumar et al., 2012). The major solitary utilization of vanillin is for flavoring. It has a role as a plant metabolite, a flavoring agent, an antioxidant, and an anticonvulsant.

3.3 OCCURRENCE, NATURE, AND BIOCHEMISTRY OF FERULIC ACID

Ferulic acid (FA, C10H10O4) is one of the most plentiful organic acids occurring extensively in fruits, vegetables, agro-industrial wastes such as sugarbeet, corn hulls, maize bran, beverages like beer and coffee, etc. (Wong et al., 2011; Rechner, 2001; D'Archivio et al., 2007). It is the most universal hydroxyl cinnamic acid consequential from phytochemical phenolic compounds. FA is widely dispersed all over the plant kingdom in cereals, spices, grains, vegetables, pulses, and fruits, their by-products such as cider oil, tea, and beverages as well as medicinal plants (Efazary et al., 2007). A wide array of biomass like maize bran, sugar cane bagasse, wheat bran, rice bran, pineapple peels, and wheat straw can be employed for the synthesis of ferulic acid (Tilay et al., 2008). Paddy straw can also serve as a competent biomass yielding ferulic acid (Salleh et al., 2011). Amongst agricultural plant products, ferulic acid is present in peak concentration (3.1% (w/w)) in maize bran (Saulnier et al., 1995). FA occurs as covalently linked structures in plant cell walls where they form ester bonds with the cell wall polysaccharides (Iiyama et al., 1994) and ether or ester bonds with the constituents of lignin. The release of ferulic acid from plants necessitates enzymatic digestion of its structures since it exists as cross-linkages between cell wall polysaccharides and lignin. (Fry, 1982). In addition, ferulic acid can as well be released from plant cell walls through the hydrolysis process by enzymes called feruloyl esterase (ferulic acid esterases) (Benoit et al., 2006). The acid serves as a renewable resource for the chemical or bio-catalytic transformation to other valuable aromatic compounds from natural agro by-products (Efazary et al., 2007).

Ferulic acid is a phenyl propenoid resultant from the cinnamic/coniferic acid (4-hydroxy-3-methoxycinnamic acid) (Figure 3.1). It exhibits two isomeric forms such as an oily yellow liquid or *Cis* and a crystalline form or *Trans*. It is named after *Ferula fetida* (Family: Umbelliferae) from which the acid was extracted for the very first time in 1866. Ferulic acid imparts an antioxidant effect because it facilitates hydrogens to the free radicals possessing phenolic-OH groups (Efazary et al., 2007).

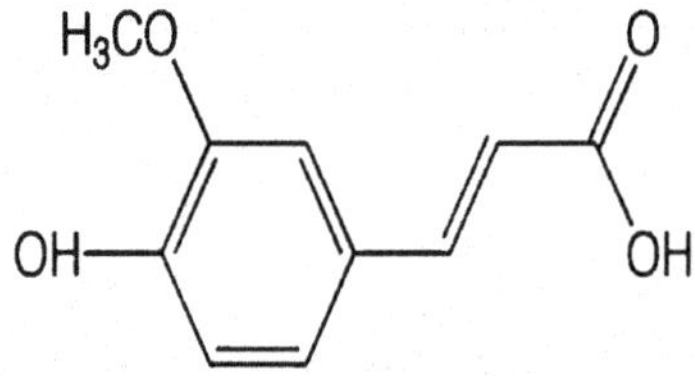

FIGURE 3.1 Structure of ferulic acid.

According to Dutt (1925), the production of ferulic acid is by virtue of its precursor features in the generation of malonic acid and vanillin. The phenolic acid can be synthesized by the phenylpropanoid pathway and it can be oxidatively coupled with other ferulic acid and derivatives, efficiently (De O Buanafina, 2009). The compound can be extracted by strong alkali treatment or enzymatic hydrolysis. Ferulic acid esterases (FAE or Feruloyl esterase) are the enzymes involved in the extraction of ferulic acid from plant cell walls by the hydrolysis process (Benoit et al., 2006). Ferulic acids are those phenolic compounds that are typically recognized and quantified by reverse-phase HPLC (Jirovsky et al., 2003).

The hydroxycinnamic acid has been described to exhibit a wide array of physiological features like anti-inflammatory, anti-cancer, therapeutic, hepatoprotective, anti-aging, anti-apoptotic, antimicrobial, anti-diabetic, anti-mutagenic, neuroprotective, photoprotective, etc., which can be mainly ascribed following its antioxidant property (Umre et al., 2018; Singh et al., 2016; Kumar and Pruthi, 2014; Zielinski et al., 2014; Fujita et al., 2013; Paiva et al., 2013; Kang et al., 2011; Dos Santos et al., 2008b, Lin et.al., 2010). Although the acid is accountable for the production of vital organic compounds like vanillic acid, vanillin, coniferyl alcohol, sinapic acid, curcumin, and ferulic acid, it primarily confers rigidity to plant cell walls. Apart from these applications, ferulic acid can also be utilized as a food preservative and display a range of pharmaceutical, biomedical and industrial applications (Kumar and Pruthi, 2014).

3.4 OCCURRENCE, NATURE, AND BIOCHEMISTRY OF VANILLIC ACID

Vanillic acid (VA, C8H8O4) is a monohydroxybenzoic acid. Conversely, it is chemically, 4-hydroxy-3-methoxybenzoic acid (Figure 3.2). Vanillic acid is a chlorogenic acid that is an oxidation product of vanillin (Dhananjaya et al., 2006). This phenolic acid possesses an important role as a plant metabolite and occurs in a variety of vanilla and various other plant extracts (Lesage-Meessen et al., 1996). The acid usually occurs in several cereal crops, fruits, whole grains, herbs, juices, green tea, wines, and beers. The phenolic acid can be observed in various plant species, mainly, *Alnus japonica*, *Fagara spp*., *Elaeagnus pungens*, *Gossypium mexicanum*, *Erica australis*, *Melia azedarach*, *Paratecoma koraiensis*, *Panax ginseng*, *Pterocarpus santalinus*, *Picrorhiza kurrooa*, *Rosa canina*, *Trachelospermum asiaticum*, *Lentinula edodes*, and *Amburana cearensis* (Itoh et al., 2009). The utmost quantity of vanillic acid observed so far in plants relates to the root of an herb indigenous to China, *Angelica sinensis*, which finds applications in customary Chinese medicine (Duke, 1992). Açaí oil which is extracted from açaí palm (*Euterpe oleracea*) fruit is a rich source of vanillic acid (*Pacheco-Palencia* et.al., 2008). VA is a prime natural phenol occurring in argan oil and is also occurs in vinegar and wine (Galvez et.al., 1994). This phenolic acid is a scent and flavoring representative that imparts a nice, creamy aroma (Lesage-Meessen et al., 1996).

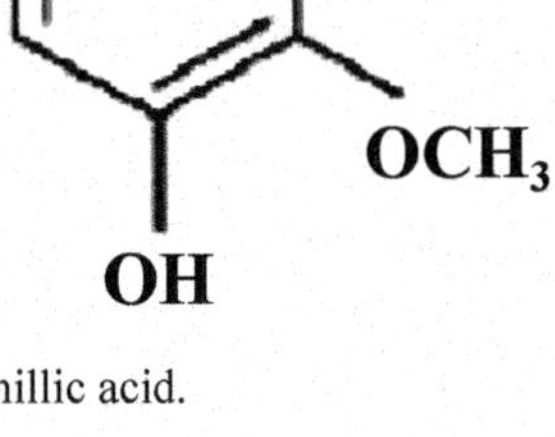

FIGURE 3.2 Structure of vanillic acid.

It is a chlorogenic acid that can be proficiently synthesized by vanillin oxidation (Dhananjaya et al., 2006). Vanillic acid is yielded as the intermediary compound in the two-phase biotransformation of ferulic acid to vanillin. (Lesage-Meessen et al., 1996). It is also used as the preparatory material in the synthesis of vanillin by chemical techniques. The acid is known to be a metabolic by-product of caffeic acid and it frequently accumulated in the urine of individuals who have consumed tea, coffee, vanilla-flavored confectionery, and chocolate (Dhananjaya et al., 2006). The acid is utilized in the production of etamivan (Kratzl and Kvasnicka,1952), an analeptic drug. The chlorogenic acid is a prime transitional in ferulic acid and lignin degeneration, and it is usually amassed in significant quantities, contrasting vanillin (Andreoni et al., 1995). Phenolic acid is utilized in diverse products, such as, in the synthesis of polyester it could be applied as a monomer or polymerized to oligomers (Bock and Anderson, 1955). 5-nitrovanillic acid as well as 5-aminovanillic acid which are the derivatives of vanillic acid depict antibacterial activity (Rai and Maurya, 1966). Vanillic acid has been employed as a flavoring agent in food and drug products, preservative, and antioxidant having useful biological activities (Stanley et al., 2011, Singh et al., 2015). The acid possesses anti-microbial, antioxidant, hepatoprotective, cardioprotective, and anti-apoptotic activities (Mourtzinos et al., 2009, Almeida et al., 2016). Vanillic acid is also described to hold neuroprotective and strong anti-inflammatory effects (Singh et al., 2015).

3.5 OCCURRENCE, NATURE, AND BIOCHEMISTRY OF VANILLIN

Vanillin (C8H8O3) is a phenolic aldehyde, an organic compound that is an important plant secondary metabolite (Figure 3.3). Chemically, 3-methoxy-4-hydroxybenzaldehyde, is a pleasant, intensely sweet-smelling aromatic compound occurring naturally in vanilla beans (Kerler et al., 2001). It is also found in *Leptotes bicolor* (Mohamad et al., 1998) roasted coffee (Sen and Grosch, 1992), and the Chinese red pine. Vanillin appears as white or very slightly yellow needles (Kumar et al., 2012). It is a fragrant aldehyde grouped under simple phenolic compounds. It structurally retains the functional groups including, ether, phenol, and aldehyde (Converti et al., 2010). The compound crystallizes in the form of colorless needles from hot water at a temperature of 81–82°C (Gildemeister and Hoffmann, 1899). Vanillin is quite a versatile flavor enjoyed by most people following its rich fragrance, landing it as the world's leading and most popular flavor (Schrader et al., 2004). It is quite often attained from the paper and pulp industry as a by-product, following the oxidative breakdown of lignin or may be extracted from the vanilla bean. (Kerler et al., 2001). It is normally considered that the phenylpropanoid pathway starting with L-phenylalanine yields vanillin as a product (Dixon, 2014). Vanillin is the end product of the two-step ferulic acid bioconversion. In other words, the phenolic acid can be extracted efficiently by oxidation (β- oxidation) of ferulic acid or synthesized from ferulic acid (Zenk,1965; Overhage et.al., 1999).

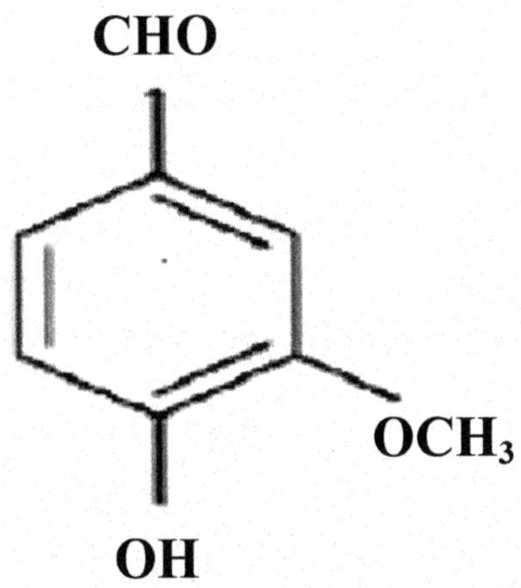

FIGURE 3.3 Structure of vanillin.

Vanillin is primarily employed in the preparation of food products, beverages, confectionery, and as a fragrance and flavor component (Ranadive, 1994). Phenolic acid acts as a useful intermediate in herbicide production, antifoaming and antimicrobial agents (Hocking, 1997). It can be suggested following the biological screening of vanillin that this compound exhibits a wide array of biological functions such as antioxidant function (Sawano and Yazama, 2011), antimicrobial activity (Naira et al., 2004), anti-diabetic function (Wang et al., 1998), anti-inflammatory function (Lim, 2004a), analgesic activity (Lim, 2004b), and anti-cancer activity (Ali et al.,2008). Vanillin is extensively utilized as a masking agent as well as a food additive in several pharmaceutical preparations (Kaur et al., 2013, Banerjee and Chattopadhyay, 2019). Vanillin also possesses the prospective to be employed as a food preservative (Sinha et al., 2008). It can also be used in various personal and home use products (Banerjee and Chattopadhyay, 2019).

3.6. BIOCONVERSION OF AGRO-WASTE BIOMASS INTO USEFUL PHENOLIC COMPOUNDS

3.6.1 FERULIC ACID BIOSYNTHESIS

Ferulic acid can be manufactured by a wide range of techniques such as release from plant cell walls, chemical synthesis, extraction from ferulic acid conjugates of low molecular weight, or through microbial fermentation. Chemical production is quite unsuitable as it produces an isomeric mixture of the compound and possesses a longer reaction time.

Importantly, the ferulic acid produced by chemical hydrolysis cannot account to be a natural source for vanillin manufacture. As per the current European Union and United States regulations, vanillin produced via microbial transformation can be branded as natural, provided the ferulic acid is obtained from a natural source like agro-waste biomass, and the retrieval technique, ideally, the biotechnological method is mild and includes the employment of enzymes with GRAS (generally regarded as safe) grade (Mathew & Abraham, 2006).

There are three pathways to manufacture ferulic acid through natural sources;

1. From ferulic acid conjugates of low molecular weight,
2. From cell walls of plants,
3. Via microbial fermentation.

In recent times, an emerging interest in the utilization of naturally occurring antioxidants consequential from dietary components has been observed. An ample interest is also given towards enzymatic methods, by use of feruloyl esterases (FAEs) manufactured by microorganisms which break the ester bonds between phenolic acid and cell wall polysaccharide of plants (Mathew & Abraham, 2004). The enzymatic treatment is currently believed to be an evolving sensational selection to generate vanillin that can be tagged as a natural product.

The degradation of lignocellulosic materials necessitates the mutual and synergistic action of several enzymes with varied functions. Such enzymes can be assembled into three classes, i.e., hemicellulose, lignin-degrading enzymes, and cellulase. The lignin-degrading enzymes include "additional enzymes," like acetyl xylan esterases, α-glucuronidase, α-L-arabinofuranosidase, and ferulic acid esterase, which assists xylanases and pectinases in the disruption of hemicelluloses present in the plant cell wall. Amongst the additional enzymes, ferulic acid esterases display a significant part in the hydrolysis of the ferulate ester groups employed in the cross-linking amongst hemicelluloses and among lignin and xylans. Ferulic acid esterases FAE; (EC 3.1.1.73) or feruloyl esterases, cinnamic acid corresponds to a subclass of carboxylic esterases (EC 3.1.1) (Wong, 2006; Fazary and Ju, 2008). The release of ferulic acid and other cinnamic acids from the polysaccharides present in the plant cell wall is made possible by these enzymes. FAE possesses industrial importance since they can act on the bond between hemicellulose present in the plant cell wall and ferulic acid, unlike the polysaccharide hydrolases (Benoit et al., 2008). Although FAE exhibits a wide array of applications but alone it is not adequate to extract ferulic acid from the agro-waste biomass matrix. Extraction of the phenolic acid requires the synergistic utilization of hemicellulases (specifically arabinofuranosidase and xylanase) and FAE along with that cellulase and protease, which are also essential (Sorensen et al., 2003; Shin et al., 2006). FAEs work synergistically with pectinases and xylanases and thereby expedite the admission of hydrolases to the backbone of cell wall polymers. (Mathew & Abraham, 2004). Ferulic acid esterases are manufactured by diverse fungal and bacterial microbes. Amongst these, *Aspergillus* species, including *Aspergillus awamori*, *Aspergillus flavipes, Aspergillus niger,* and *Aspergillus oryzae*, are the most dynamic manufacturers of FAEs. Under submerged cultivation, *Aspergillus niger* (Johnson et al., 1989) and *Aspergillus flavipes* (Mathew and Abraham, 2005) strains were observed to be the most active manufacturers employing lignocellulose-derived carbon sources like de-starched wheat bran and maize bran. A wide array of feruloyl esterases have been isolated/refined and described from various microorganisms, like *Clostridium thermocellum* (Blum et al., 2000), *Clostridium stercorarium* (Donaghy et al., 2000), *Streptomyces thermophile* (Topakas et al., 2004), *Aspergillus awamori* (Shin and Chen, 2007), *Streptomyces olivochromogenes* (Topakas and Christakopoulos, 2004), *Fusarium proliferatum* (Shin and Chen, 2006), *Fusarium oxysporum* (Topakas and Christakopoulos, 2004; Faulds and Williamson, 1991). But despite a lot of extensive research, until now the commercial availability of refined FAE is not possible. This explains the prime reason behind the dearth of industrial technology built on the reclamation of ferulic acid from agro-waste biomass like wheat bran (Fazary and Ju, 2008). Ferulic acid esterase catalyzes the process conversion, optimization, and enzymatic activities of several other microorganisms and may further increase the amount of ferulic acid extraction. Also, the other metabolite that was detected along with FA could be some other phenolic compounds like vanillic acid or vanillin that are the breakdown products of ferulic acid. Extensive studies were conducted among various fungi and bacteria species. But there are few reports on the FAE activities by probiotic LAB (Liu et al., 2016). The release of ferulic acid from agricultural crop residues via alkaline hydrolysis needs a shorter time The crop biomass is subjected to a dilute NaOH solution at a temperature of 50–70°C (Barberousse et al., 2009).

3.7 VANILLIN BIOSYNTHESIS

3.7.1 VANILLIN BIOSYNTHESIS IN VANILLA PLANIFOLIA

The vanilla extract comprises more than 200 components in its flavor profile. Out of these, the characteristic vanilla fragrance is primarily due to vanillin which occurs as the peak abundant organic compound. The concentration of vanillin ranges from 1.0–2.0% w/w of dry weight in treated *Vanilla* pods where it is gathered mainly in coupled form, predominantly as the β-D-glucoside. The *vanilla* beans do not display any trace of vanilla flavor at six to eight months following pollination when the green beans are yielded (Walton et al., 2003). Fermentation of *vanilla* pods is popularly called "curing." It causes the hydrolysis of glucovanillin, glucoside of vanillin, as well as related β-D-glucosides by the action of the enzyme β-D-glucosidase. This results in the release of free vanillin and associated substances (particularly 4-hydroxybenzaldehyde) thus producing a pleasant aroma (Odoux et al., 2003).

The production of vanillin in the course of curing is quite a simple technique as compared to the preliminary pathway of synthesis of vanillin β-D-glucoside. Although numerous biosynthetic protocols have been devised, the comprehensive route of vanillin production is yet to be solved.

As per the literature point of view, there is an overall acceptance that vanillin is derived from the shikimic acid route. There exist two key interpretations to suggest in what way a phenylpropanoid precursor is transformed to vanillin (Havkin-Frenkel & Belanger, 2008). The first view termed the 'ferulate pathway' (Figure 3.4A) suggested by Zenk (1965) involves hydroxylation as well as methylation of the phenylpropoanoid compound to generate ferulic acid. This is followed by β-oxidation of the acid-producing vanillin. Whilst, the other viewpoint is designated the 'benzoate pathway' (Figure 3.4B) proposed by Podstolski et al., (2002). It involves chain shortening and then hydroxylation as well as methylation of the phenylpropanoid compound to generate vanillin.

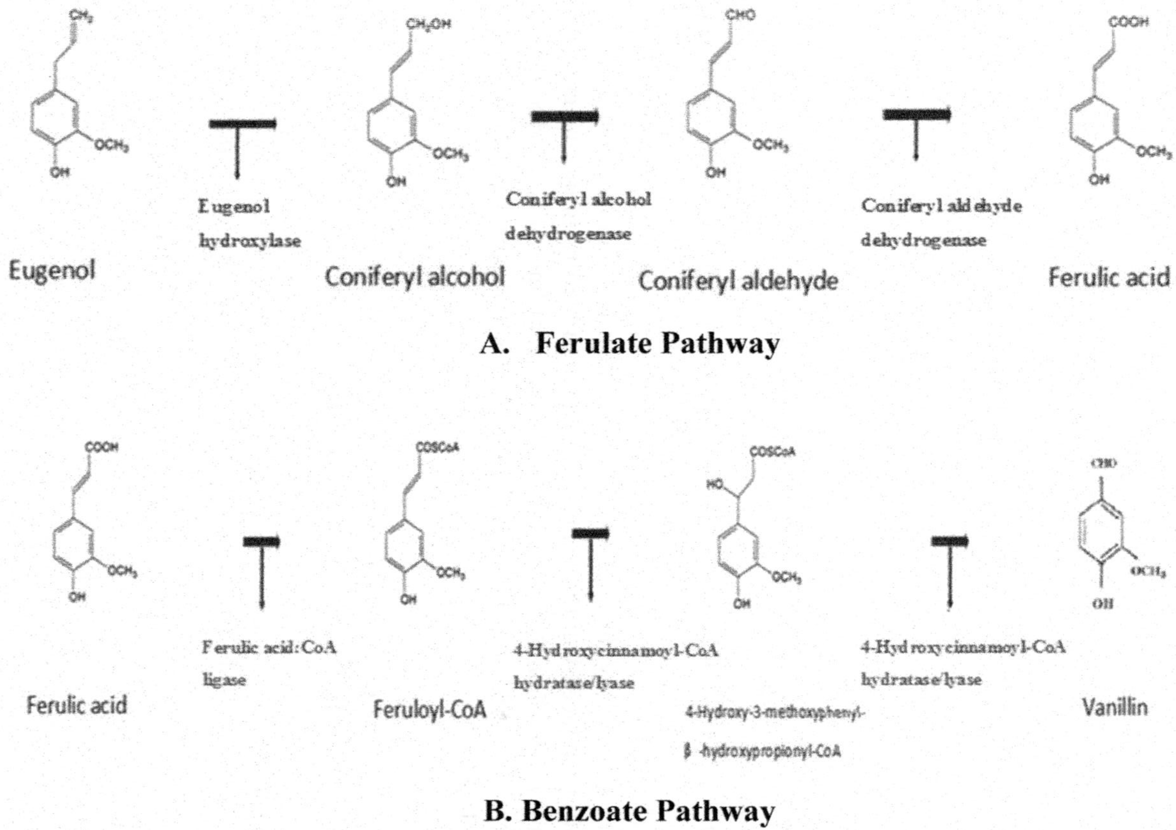

FIGURE 3.4 Metabolic routes yielding vanillin: (A) ferulate pathway and (B) benzoate pathway.

3.8 VANILLIN PRODUCTION THROUGH BIOTECHNOLOGICAL APPROACHES

3.8.1 USE OF ENZYMES

Vanillin can be synthesized via in vitro enzyme-based systems. For designing such systems, insight on the vanillin biosynthetic route and the associated enzymes, which catalyze sequential phases in the process is essential. Production of vanillin or else vanillin intermediates can be made possible by controlling some of the distinct stages in the synthetic process by employing biotechnology to clone genes for appropriate enzymes. Enzymatic preparations containing β-glucosidase can be employed for the extraction of vanillin from its pods since the enzyme facilitates the release of the compound from its fruit.

Vanillin can be extracted from vanilla pods by employing enzymatic preparations comprising β-glucosidase which facilitates the discharge of vanillin from pods. This technique can serve as a substitute for traditional curing (Dignum et al., 2001a; Ruiz-Terán et al., 2001; Odoux & Havkin-Frenkel, 2005). Vanillin can also be produced by biotransformation from plant material by the utilization of enzymes. Ex. Van den Heuvel et al., (2001) produced vanillin via bioconversion of vanillylamine and cresol by using *Penicillium* flavoenzyme. Similarly, vanillin can be produced from esters of coniferyl alcohol by utilizing Soybean lipoxygenase (Markus et al., 1992).

3.8.4 USE OF PLANT TISSUE CULTURE

This technique is being employed for a long time to produce vanillin. It is advantageous because following vanillin manufacture, synthesis of some varied compounds occurring in pods of vanilla could be possible. For example, the cells of vanilla and *Capsicum frutescens* and the organs of vanilla have been effectively cultured then observed to mount up vanillin with additional metabolites (Ramachandra Rao and Ravishankar, 2000b; Dignum et al., 2001b). But the quantity of vanillin yielded using this technique is usually low which appears to be the main disadvantage of using this technique. Although steps were undertaken to raise the harvest of vanillin by these methods such as the use of hormones or elicitors, feeding of putative precursors, adjustment of culture conditions of the environment, and utilization of adsorbent like charcoal (Walton et al., 2003). But none have been proven to be worthy enough.

Genetic engineering utilization provides a new and feasible strategy of vanillin production from plants by the introduction of a pathway or an enzyme to produce vanillin as a typical intermediate of the phenylpropanoid route in plants. Walton et al., (2003) proposed some rise in the potentials of vanillin manufacture following isolation of the enzyme HCHL (4-hydroxycinnamoyl-CoA lyase/ hydratase). Any vanillin produced by genetic engineering is bound to be transformed to its β-D-glucoside form and can rarely remain in its free state. But the vanillin once generated is bound to undergo reduction and oxidation. Further, following the inadequate understanding of the biosynthetic pathways of vanillin and other associated enzymes, the genetic engineering of *Vanilla* plants is quite challenging (Walton et al., 2003; Havkin-Frenkel and Belanger, 2008).

3.8.5 APPLICATION OF MICROORGANISMS

A wide range of compounds, including lignin, ferulic acid, isoeugenol, eugenol, vanillic acid, aromatic amino acid, glucose, and phenolic stilbenes, have been employed for the laboratory-scale manufacture of vanillin with the help of a wide array of microorganisms mainly fungi and bacteria.

3.8.5.1 BIOCONVERSION OF LIGNIN

Lignin is an essential component of the plant cell wall and it exists as one of the utmost prolific natural reservoirs of flavoring compounds. It is a complex aromatic polymer and accommodates vanillin subcomponents in its polymeric organization. Lignin is produced by the dehydrogenative polymerization of three monolignols (cinnamyl alcohols) namely, coniferyl, sinapyl alcohol, and *p*-coumaryl. Lignin is the precursor for vanillin production by chemical oxidation and is also a profuse derivative of the paper production house. Despite this, meager information about microbial production of vanillin from lignin is available. (Priefert et al., 2001).

Six independent lignin degeneration pathways were identified based upon the scientific literature. But two out of those routes namely, ferulate catabolic pathways and the β-aryl ether cleavage, were given much emphasis in the microbial society because the intermediate metabolite was observed as vanillin.

β-aryl ether cleavage pathway is illustrated in Figure 3.6 (*Rhodococcus* sp., and *Delftia acidovorance* (Masai et al., 2002), *Sphingomonas paucimobilis*) but presently such studies are limited to scientific importance only.

3.8.5.2 BIOTRANSFORMATION OF FERULIC ACID

The microbial transformation of ferulic acid to vanillin exists as one of the most favorable and comprehensively studied methods for the viable manufacture of biovanillin. Ferulic acid (FA), the major phenolic compound occurring in monocot plant cell walls and therefore derived from agro-industrial derivatives like maize bran (30 g/kg), corn hulls (31.0 g/kg), sugarbeet (5–10 g/kg), wheat (6.6 g/kg), barley grains (1.4 g/kg) as well as rice endosperm cell wall (9 g/kg). Isolation of ferulic acid can occur by enzymatic hydrolysis or by treatment with a strong alkali. The former remains to be the ideal selection for the generation of vanillin that can be branded as "natural" (Hasyierah et al., 2008; Mathew and Abraham, 2006). Furthermore, ferulic acid is supposed to be the prime appropriate aspirant for biovanillin manufacture, following its minimal toxicity among the full range of examined precursors.

3.8.5.2.1 Catabolic Pathways of Ferulic Acid

Based on scientific literature for the initial reaction of ferulic acid degradation four main catabolic pathways have been proposed;

- **Side chain reduction pathway** which encompasses the anaerobic degradation or side-chain reduction of ferulic acid. The pathway is instigated by an isomerization reaction comparable to the decarboxylation reaction (Rosazza et al.,1995) but yields in the formation of dihydroferulic acid under aerobic conditions.
- Non-oxidative decarboxylation pathway encompassing preliminary isomerization of ferulic acid to 4-vinylguaiacol, which further decarboxylates spontaneously. The reaction is accelerated by the enzyme ferulic acid decarboxylase (Huang et al., 1993). Apart from 4-vinylguaiacol, the existence of other metabolic products, like vanillin, dihydroferulic acid, vanillic acid, protocatechuic acid, and vanillyl alcohol have also been recognized in some fungal and bacterial strains.

- Coenzyme-A-independent deacetylation pathway which includes the preliminary removal of an acetate group from the unsaturated ferulic acid side-chain, which leads to the direct synthesis of vanillin (Sutherland et al., 1983; Nazareth and Mavinkurve, 1986; Jurková and Wurst, 1993). Rosazza et al., (1995) proposed the pathway for the acetate cleavage from ferulic acid.
- Coenzyme-A-dependent deacetylation pathway which involves the thioclastic cleavage of 4-hydroxy-3-methoxyphenyl-β-ketopropionyl-CoA to generate Acetyl-CoA and vanillyl- CoA, in the presence of enzyme β- ketothiolase.

3.9 CONCLUSION

Lignocellulosic industrial waste is the most inexpensive and easily accessible form of carbohydrates for valorization and consequent value addition. They account for a renewable reserve from which various valuable chemical and biological products can be obtained. Both fungal and bacterial strains can be organized for the manufacture of various phenolics using agricultural wastes. The abilities of agro-waste biomass in the production of biochemicals and commercially valued products are not well understood. Bioconversion is an alternative technology for the transformation of lignocellulosic biomass to generate value-added materials such as ferulic acid vanillin, vanillic acid, and numerous other biochemicals. Feruloyl esterase was observed capable to isolate ferulic acid from agro-waste biomass. Novel bioconversion technologies employing recombinant microorganisms can go a long way in the manufacture of vital phenolic substances thus augmenting renewable energy sources and food security complications in the future. Conversely, such technologies can result in a reduction of environmental deterioration as well. Diverse biocatalytic pathways during the biodegradation technologies can also be observed for the development of novel compounds possessing universal significance and addressing diverse problems in food sectors as well as pharmaceutical industries. An extensive exploration of several studies on the utilization of inexpensive media components for biochemical production suggests that lignocellulosic waste shows much potential and can be employed as a prime carbon source in mainstream upscale fermentation processes.

KEYWORDS

- **agro-industrial wastes**
- **biological degradation**
- **ferulic acid**
- **lignocellulose**
- **microorganisms**
- **phenolic compounds**
- **vanillin**

REFERENCES

Ali, M. M., Jesmin, M., Sarker, M. K., Salahuddin, M. S., Habib, M. R., & Khanam, J. A., (2008). Antineoplastic activity of N-salicylideneglycinato-di-aquanickel (II) complex against Ehrlich ascites carcinoma (EAC) cells in mice. *Int. J. Biol. Chem. Sci., 2*, 292–298.

Ali, S. M. M., Azad, M. A. K., Jesmin, M., & Ahsan, S., (2010). *In vivo* anticancer activity of vanillin semicarbazone. *Asian Pac. J. Trop. Biomed., 12*, 1–5.

Babenko, L. M., Smirnov, O. E., Romanenko, K. O., Trunova, O. K., & Kosakivska, I. V., (2019). Phenolic compounds in plants: Biogenesis and functions. *Ukr. Biochem. J., 91*(3), 5–18.

Banerjee, G., & Chattopadhyay, P., (2019). Vanillin biotechnology: The perspectives and future. *J. Sci. Food Agric., 99*(2), 499–506.

Barberousse, H., Kamoun, A., Chaabouni, M., Giet, J. M., Roiseux, O., Paquot, M., Deroanne, C., & Blecker, C., (2009). Optimization of enzymatic extraction of ferulic acid from wheat bran, using response surface methodology, and characterization of the resulting fractions. *J. Sci. Food Agric., 89*(10), 1634–1641.

Benoit, I., Danchin, E. G. J., Bleichrodt, R., & Vries, R. P., (2008). Biotechnological applications and potential of fungal feruloyl esterases based on prevalence, classification and biochemical diversity. *Biotechnol. Lett., 30*, 387–396.

Benoit, I., Navarro, D., Marnet, N., Rakotomanomana, N., Lesage-Meessen, L., Sigoillot, J. C., Asther, M., & Asther, M., (2006). Feruloyl esterases as a tool for the release of phenolic compounds from agro-industrial by-products. *Carbohydr. Res., 341*(11), 1820–1827.

Blum, D. L., Kataeva, I. A., Li, X. L., & Ljungdahl, L. G., (2000). Feruloyl esterase activity of the *Clostridium thermocellum* cellulosome can be attributed to previously unknown domains of XynY and XynZ. *J. Bacteriol., 182* 1346–1351.

Bock, L. H., & Anderson, J. K., (1955). Linear polyesters derived from vanillic acid. *J. Polym. Sci., 17*, 553–558.

Bugg, T. D. H., Ahmad, M., Hardiman, E. M., & Singh, R., (2011). The emerging role for bacteria in lignin degradation and bio-product formation. *Curr. Opin. Biotechnol., 22*(3), 394–400.

Converti, A., Aliakbarian, B., Domínguez, J. M., Vázquez, G. B., & Perego, P., (2010). Microbial production of biovanillin. *Braz. J. Microbiol., 41*, 519–530.

D'Archivio, M., Filesi, C., Di Benedetto, R., Gargiulo, R., Giovannini, C., & Masella, R., (2007). Polyphenols, dietary sources and bioavailability. *Ann. Ist. Super. Sanita., 43*(4), 348–361.

De la Rosa, L. A., Moreno-Escamilla, J. O., Rodrigo-Garcia, J., & Alvarez-Parrilla, E., (2019). *Phenolic compounds. Postharvest Physiology and Biochemistry of Fruits and Vegetables* (1st edn., pp. 253–271) Wood head Publishing Co. Inc., Sawston, UK.

De O Buanafina, M. M., (2009). Feruloylation in grasses: Current and future perspectives. *Mol. Plant, 2*(5), 861–872.

Dhananjaya, B. L., Nataraju, A., Gowda, C. D. R., Sharath, B. K., & D'Souza, C. J. M., (2009). Vanillic acid as a novel specific inhibitor of snake venom 5'-nucleotidase: A pharmacological tool in evaluating the role of the enzyme in snake envenomation. *Biochem. Moscow., 74*, 1315–1319.

Dhananjaya, B. L., Nataraju, A., Rajesh, R., Raghavendra, G. C. D., Sharath, B. K., Vishwanath, B. S., & D'Souza, C. J., (2006). Anticoagulant effect of naja venom 5'nucleotidase: Demonstration through the use of novel specific inhibitor, vanillic acid. *Toxicon., 48*(4), 411–421.

Dignum, M. J. W., Kerler, J., & Verpoorte, R., (2001a). Vanilla production: Technological, chemical, and biosynthetic aspects. *Food Rev. Int., 7*(2), 119–120.

Dignum, M., Kerler, J., & Verpoorte, R., (2001b). Alpha-glucosidase and peroxidase stability in crude enzyme extracts from green beans of *Vanilla planifilolia* Andrews. *Phytochem. Anal., 12*, 174–179.

Dixon, R. A., (2014). *Vanillin Biosynthesis – Not as Simple as it Seem?* (Vol. 18, pp. 292–298). Hoboken, New Jersey.

Donaghy, J. A., Bronnenmeier, K., Soto-Kelly, P. F., & McKay, A. M., (2000). Purification and characterization of an extracellular feruloyl esterase from the thermophilic anaerobe *Clostridium stercorarium. J. Appl. Microbiol., 88*, 458–466.

Dos, S. W. D., Ferrarese, M. L. L., Nakamura, C. V., Mourao, K. S. M., Mangolin, C. A., & Ferrarese-Filho, O., (2008b). Soybean (*Glycine max*) root lignifications induced by Ferulic acid. The possible mode of action. *J. Chem. Ecol., 34*(9), 1230–1241.

Duke, J. A., (1992). *Handbook Of Phytochemical Constituents of GRAS Herbs and Other Economic Plants* (999th edn.). CRC Press. ISBN 978-0-8493-3865-6.

Dutt, S., (1925). General synthesis of (α)-unsaturated acids from malonic acid. *Quart. J. Chem. Soc., 1*, 297–301.

Efazary, A. E., Yi-Hsu, J., et al., (2007). Feruloyl esterases as biotechnological tools; current and future perspectives. *Acta Biochem. Biophys. sin., 39*(11), 811–828.

Faulds, C. B., & Williamson, G., (1991). The purification and characterization of 4-hydroxy-3-methoxycinnamic (ferulic) acid esterase from *Streptomyces olivochromogenes. J. Gen. Microbiol., 137*, 2337–2345.

Fazary, A. E., & Ju, Y. H., (2008). The large-scale use of feruloyl esterases in industry. *Biotechnol. Mol. Biol. Rev., 3*(5), 095–110.

Fry, S. C., (1982). Phenolic components of the primary cell wall. *Biochem. J., 203*, 493–504.

Fujita, A., Borges, K., Correia, R., De Melo, F. B. D. G., & Genovese, M. I., (2013). Impact of spouted bed drying on bioactive compounds, antimicrobial and antioxidant activities of commercial frozen pulp of camu–camu (*Myrciariadubia* Mc. Vaugh). *Food Res. Int., 54*, 495–500.

Gálvez, M. C., Barroso, C. G., & Pérez-Bustamante, J. A., (1994). Analysis of polyphenolic compounds of different vinegar samples. *Z Lebensm Unters Forch, 199*, 29–31.

Gildemeister, E., & Hoffmann, F. R., (1899). *Die Ätherischen Öle, 3*(3), 288. Julius Springer.
Graf, E., (1992). Antioxidant potential of ferulic acid. *Free Radic. Biol. Med., 13*(4), 435–448.
Hasyierah, N., & Salleh, M. (2008). *Ferulic Acid from Lignocellulosic Biomass: Review*. Malaysian Universities Conferences on Engineering and Technology (MUCET2008).
Havkin-Frenkel, D., & Belanger, F., (2008). *Biotechnology in Flavor Production* (pp. 86, 87). Blackwell Publishing Ltd.
Herrero, M., Castro-Puyana, M., Ibáñez, E., & Cifuentes, A., (2013). compositional analysis of foods. In: *Liquid Chromatography: Application* (pp. 295–317). Elsevier, Waltham, USA.
Hocking, M. B., (1997). Vanillin: Synthetic flavoring from spent sulfite liquor. *J. Chem. Educ., 74*, 1055–1059.
Huang, Z., Dostal, L., & Rosazza, J. P. N., (1993). Microbial transformations of ferulic acid by *Saccharomyces cerevisiae* and *Pseudomonas fluorescens*. *Appl. Environ. Microbiol., 59*, 2244–2250.
Iiyama, K., Lam, T. B. T., & Stone, B. A., (1994). Covalent cross-links in the cell wall. *Plant Physiol., 104*(2), 315–320.
Itoh, A., Isoda, K., Kondoh, M., Kawase, M., Kobayashi, M., Tamesada, M., & Yagi, K., (2009). Hepatoprotective effect of syringic acid and vanillic acid on concanavalin a-induced liver injury. *Biol. Pharm. Bull., 32*, 1215–1219.
Itoh, A., Isoda, K., Kondoh, M., Kawase, M., Watari, A., Kobayashi, M., Tamesada, M., & Yagi, K., (2010). Hepatoprotective effect of syringic acid and vanillic acid on CCl4-induced liver injury. *Biol. Pharm. Bull., 33*, 983–987.
Jirovsky, D., Horáková, D., Kotouček, M., Valentová, K., & Ulrichova, J., (2003). Analysis of phenolic acids in plant materials using HPLC with amperometric detection at a platinum tubular electrode. *J. Sep. Sci., 26*(8), 739–742.
Johnson, K. G., Silva, M. C., MacKenzie, C. R., Schneider, H., & Fontana, J. D., (1989). Microbial degradation of hemicellulosic materials. *Appl. Biochem. Biotechnol., 20, 21*, 245–258.
Jurková, M., & Wurst, M., (1993). Biodegradation of aromatic carboxylic acids by *Pseudomonas mira*. *FEMS Microbiol. Lett., 111*, 245–250.
Kang, J., Xie, C., Li, Z., Nagarajan, S., Schauss, A. G., Wu, T., & Wu, X., (2011). Flavonoids from acai (*Euterpe oleracea* Mart.) pulp and their antioxidant and anti-inflammatory activities. *Food Chem., 128*(1), 152–157.
Kaur, B., & Chakraborty, D., (2013). Biotechnological and molecular approaches for vanillin production: A review. *Appl. Biochem. Biotechnol., 169*(4), 1353–1372.
Kumar, N., & Pruthi, V., (2014). Potential applications of ferulic acid from natural sources. *Biotechnol. Rep., 4*, 86–93.
Kumar, R., Sharma, P. K., & Mishra, P. S., (2012). A review on the vanillin derivatives showing various biological activities. *Int. J. PharmTech. Res., 4*(1), 266–279.
Kvasnicka, E., & Kratzl, K. U.S. Patent 2,641,612 (1952 to Chemie Linz Ag).
Lesage-Meessen, L., Delattre, M., Haon, M., Thibault, J. F., Ceccaldi, B. C., Brunerie, P., & Asther, M., (*1996*). A two-step bioconversion process for vanillin production from ferulic acid combining *Aspergillus Niger* and *Pycnoporus cinnabarinus*. *J. Biotechnol., 50*(2, 3), 107–113.
Lim, J. C., (2004a). Anti-antigenic, anti-inflammatory and anti-nociceptive activities of vanillyl alcohol. *Arch. Pharm. Res., 10*, 1275–1279.
Lim, J. C., (2004b). Anti-inflammatory and anti-nociceptive activities of vanillyl alcohol. *Arch. Pharm. Res., 31*, 1275–1279.

Lin, C. M., Chui, J. H., Wu, I. H., Wang, B. W., Pan, C. M., & Chen, Y. H., (2010). Ferulic acid augments angiogenesis via VEGF, PDGF and HIF-1α. *J. Nutr. Biochem., 21*(7), 627–633.
Liu, S., Bischoff, K. M., Anderson, A. M., & Rich, J. O., (2016). Novel feruloyl esterase from *Lactobacillus fermentum* NRRL B-1932 and analysis of the recombinant enzyme produced in *Escherichia coli. Appl. Environ. Microbiol., 82*(17), 5068–5076.
Markus, P. H., Peters, A. L. J., & Roos, R., (1992). *Process for the Preparation of Phenylaldehydes*. Eur Patent Appl EP 0 542 348 A2.
Masai, E., Harada, K., Kitayama, H., Peng, X., Katayama, Y., & Fukuda, M., (2002). Cloning and characterization of the ferulic acid catabolic genes of *Sphingomonas paucimobilis* SYK-6. *Appl. Environ. Microbiol., 68*(9), 4416–4424.
Mathew, S., & Abraham, T. E., (2004). Ferulic acid: An antioxidant found naturally in plant cell walls and feruloyl esterases involved in its release and their applications. *Crit. Rev. Biotechnol., 24*(2, 3), 59–83.
Mathew, S., & Abraham, T. E., (2005). Studies on the production of feruloyl esterase from cereal brans and sugar cane bagasse by microbial fermentation. *Enzyme Microb. Technol., 36*, 565–570.
Menon, V., & Rao, M., (2012). Trends in bioconversion of lignocellulose: Biofuels, platform chemicals & biorefinery concept. *Prog. Energ. Combust., 38*(4), 522–550.
Mohamed, A., Rene, J., Heijden, A., & Udo, A., (1998). Trace-level determination of polar flavor compounds in butter by solid-phase extraction and gas chromatography–mass spectrometry. J. *Chromatography, 1*, 295–305.
Mourtzinos, I., Konteles, S., Kalogeropoulos, N., & Karathanos, V. T., (2009). Thermal oxidation of vanillin affects its antioxidant and antimicrobial properties. *Food Chem., 114*, 791–797.
Mussatto, S. I., & Teixeira, J. A., (2010). Lignocellulose as raw material in fermentation processes. In: Méndez-Vilas, A., (ed.), *Current Research, Technology and Education Topics in Applied Microbiology and Microbial Biotechnology* (pp. 897–907). Badajoz: Formatex.
Naira, R., Sonib, M., Balujab, S., & Chanda, S., (2004). Synthesis, structural determination and antibacterial activity of compounds derived from vanillin and 4-aminoantipyrine. *J. Serb. Chem., 12*, 991–998.
Nazareth, S., & Mavinkurve, S., (1986). Degradation of ferulic acid via 4-vinylguaiacol by *Fusarium solani* (Mart) sacc. *Can. J. Microbiol., 32*, 494–497.
Odoux, E., & Havkin-Frenkel, D., (2005). *Hydrolysis of glucovanillin by β-glucosidase during curing of vanilla bean (Vanilla planifolia Andrews)*. In: *Vanilla: The First International Congress* (pp. 95–100). Allured, Carol Stream.
Odoux, E., Escoute, J., Verdeil, J. L., & Brillouet, J. M., (2003). Localization of β-D-glucosidase activity and glucovanillin in vanilla bean (*Vanilla planifolia* Andrews). *Ann. Bot., 92*, 437–444.
Overhage, J., Priefert, H., & Steinbüchel, A., (1999a). Biochemical and genetic analyses of ferulic acid catabolism in *Pseudomonas* sp. strain HR199. *Appl. Environ. Microbiol., 65*, 4837–4847.
Overhage, J., Priefert, H., Rabenhorst, J., & Steinbuchel, A., (1999b). Biotransformation of eugenol to vanillin by a mutant of *Pseudomonas* sp. strain HR199 constructed by disruption of the vanillin dehydrogenase (*vdh*) gene. *Appl. Microbiol. Biotechnol., 52*, 820–828.
Pacheco-Palencia, L. A., Mertens-Talcott, S., & Talcott, S. T., (*2008*). Chemical composition, antioxidant properties, and thermal stability of a phytochemical enriched oil from açaí (*Euterpe oleracea* Mart.). *J. Agric. Food Chem., 56*(12), 4631–4636.
Paiva, L. B., Goldbeck, R., Santos, W. D., & Squina, F., (2013). Ferulic acid and derivatives: Molecules with potential application in the pharmaceutical field. *Braz. J. Pharm. Sci., 49*(3), 395–411.

Podstolski, A., Havin-Frenkel, D., Malinowski, J., Blount, J. W., Kourteva, G., & Dixon, R. A., (2002). Unusual 4-hydroxybenzaldehyde synthase activity from tissue cultures of the vanilla orchid *Vanilla planifolia. Phytochemistry, 61*, 611–620.
Priefert, H., Rabenhorst, J., & Steinbuchel, A., (2001). Minireview: Biotechnological production of vanillin. *Appl. Microbiol. Biotechnol., 56*, 296–314.
Rai, R. P., & Maurya, M. S., (1966). Synthesis and evaluation of antibacterial activity of vanillin derivatives. *J. Sci. Technol. India, 4*, 275–276.
Ramachandra, R. S., & Ravishankar, G. A., (2000b). Biotransformation of protocatechuic aldehyde and caffeic acid to vanillin and capsaicin in freely suspended and immobilized cell cultures of *Capsicum frutescens*." *J. Biotechnol., 76*, 137–146.
Ranadive, A., (1994). *Vanilla--Cultivation, Curing, Chemistry, Technology and Commercial Products.* Developments in food science.
Rechner, A. R., Pannala, A. S., & Rice-Evans, C. A., (2001). Caffeic acid derivatives in artichoke extract are metabolized to phenolic acids *in vivo. Free Radic. Res., 35*(2), 195–202.
Rosazza, J. P. N., Huang, Z., Dostal, L., Volm, T., & Rousseau, B., (1995). Review: Biocatalytic transformations of ferulic acid: An abundant aromatic natural product. *J. Ind. Microbiol., 15*, 457–471.
Ruiz-Terán, F., Perez-Amador, I., & López-Munguia, A., (2001). Enzymatic extraction and transformation of glucovanillin to vanillin from vanilla green pods. *J. Agric. Food Chem., 49*(11), 5207–5209.
Sadh, P. K., Duhan, S., & Duhan, J. S., (2018). Agro-industrial wastes and their utilization using solid state fermentation: A review. *Bioresour. Bioprocess, 5*, 1.
Salleh, N. H. M., Daud, M. Z. M., Arbain, D., Ahmad, M. S., & Ismail, K. S. K., (2011). Optimization of alkaline hydrolysis of paddy straw for ferulic acid extraction. *Ind. Crops Prod., 34*(3), 1635–1640.
Saulnier, L., Vigouroux, J., & Thibault, J. F., (1995). Isolation and partial characterization of feruloylated oligosaccharides from maize bran. *Carbohydr. Res., 272*, 241–253.
Sawano, T., & Yazama, F., (2011). Evaluation of antioxidant activity of vanillin by using multiple antioxidant assays. *J. Biophys. Acta,* (2), 170–700.
Schrader, J., Etschmann, M. M. W., Sell, D., Hilmer, J. M., & Rabenhorst, J., (2004). Applied biocatalysis for the synthesis of natural flavor compounds – current industrial processes and future prospects. *Biotechnol. Lett., 26*(6), 463–472.
Sen, A., Grosch, W., (1992). Potent odorants of the roasted powder and brew of Arabica coffee. Zeitschrift für lebensmittel-untersuchung und. hung "*Journal de Pharmacie et de Chimie,'3*, 239–245.
Shin, H. D., & Chen, R. R., (2006). Production and characterization of a type B feruloyl esterase from *Fusarium proliferatum* NRRL 26517. *Enzyme Microb. Technol., 38*, 478–485.
Shin, H. D., & Chen, R. R., (2007). A type B feruloyl esterase from *Aspergillus nidulans* with broad pH applicability. *Appl. Microbiol. Biotechnol., 73*, 1323–1330.
Shin, H. D., McClendon, S., Le, T., Taylor, F., & Chen, R. R., (2006). A complete enzymatic recovery of ferulic acid from corn residues with extracellular enzymes from *Neosartorya spinosa* NRRL185. *Biotechnol. Bioeng., 95*, 1108–1115.
Singh, J. C. H., Kakalij, R. M., Kshirsagar, R. P., Kumar, B. H., Komakula, B. S. B., & Diwan, P. V., (2015). Cognitive effects of vanillic acid against streptozotocin-induced neurodegeneration in mice. *Pharm. Biol., 53*(5), 630–636.

Singh, J. P., Kaur, A., Singh, N., Nim, L., Shevkani, K., Kaur, H., & Arora, D. S., (2016). *In vitro* antioxidant and antimicrobial properties of jambolan (*Syzygium cumini*) fruit polyphenols. *LWT-Food Sci. Technol., 65*, 1025–1030.

Sinha, A. K., Sharma, U. K., & Sharma, N., (2008). A comprehensive review on vanilla flavor: Extraction, isolation and quantification of vanillin and others constituents. *Int. J. Food Sci. Nutr., 59*(4): 299–326.

Sørensen, H. R., Meyer, A. S., & Pedersen, S., (2003). Enzymatic hydrolysis of water-soluble wheat arabinoxylan. 1. Synergy between α-L-arabinofuranosidases, endo-1,4-β-xylanases, and β-xylosidase activities. *Biotechnol. Bioeng., 81*(6), 726–731.

Srinivasan, M., Sudheer, A. R., & Menon, V. P., (2007). Ferulic Acid: Therapeutic potential through its antioxidant property. *J. Clin. Biochem. Nutr., 40*(2), 92–100.

Stanely, P., Rajakumar, S., & Dhanasekar, K., (2011). Protective effects of vanillic acid on electrocardiogram, lipid peroxidation, antioxidants, proinflammatory markers and histopathology in isoproterenol induced cardiotoxic rats. *Eur. J. Pharmacol., 668*, 233–240.

Sutherland, J. B., Crawford, D. L., & Pometto, III. A. L., (1983). Metabolism of cinnamic, *p*-coumaric and ferulic acids by *Streptomyces setonii*. *Can. J. Microbiol., 29*, 1253–1257.

Tilay, A., Bule, M., Kishenkumar, J., & Annapure, U., (2008). Preparation of ferulic acid from agricultural wastes: Its improved extraction and purification. *J. Agric. Food Chem., 56*, 7644–7648.

Topakas, E., & Christakopoulos, P., (2004). Production and partial characterization of alkaline feruloyl esterases by *Fusarium oxysporum* during submerged batch cultivation. *World J. Microb. Biot., 20*, 245–250.

Topakas, E., Stamatis, H., Biely, P., & Christakopoulos, P., (2004). Purification and characterization of a type B feruloyl esterase (StFAE-A) from the thermophilic fungus *Sporotrichum thermophile*. *Appl. Microbiol. Biotechnol., 63*, 686–690.

Umre, R., Ganeshpurkar, A., Ganeshpurkar, A., Pandey, S., Pandey, V., Shrivastava, A., & Dubey, N., (2018). *In vitro, in vivo* and *in silico* antiulcer activity of ferulic acid. *Future J. Pharma. Sci., 4*(2), 248–253.

Van, D. H. R. R. H., Fraaije, M. W., Laane, C., & Van, B. W. J. H., (2001). Enzymatic synthesis of vanillin. *J. Agric. Food Chem., 49*, 2954–2958.

Walton, N. J., Mayer, M. J., & Narbad, A., (2003). Vanillin. *Phytochemistry, 63*, 505–515.

Wang, Y., Guanzhong, W., & Jiangchuan, L., (1998). Design, synthesis and hypoglycemic activity of 3-methyl-1-phenyl-4-{4-[(5-methyl-2-phenyloxazol-4 yl)methoxy] benzyl (benzyl)}-2-pyrazol-5-one. *Chem. Mater. Sci., 3*(1), 118–123.

Wong, D. W. S., (2006). Feruloyl esterase. *Appl. Biochem. Biotechnol., 133*(2), 87–112.

Wong, D. W., Chan, V. J., Batt, S. B., Sarath, G., & Liao, H., (2011). Engineering *Saccharomyces cerevisiae* to produce feruloyl esterase for the release of ferulic acid from switchgrass. *J. Ind. Microbiol. Biotechnol., 38*(12), 1961–1967.

Zenk, M. H., (1965). Biosynthese von vanillin in *Vanilla planifolia* Andr. *Z. Pflanzenphysiology, 53*, 404.

Zielinski, A. A. F., Ávila, S., Ito, V., Nogueira, A., Wosiacki, G., & Haminiuk, C. W. I., (2014). The association between chromaticity, phenolics, carotenoids, and *in vitro* antioxidant activity of frozen fruit pulp in Brazil: An application of chemometrics. *J. Food Sci., 79*, C510–C516.

CHAPTER 4

Multifaceted Utilization of Microalgal Biomass Towards Industrial Applications

O. N. TIWARI,[1] DIPANKAR GHOSH,[2] SHRESTHA DEBNATH,[2] MINAKSHI SAHU,[1] and KONDI VANITHA[3]

[1]*Centre for Conservation and Utilization of Blue Green Algae, Division of Microbiology, ICAR-Indian Agricultural Research Institute, New Delhi-110012, India, E-mail: ontiwari1968@gmail.com (Dr. O.N. Tiwari)*

[2]*Microbial Engineering and Algal Biotechnology Laboratory, Department of Biotechnology, JIS University, Kolkata, West Bengal, 700109, India, E-mail: dghosh.jisuniversity@gmail.com or d.ghosh@jisuniversity.ac.in (Dr. Dipankar Ghosh)*

[3]*Department of Pharmaceutics, Vishnu Institute of Pharmaceutical Education and Research, Narsapur, Medak, Telangana, 502313, India*

ABSTRACT

Microalgae are the broad spectrum of photosynthetic autotrophs which have been found in marine as well as fresh water bodies in all around the globe. Microalgae have salient features, i.e., shorter generation time, higher growth rate, higher biomass productivities, higher substrate conversion efficiency, higher light conversion efficiency and higher productivity of high value added products through carbon dioxide sequestration. It has also been observed that microalgae biomass is very potential microbial cell factories to biosynthesize diverse ranges of value-added biomolecules generations, i.e., food, bioenergy, food supplements, therapeutic compounds, biopigments, biofertilizers, animal feeds, and cosmetics, etc. To this end it seems

Biotechnology for Waste Biomass Utilization. Prakash Kumar Sarangi & Latika Bhatia (Eds.)

microalgal biomass have multifaceted utility towards sustainable industrial and biorefinery implications. Based on these current facts and figures, current chapter deals with detail literature survey on aforementioned scientific areas to highlight microalgal potential towards community services in near future.

4.1 INTRODUCTION

Microalgae are used as sustainable microbial cell factories to biosynthesis of promising high potential biomaterials likely bioactive compounds, biobased fuels, feed supplements including many more utilizing cheaper raw materials, i.e., waste biomass. Microalgae have been used as alternative resources over unsustainable overexploitation of various natural constituents which gives us various high value-added biomolecules. Broad spectrum of market analysis has been shown that diverse ranges of microalgae are most potential industrial biocatalysts. Micro algal biomass generation has been paid great attention due its economic feasibility and ability to grow in diverse spectrum of extreme environmental conditions (Kumar et al., 2014; Gonzalez et al., 2016). Microalgae predominantly belong to clades of prokaryotes and eukaryotes either in unicellular or multicellular form. These diverse ranges of micro algae have been fixed atmospheric greenhouse gases, i.e., carbon dioxide and transformed it into greener gas, i.e., oxygen through photosynthesis. Broad spectrums of micro algal regimes are tolerant to harsh environmental stress factors. *Nostoc* sp., is most prevalence micro algal species which has been dramatically being used since several decades. Moreover, microalgae can able to generate huge amount of biomass through the utilization of minute amount of nutrient and water. Even micro algae do have higher potential towards utilization of industrial waste water as cheaper feedstocks to fetch out high value added sustainable products for community services. A statical survey has shown that 1 kg of algal biomass generation requires approximately 333 liters of water. In contrary, agriculture crop production, i.e., 1 kg of soy bean production requires 2204 liters of water towards its cultivation (Liu et al., 2016; Bhalamurugan et al., 2018). Broad ranges of micro algae species exist in nature including *Haematococcus pluvialis, Chlorella vulgaris, Dunaliella species, Nostoc* spp., *Arthrospira (Spirulina)* have been utilized for valuable products generations (Sharma and Sharma, 2017). *Arthrospira (Spirulina)* and *Chlorella sp.* are one of the most usable microalgal communities in food and bio-neutraceuticals industries. Even there are several micro algae, i.e., *I. galbana, N. opsisoculata, C. muelleri, C. gracilis, P. Tricornutum,* which are potential cell factories to generate therapeutic biomolecules concerning antioxidant, anti-inflammatory, antitumor, and anticancer, anti-aging, anti-tanning attributes (Liu et al., 2016). Micro algae harvesting approaches are ecofriendly and economically feasible as it recycles diverse ranges of waste materials (Wolkers et al., 2011). Microalgae are the most potent reservoirs of rich commodity biochemicals including biopigments, essential proteins, polysaccharides, fatty acids and lipids, etc. These micro algae-based bio active molecules have been utilized several decades for various industrial productions of cosmetics, feed additives, aquaculture foods, therapeutic agents, bioenergy production. However, current industrial implementation of micro algae based biorefinery approaches have been facing few major hindrances to attain highest commercialization due to higher production extraction cost, lacking of purification methodologies, improper biocatalytic processes towards value added production biosynthesize utilizing wastes (Ventura et al., 2017). Elevated micro algal biomass productivity has been predominantly derived through the assimilation of carbon, nitrogen and phosphorous content in the feedstocks (Wijffels and Barbosa, 2010; Chandra et al., 2016). Day by day increase in global warming, environmental pollution, food crisis and energy demands are the most common issues in today's world. Moreover, worldwide mass human population have been expected to reach up to 9 billion by 2050. To this end, food and energy crisis will be spontaneously growing as human communities expand. Microalgae are the pivotal key role players for high value-added biomolecules production through carbon dioxide sequestration to minimize the acceleration of global warming and environmental pollution rate. It has been reported that 1.8 kgs of carbon dioxides will be fixed through the generation of 1 kg of micro algal biomass. Even micro algae-based bioenergy generations are highly admirable than do the alternative fossil fuels (Uduman et al., 2010; Medipally et al., 2015; Raheem et al., 2018; Sharma and Sharma, 2017). Moreover, over the past 50 years, micro algae-based biomass cultivation approaches have been paid accelerated attention towards high value-added products generations (Figure 4.1). Therefore, the current chapter deals with extensive literature survey on most of the industrial and commercial utilization of micro algal biomass to motivate the production of next generation value added bio commodities production to attain economic sustainability of industries and resolve environmental perspectives in near future.

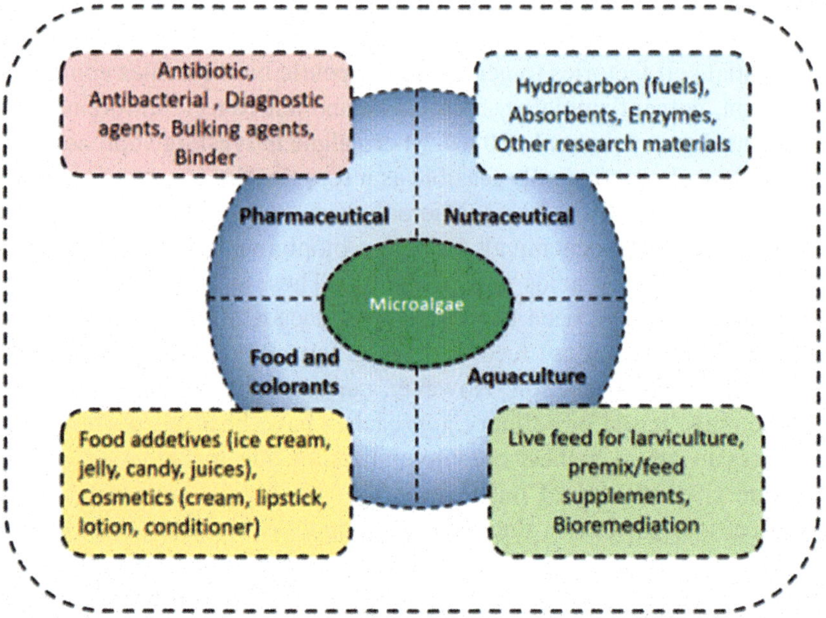

FIGURE 4.1 Commercial application of microalgae for various approaches (Begum et al., 2016).

4.2 SUSTAINABLE UTILIZATION OF MICRO-ALGAL BIOMASS TOWARDS SUSTAINABILITY

4.2.1 MICROALGAE AS POTENTIAL BIOENERGY PRODUCERS

Energy resources are one of the most unavoidable essential requirements to sustain the universal economic growth towards enhancing quality of living. Global population and consumption of fossil fuel reserves have been increasing in rapid pace along with global depletion of fossil fuel reserves in a regular basis. Hence, the bioenergy generations from micro algal biomass is entirely a new platform having innovation to sustain global mass population demands (Ratha and Prasanna, 2012). Algae is the most probable energy reservoir to pave an alternative way out to decline fossil fuel usage, maintain the ecological balance, food safety and secure regular energy domain, etc. (Sharma and Sharma, 2017). Therefore, various commercial application tends to microalgae become a most potential future platform towards value added biomolecule biosynthesis bioremediation of waste biomass (Ratha and Prasanna, 2012).

4.2.1.1 BIOFUELS

Biofuels are energy sources which exist as solid, gaseous and liquid forms. Biofuels have been derived from diverse ranges of cheaper raw materials including soybean, corn, rapseed, sunflower, palm without interfering the food chain (Sharma and Sharma, 2017; Odjadjare et al., 2017). Micro algae-based biomass has been recognized as a prospect staple to generate biofuels using various bioprocess engineering approaches. Micro algae-based biofuels production is beneficial due to its higher biomass yield, high natural oil production capability, nontoxicity, economic sustainability, CO_2 fixation efficacies; higher efficiency to triaceyl glycerol (TAG) accumulation. Microalgal biomass can show higher productivities of aforementioned biomolecules which ranges from 1000–4000 gallon/acre/yr compared to any other cheaper raw feedstock utilizations. Broad spectrums of avenues are in trial for biotransformation of microalgal biooils into value added energy generations. These diverse ranges of approaches include transesterification, green diesel and gasoline replacement method, hydrothermal conversion, fermentation, thermochemical conversion methods, gasification methods for syn gas production, thermochemical liquefaction, pyrolysis, and anaerobic digestion. Microalgae are capable of rapidly synthesis the photosynthetic energy. Microalgal photosynthesis transforms 9–10% of solar energy (e.g., synthesis of electron, proton for bio hydrogen production, oils for biodiesel production, starch, sugar for bioethanol production, biomass for biomethane production, etc.) and produce biomass 77 g/m^2 /day as well as 280 ton/ha/year. Some microalgal species *T. suecia, B. braunii, C. vulgaris, C. cohnii, Cylindrotheca* sp., *Nitzschia* sp., *Schizochytrium* sp., *P. Tricornutum* carry higher quantities of biooils per unit dry weight (i.e., 15–75%) which are potential sources for any bioenergy production (Suganya et al., 2016; Priyadarshani and Rath, 2012; Khan et al., 2018; Gouveia, 2011; McGinn et al., 2011). During the cell disruption methods ionic liquids and proteins are extracted from microalgae which are essential for consumption of energy (Koopmans et al., 2013). Moreover, different stress factors induce intracellular microalgal production of triglycerides through enormous shifts to cellular lipid and synthesize oleic acids, palmitoleic acids, palmitic acids, zeolites and silica alumina used as new generation catalyst for raising the fuel production from triglycerides (Zhao et al., 2013). Few scientific reports have shown that microalgae are also potential for reduction of greenhouse gas (GHG) emissions, which is 25% less than regular diesel combustion (Kumar et al., 2010). *Botryococcus braunii* produces higher quantities of lipids which carry considerable amount of monounsaturated, polyunsaturated fatty acids up to 75% (w/w), which are used as reliable sources of biofuel (e.g., biodiesel) biocatalysis (Koller et al., 2012; Wijffels et al., 2010; Tiwari et al., 2020). Microalgae *S. obliquus, C. reinhardtii, C. vulgaris* are good energy carrier towards generation of biogases (Koller et al., 2014). Many scientific literatures have shown that dried microalgal cellular biomass (DMCB) of *Nannochloropsis sp, Scenedesmus obliquus, Lyngbya majuscule, Chlorella vulgaris, Chlorella sorokiniana, Auxenochlorella protothecoides, Microspora* species enhance the generations of biogases by 40 percent. Aforementioned potential microalgal species are also capable to generate bioethanol, biohydrogen, biomethane generation under permissive physico chemical conditions. *Microspora sp.* and *Chlorella variabilis* are one of the efficient microalgal species which produce 426.26 and 446.02 $mL.d^{-1}$ yielded biogas, respectively. *Chlorella vulgaris* produces 11.7 g/L ethanol using lignocellulosic biomass under mixotrophic or heterotopic conditions. *Chlorella vulgaris* also produces extensive amount of methane. It also produce highest amount of lipid (22.9 $mg.L^{-1}.L.d^{-1}$) and biomass (0.074 $g.L^{-1}.L.d^{-1}$) during the waste water treatment (Maurya et al., 2016; Khan et al., 2018; Varfolomeeva and Wassermanb, 2011; Guldhe et al., 2017). At present, bioprocess industries have been accommodating few novel approaches, i.e., some upstream strategies (enhance lipid productivity, enhance growth for improving biomass quantity, etc.), biorefinery processes for reduce the cost and to get more profit from the microalgal energy production (Garciaa et al., 2017; Zhu and Hiltunen, 2016). *Nannochloropsis* sp. and *Dunaliella salina* produce around 68% and 66% of biodiesel per dry biomass weight of microalgal regimes (Bhalamurugan et al., 2018). *Scenedesmus obliquus* has depicted already having higher potentiality for biofuels generations using cheaper feedstocks. It contains 40–55% lipid per dry weight biomass. With the dry method it increases the lipid content 18.95%. During the cultivation in photobioreactor it enhances the triaceylglycerol content 57 ± 0.2% DW (Pignolet et al., 2013; Shubaa and Kifle, 2018; Kapoore et al., 2018). Some studies have shown that reduction the particular size of microalgal cell increases lipid recovery and enhance the lipid production. Small amount of microalgal dried powder have enormous effectivities for re-adsorption large amount of lipid biomolecules. By the extraction of supercritical carbon dioxide from very small amount of microalgal biomass recovers very high amount of biooils (Klinthong et al., 2015). *Neochloris oleoabundans* HK-129 microalgal strain synthesizes of 1.2–1.7 g/L biomass with increase in light intensity ranges from 50–200 $\mu mol/m^2/s$. *C. zofingiensis* produces 58% dry weight lipid (potential precursor for biodiesel generation) upon higher light intensities and nitrogen deprivation (Show et al., 2017). Physical independent process variable, i.e., temperature has shown great positive impact on lipid biosynthesis and accumulation in *Nannochloropsis oculata, Chlorella protothecoides, Haematococcus pluvilalis, Prophyridium cruentum,* and *Isochrysis galbana* (Katiyar et al., 2017). *Scenedesmus dimorphus* brings forth 53.7 w/w carbohydrates by the hydrolysis of 80% fermentable accessible residual sugars for bioethanol production (Khan et al., 2018). *Nannochloropsis* sp., *Nostoc* sp., *D. salina, G. partita,* and *Cosmarium* sp, are potential resources for acetone, ethanol, butanol production via the fermentation process (Rodrigues et al., 2015). *Thalassiosira pseudonana, Nitzschia frustulumm, Skeletonema costatum, Micromonas pusilla, Microcystis aeruginosa, Cyclotella cryptica, Chaetoceros gracilis, Aphanothece microscopica Nageli,* and *Porphyridium cruentum* microalgal species are potential for syngas and biohydrogen production via gasification process (Picardo et al., 2013). Microalgal transesterification produce 35–41(MJ/kg) biodiesel during the fermentation. In contrary, hydrothermal treatment, anaerobic digestion, biological production process using micro algae produce 23.4 (MJ/kg), 37.2 (MJ/kg), 33–39 (MJ/kg), 144 (MJ/kg) of bioethanol, biogas, bio-oil, biohydrogen, respectively (Usher et al., 2014). *Dunaliella salina, Dunaliella tertiolecta,* and *Dunaliella acidophila* species contain higher amount of hydrocarbons, unsaturated fatty acids, lipids towards bioconversion of liquid biofuels (e.g., ethanol 11.0 $mg.g^{-1}$) which have enormous sustainability towards industrialization. *Chlorococum sp.* produce 3.83 $g.L^{-1}$ ethanol, *Neochlorosis oleabundans* produce 56.0 $g.g^{-1}$ biodiesel fuels, *Chlamydomonas reinhardtii* produce 2.5 $mL.h^{-1}$ biohydrogen, *and Spirulina platensis* produce 1.8 $mol.mg^{-1}$ biohydrogen, *S. platensis UTEX 1926* produces 0.4 $m^3.kg^{-1}$ biomethane, *Spirulina sp.* (LEB 18) produces approximately 0.8 $g.L^{-1}$ methane (Sudhakar and Premalatha, 2015; Costa and Morais, 2011). Very recently algae industry magazine, USA and Oilgae (algal fuel industry, Tamilnadu) in India has been continuously working on algal biofuels generations along with algal co-products biosynthesis (Han et al., 2015).

4.2.1.2 BIODIESEL

Currently microalgae are used as potential biocatalysts for biodiesel generation using diverse ranges of waste materials under different environmental accessibility zone. However, major pitfall of micro algae driven biomolecule generations is higher production cost especially downstream processing expenditure. Although some industries adopt bioenergy production processes to minimize the gross production cost as much as possible. There are various steps have been involved for micro algae-based biodiesel production (e.g., transesterification, lipid rich biomass production, harvesting, microalgal cell disruption, extraction, drying from oil bodies, etc.) (Chunga et al., 2017; Katiyar et al., 2017). At present an extensive quantities of biodiesel has been produced using soybean oil, palm oil, rapeseed oils, etc. various seed oils as major feedstock or initial substrates. However, this approach has been interfering with the food chain in the human community. Microalgae based biodiesel production has not been competing with food chain. This salient feature of microalgal biodiesel generation provides additional merit to it compared to chemical approaches. As per chemical reaction lipids or triacylglycerols or free fatty acids have been converted into biodiesel through microalgae based transesterification reactions. Micro algal crude oil does have higher viscosity. Moreover, high molecular weight free fatty acids need to be transformed into low molecular weight by transesterification process. In this transesterification bioprocess, crud algal oils react with C1 alcohol likely methyl alcohol in association with a chemical accelerator to generate precursor of fatty acid methyl ester (FAME) which ultimately converts in to biodiesel along with glycerol as major by product. Microalgae, i.e., *C. vulgaris* and *C. protothecoides* contain elevated amount of microbial fatty acid based oils to produce higher biodiesel yield. The amounts of oils 16 to 68% dry weight have been produced by microalgal biomass. Also, microalgae contain 70% dry weight lipids (triglycerides) under the nitrogen deprivation condition. *Chlorella protothecoides* contains 55.2% dry weight crude lipids where KOH utilized as chemical accelerator during transesterification. *Euglena* sp., *Spirogyra* sp., *Scenedesmus* sp., *Pseudokirchneriella* sp., *Phormidium* sp, *Nitzschia* sp., and *Desmodesmus* sp., contain elevated amount of unsaturated fatty acid and fatty alcohols. These unsaturated fatty acid-based precursors are major backbone for production of biodiesel. Microalgae are not only the major contributor as biodiesel producers as well microalgae reduce carbon dioxide level in atmosphere for minimizing the impact of global warming and air pollution. *Chlorella vulgaris* can sequester 74% of the CO_2 during the photoautotrophic growth in a proper simulated physico chemical condition with a photobioreactor. *Chlorella vulgaris* is a novel microalgal species for biodiesel production as well as CO_2 fixation (Sharma and Sharma, 2017; Chu, 2012; Abdelaziz et al., 2014). *C. vulgaris* contains high amount of starch (~37% dry wt), saturated fatty acids (22.65 %), unsaturated fatty acids (~77%), C14-C18 chain length of fatty acids, and ethanol (65 %). It is a good biocatalytic microorganism towards biodiesel production. *Chlorella vulgaris* produces 3726 kg-water/kg biodiesel during the open pond systems (Maeda et al., 2018; Choi, 2016). *Scenedesmus* sp. and *Botryococcus* sp. produce ~10–28 g.L^{-1} total lipids, respectively. The mixture of two microalgal species (*Phaedoactylum tricornutum* and *Porphyridium purpuruem*) shows great potential towards algal-diesel biosynthesis. *Nannochloropsis* sp. bears ~86% (w/w) neutral lipids through hexane extraction. *Euglena gracilis, Monochrysis lutheri, Dunaliela salina* synthesize di-galactosyl-diglycerides, polyunsaturated C16-C18 fatty acids, sphingolipids, phosphoglyceride, under impact of broad spectrum of light intensity and temperature (Odjadjare et al., 2017). *Chlamydomonas pitschmannii* contains 51% of lipid (Singh et al., 2014). *Chlorella sorokiniana* is predominant biocatalyst as good quality of algal diesel producer. *S. obliquus* contains comparatively elevated amount of 56.4% saturated and 43.6% unsaturated esters, and 11.42% of linoleate. The saturated esters have higher potentiality towards biodiesel production. *Scenedesmus dimorphus* biosynthesizes 44% of free fatty acid methyl esters along with alcohol based biomolecules, i.e., methanol or ethanol or propanol after the transesterification. *Scenedesmus bijuga* is another biodiesel producer utilizing diverse ranges of waste waterbodies. The highest productivity of *Scenedesmus bijuga* is dealing with 24.66 mg. $L^{-1}.d^{-1}$. *Scenedesmus acuminatus* has been considered as an economically feasible microbial cellular factory for biodiesel production. *Dunaliella sp.* and *Nannochloropsis sp.* produce biodiesel production yield of 66.6% and 68.5% (dry basis) per unit biomass content. *D. salina* contains 1:2 (35%: 65% dry basis) of saturated and unsaturated fatty acids. Even *D. salina* produces higher amount of FAME (fatty acid methyl ester) including linolenate and linoleicate major precursors as algae based diesel producer. Moreover, *D. salina* has higher oxidative stability to combat different stress conditions during biodiesel generation. *Nannochloropsis gaditana, N. granulate, N. salina, N. ocenaica, N. oculata* has been considered as excellent microalgal strains towards algal diesel generation having its better lipid generation rates (158.76 ± 13.83 mg.$L^{-1}.d^{-1}$) (Bhalamurugan et al., 2018; Taleb et al., 2015). Algal cellular mass, fatty acids and biodiesel generation depend on different extraction methods, i.e., extraction of fatty acid based bio-oils, physical extraction, and chemical extraction, etc. (Bux, 2013). Few comparative scientific reports have shown that microalgae is giving 70% oil per unit biomass 136 Oil titer (L per hector) and 1,21,104 Kg biodiesel per hector per year, 30% oil in biomass 58 Oil yield L.ha^{-1} and 51,927 Kg biodiesel.ha^{-1}.$year^{-1}$, 50% oil in biomass (86,515 Kg biodiesel.ha^{-1}.$year^{-1}$), there corn, soybean, palm oil give 169,443,5938 Oil yield L.ha^{-1} and

152,562,4747 Kg biodiesel per hector per year, respectively (Frac et al., 2010; Patel et al., 2016; Aguirre et al., 2013; Farieda et al., 2017). During transesterification process, acid-alkali both is used as catalyst but some study has been shown that alkali is not suitable for microalgal lipid transesterification. *Spirogyra sp, Oedogonium sp.* have produced a good amount of biodiesel along with sodium hydroxide catalyst and hexane as co-solvent. *Schizochytrium limacinum* produces 50% Wt. biodiesel by using chloroform as solvent. *Nannochloropsis oculata* shows highest activity with the using of CaO and MgO catalysts which produce the highest 97.5% biomass yield. With the enzymatic reaction (lipase) microalgae *Chlorella protothecoides* produces higher yielded biodiesel (98 percent) using of lipase enzyme of *Candidiasis sp. Chaetoceros gracilis* microalgal species contain longer chain of primary alcohols likely bioethanol, butanol, 3-methyl 1-butanol, 2-methyl 1-propanol, etc. those are capable to produce biodiesel through direct transesterification process (Hidalgo et al., 2013). *Cynophyceae sp., Nannochloropsis* sp., *S. platensis* contain elevated quantity of fatty acids, lipids, palmitate such as C18:0, C16:1, C20:0 respectively which influence to direct transesterification for biodiesel production. Some microalgal species and their lipid productivity accelerate production of biodiesel (e.g., *Ankistrodesmus falcatus* (56.07 ± 1.75 mg/L/day), *Chaetoceros muelleri* (21.8 mg/L/day), *Chaetoceros calcitrans* (17.6 mg/L/day), *Chlorella emersonii* (10.3–50.0 $mg.L^{-1}.day^{-1}$), *C. sorokiniana* (44.7 $mg.L^{-1}.day^{-1}$), *C. vulgaris* (11.2–40.0 $mg.L^{-1}.day^{-1}$), *Chlorococcum sp.* (53.7 $mg.L^{-1}.day^{-1}$), *Dunaliella salina* (116.0 $mg.L^{-1}.day^{-1}$), *Ellipsoidion sp.* (47.3 $mg.L^{-1}.day^{-1}$), *Isochrysis sp.* (37.8 mg/L/day), *Monodus subterraneus* (30.4 mg/L/day), *Nannochloris sp.* (~61–77 $mg.L^{-1}.day^{-1}$), *Nannochloropsis oculata* (84.0–142.0 mg/L/day), *Pavlova salina* (49.4 mg/L/day), *Pavlova lutheri* (40.2 mg/L/day), *Phaeodactylum tricornutum* (44.8 mg/L/day), *Porphyridium cruentum* (34.8 mg/L/day), *S. quadricauda* (35.1 $mg.L^{-1}.day^{-1}$), *Scenedesmus sp.* (40.8–53.9 mg/L/day), *Tetraselmis suecica* (27.0–36.4 $mg.L^{-1}.day^{-1}$), *Tetraselmis sp.* (43.4 mg/L/day). Microalgae *Haematococcus pluvialis* can abide 34% CO_2 fixation rate of ~0.14 $gm.L^{-1}.d^{-1}$. Moreover, *Chlorella sp., Scenedesmus sp., Dunaliella sp.*, etc. are also capable of fixing the CO_2 and can be applied in the industrial purposes (Farieda et al., 2017; Patel et al., 2016; Kiran et al., 2014). Some experimental evidence has been depicted that the close system photobioreactor produce more microalgal biomass than open system photobioreactor which is quite suitable for biodiesel production. *Neochloris oleoabundans* produce 0.4188 ($gm.L^{-1}.d^{-1}$) DW biomass comparatively higher than others microalgal species (Arenas et al., 2016). *Haematococcus pluvialis*, *Neochlorosis oleabundans*, *Chlorella protothecoides*, and *Chlorococcum sp.* have shown biodiesel productivities of 420 GJ/ha/year, 56.0 gm.gm^{-1}, 15.5 gm.L^{-1}, and 10.0 gm.L^{-1} (Farieda et al., 2017). *Neochloris oleoabundans, Isochrysis zhangjiangensis, Chlorella sorokiniana, Pseudochlorococcum sp.* produce lipids contents of about 53.8%, 53%, 44%, and 52.1% per unit cell dry weight biomass, respectively under nitrogen starved condition. *Chlamydomonas reinhardtii, Monodus subterraneus, Chlorella vulgaris* produce higher lipids under phosphorus starvation condition. In another study, it has been shown that *Chlorella vulgaris generates* maximum lipid productivity of around 58.39 mg/L/day. *Heterochlorella luteoviridis* produces 83% lipid following the ethanol pretreatment. *Scenedesmus sp.* yields ~28% of lipid (gm^{-1} of dry biomass) through microwave method. *Nannochloropsis oculata* produces 0.142 g/L/d lipid with the 0.480 g/L/d maximum biomass which are effective for biodiesel production. *Nannochloropsis gaditana* produces 97.8% weight FAEE (fatty acid ethyl ester) and 99.5 weight % FAME with highest algal diesel yields following the transesterification (Goha et al., 2019; Kiran et al., 2014). Most potent avenue, i.e., transesterification process induces microalgae to produce high amount of biodiesel yields in different microalgal regimes likely *Chlorella protothecoides* (80% weight), *Nannochloropsis oculata* (97.5% weight), *Chlorella protothecoides* (98.15% weight), *Chlorella minutissima* (82% weight), *Chlorella emersonii* (88% weight) (Bahadar and Khan, 2013). *Scenesdmus* sp. is one of the most potential biocatalytic microbial communities for biodiesel production. Few scientific studies have shown that the surface level in those species contains lipid layer, alcoholic groups, carboxyl groups which are capable for biodiesel production (Sudhakar and Premalatha, 2015). *Chlamydomonas* sp. and *Desmodesmus* sp. are biosynthesizing high yielding strains for lipids via CO_2 sequestration and lower nitrogen concentrations during wastewater treatment. *Chlamydomonas* sp. shows lipid productivity of maximum 0.31 (g/L) during CO_2 concentration of 0.34 g/L during nitrogen starvation along with under sodium salt stress conditions which are the most promising platform of biodiesel production (Wu et al., 2012). Some solid acid catalyst (NiO, MoO_2/Al_2O, H-ZSM-5, (HY-340) niobium oxide) act as inducer of microalgal biodiesel producing in following micro algal species, i.e., *Monoraphidium contortum*, and *Nannochloropsis gaditana*. Usage of these acid-based catalysts reduces the overall production cost of about 25% and enhances biomass yield of about 77% towards 99% of biodiesel conversion efficiency in presence of Ni-Mo catalyst using microalgal bio-oil as precursor (Sani et al., 2013).

4.2.1.3 BIOHYDROGEN

Biohydrogen has highest heat content (143 GJ.ton^{-1}) compared to any other available fuels till date. It is only one fuel which is not bound with any carbon atom. Biohydrogen production is mainly derived through the involvement of biologically mediated acidogenesis processes where organic compounds has been converted into high value-added bio products likely hydrogen, volatile fatty acids, end product of glycolysis, i.e., pyruvate. Pyruvate is further oxidized to acetyl CoA following acetyl phosphate and ATP. The pyruvate oxidation process induces reduction of ferredoxin using hydrogenase enzymes. Then reduced ferredoxin releases electrons and molecular hydrogen which combines with proton to develop biohydrogen gas (Venkata Mohan et al., 2016; Ghosh and Hallenbeck, 2015). Biohydrogen production has been categories as follows: Direct photolysis, indirect photolysis, and ATP driven pathways. Hydrogenase as well as Nitrogenase enzymes act as catalyst during these processes (Sharma and Sharma, 2017; Ghosh and Hallenbeck, 2015). Most common hydrogen evolving microalgal species include *Chlamydomonas reinhardtii, Platymonas subcordiformis, S. obliquus*, *Chlorella sorokiniana, Chlorella fusca,* and *Chlorococcum littorale. S. obliquus* has been first considered as microalgal hydrogen producers however it produces very lower amount of biohydrogen. Light or photons are the most important process parameter for generation of micro algal biohydrogen gas either in autotrophic or mixotrophic processes (Bhalamurugan et al., 2018). *Chlamydomonas reinhardtii* produces hydrogen through the activities of hydrogenase enzyme (Enamalaa et al., 2018; Ghosh and Hallenbeck, 2015). However, large scale production of biohydrogen by *Chlamydomonas reinhardtii* have not yet commercially reliable due to its various process parameter optimization and bioprocess engineering issues. The sulfur limitation also induces biohydrogen yields in *Chlorella sorokiniana.* Metabolically engineered two *chlorella species* have been reported to generate higher amount of molar yield of biohydrogen (e.g., *Chlorella sp., TISTR 8262* produce 13.03% yielded highest biohydrogen and *Chlorella ellipsoidea TISTR 8260* produce 3.05% yielded biohydrogen). Some scientific studies have also depicted that genetically modified microalgal strains have more potentiality to produce biohydrogen than single strains (Bhalamurugan et al., 2018). Nutrient deprivation is one of the main bottlenecks for lower biohydrogen productivity. Two step continuous processes should be suitable for ameliorating biohydrogen production (Koller et al., 2012). About 70 microalgal species more than 30 genera have great potentiality to produce biohydrogen (e.g., *Chlorococcum* sp, *Dunaliella* sp,

Micromonas sp, *Parachlorella kessleri,* and *Scenedesmus*, etc.) (Skjånes et al., 2013). Various ranges of algal species (e.g., *Ulva lactua, Prymnesium parvum, Dunaliella* sp., *Chlorella* sp., *Chlamydomonas* sp., *Gracilaria* sp., *Arthrospira* sp., *Sargassum* sp., *Spirulina* sp., *Euglena gracilis,* and *Scenedesmus* sp.) produce hydrogen by fermentation process with the help of yeasts. *Ulva lactuca* produces high amount of hydrogen after hydrothermal pre-treatment ~0.14 $gm.gm^{-1}$ biomass yield (Sharma and Sharma, 2017; Suganya et al., 2016).

4.2.1.4 BIOMETHANE

Microalgae are a potential microbial regime which can catalyze the production of biomethane. In worldwide market its received attention due to its high productivity of valuable by products likes biogas. Biogas mainly composes the mixture of biogas specially biomethane (~55–75%) and CO_2 (~25–45%) in anoxygenic condition. Biomethane is being utilized a biogas which has been further transformed into bioelectricity. Microalgal species *Ulva* sp. produces volatile solid based 180 mL/g methane under mesophilic condition (35°C) during anaerobic digestion. *Laminaria sp.* yields 0.26–0.28 $m^3\ kg^{-1}$ of biomethane. Biomethane yield is comparatively higher considering micro algal biomass rather than biomass of grass, wood, etc. Moreover, currently mixed microbial consortia approaches have been initiated to cultivate two or multiple micro algal species for biomethane generation utilizing waste water treatment plant effluents which makes the entire approach economically and industrially viable and sustainable (Venkata Mohan et al., 2016; Harun et al., 2010).

4.2.1.5 BIOETHANOL

Microalgae are the most potential cell factory to biosynthesize bioethanol due to its certain salient features, i.e., thin cellulosic cellular wall, broad spectrum of metabolic diversity, higher photosynthetic capacity, higher amount of biomass productivity, variable chemical cellular constituents including higher carbon dioxide sequestration. Microalgae include majorly two metabolic pathways likely fermentation and gasification towards bioethanol generations. However, microalgal starch enriched cell wall is the most potent feedstock for bioethanol generation in association with yeast *Saccharomycess cerevisiae.* Microalgal biomass has been subjected to acid, alkali, and enzyme and or whole cells of yeast treatment which accelerates to release simple residual sugars which can further be biotransformed to ethanol through yeast fermentation. Microalgae *Chlorococcum littoraleto* produces ethanol through the dark fermentation process in absence of light sources through the generation of 27% of starch material release from the cell wall of microalgae. Moreover, ethanol productivities through microalgae has been entirely dependent on mode of fermentation and hydrolytic processes. In similar fashion, *Chlorococcum, Chlorella, Chlamydomonas, Scenedesmus, Tetraselmis* species contains high amount of carbohydrates which are effective for bioethanol production following aforementioned conceptualization (Rodrigues et al., 2015). Different studies show that some microalgal species produce an adequate biomass yield after the pretreatment process with various microorganisms' fermentation. (e.g., microalge *Dunaliella* sp produce 0.011 gram weight ethanol from total dry weight of biomass with the pretreatment of Glucoamylase with *Saccharomyces cerevisiae IAM 4140*, *Chlorella vulgaris IAMC-534* produces 0.123 gram weight ethanol from total dry weight biomass with the pretreatment of a-Amylase & glucoamylase with *Saccharomyces cerevisiae IFO 039*, *Chlorococcum humicola* produces 0.520 gram weight ethanol from total dry weight biomass with the pretreatment with Diluted acid-heat with *S. cerevisiae, C. vulgaris* produce 0.4 gram weight ethanol from total dry weight biomass with the pretreatment of Diluted acid-heat, to activate cellulase & cellobiase in *Escherichia coli* SJL2526 microorganism). Marine microalgal genetically modified strains *C. reinhardtii* UTEX2247 and *Chlamydomonas sp.* YA-SH-1 produce 0.08 and 0.154 dry weight ethanol yield by dark fermentation process. *Chlorella vulgaris* microalgal species consider as a one of the highest ethanol productive microalgae (238 $dm^3 . mg^{-1}$) (Doan et al., 2012). *S. abundans* utilizes raw substrate towards bio-ethanol generation. Treatment of cellulose with diluted acid produce ~0.1 gm bio-ethanol per gm of dry biomass of microalgae. Even when it has been treated with the sulfuric acid using by the pretreatment of 15 g/L of microalgae in 140°C temperature then the bioethanol concentration yield was 7.20 g/L (Odjadjare et al., 2017). *Spirulina platensis* produces simultaneously 16.32% and 16.27% bioethanol with the treatment of ~0.5 N strong acids likely nitric and sulfuric acid in 100°C temperature. *Chlorococcum humicola* increases the bioethanol production 16% to 52% with the pretreatment of 1% v/v sulfuric acid in 140°C temperature (Bhalamurugan et al., 2018). *Scenedesmus dimorphus* contains 53.7 w/w carbohydrates with the hydrolysis of 80% fermentable sugar, is a feasible stock of bioethanol (Khan et al., 2018). *Chlorococcum* sp. and *Spirulina* sp. produce (3.83 g/l) bioethanol under the 30°C temperature (Varfolomeeva and Wassermanb, 2011). Genetically engineered microalgal strain *Chlamydomonas fasciata Ettl 437* produces 80% yielded ethanol following the fermentation of *S. cerevisiae* AM12 with 98% extracted carbohydrates. *Chlorococcum humicola* is also a potential microalgal species for ethanol fermentation. *Chlorococcum humicola* produces 7.2g.L^{-1} ethanol considering a mixed consortium approach in association with *S. cerevisiae*. *Tribonema sp.* produces ~70% ethanol with the ~80% algal-biomass yield in association of fermentation with *S. Cerevisiae* (Eduardo and Alberto, 2016). *Chlorococcum infusionum* microalgae have been achieved highest yield of ethanol by the alkaline hydrolysis process. It contains 26.1% (g ethanol/g algae) (Phwana et al., 2018). *Chlorococcum sp.* and *Chlorella vulgaris* have extensively applied to generate bioethanol. It produces ~3.8 gm.L^{-1} of bio-ethanol utilizing 10 gm.L^{-1} of microalgal-lipid extraction (Bahadar and Khan, 2013).

4.2.1.6 BIOBUTANOL

Over the ten decades, biobutanol has been applied as potential transportation fuel having its lower vapor pressure, higher energy density. Biobutanol is the better alternative to ethanol, petroleum as biofuel additives though its economic production rates are as same as ethanol. In the production process of butanol bacteria utilizes the cellulose contain in algal biomass, also digest starch, reducing residual sugars. Microalgae *Ulva lactuca* generates biobutanol in association with *Clotridium species.* Biobutanol production yield lies to ~0.16 gm.gm^{-1} which is lower compared to ethanol molar yield. Brown algae *Saccharina* sp. ferments biobutanol after acid treatment of organic matters. However, biobutanol production yield lowers up to 0.12 g g^{-1}. Some studies show that after ionic liquid extraction from microalgal derive carbohydrate compare with Ionic Liquid Extracted Algae (ILEA) and Hexane Extracted Algae (HEA) for ABE (bio-acetone, bio-ethanol, and bio-butanol) fermentative avenues. Biobutanol fermentative avenues by ILEA and HEA produce 4.99 and 6.63 g.L^{-1} yield biobutanol. Also, biobutanol and biodiesel production cost have been reducing after using cheaper feedstocks in aforementioned biorefinery processes (Sharma and Sharma, 2017).

4.2.1.7 BIO-OILS

Bio-oils have been generated through thermochemical conversion approach. In the high temperature absence of oxygen along the mixture with char and gas the microalgal biomass convert into oil. This process ordains through another two fermentation processes, those are pyrolysis and thermochemical liquefaction bio-oil formation process. Pyrolysis process fermented through higher temperature (~350–530°C). Solid, gaseous, liquid parts are produced. A non-aqueous and liquid aqueous part is referred to as bio-oil or tar. In thermochemical fermentation process microalgal biomass hold sway higher pressure and minimized temperature of 10 MPa and 300°C approximately. Various microalgal species are used for bio-oil fermentation and their bio oil producing yield is approvable. *Dunaliella sp* ~37%, *Spirulina* up to ~41%, *Scenedesmus* ~24%–45%, *B. braunii* ~50%, *Chlorella sp.* ~30%, *C. cohnii* ~20%, *Cylindrotheca sp.* ~26%, *D. primolecta* ~22%, *Isochrysis sp.* ~29%, *M. Salina* >20%, *Nannochloris sp.* ~30%, *Nannochloropsis sp.* ~50% *N. oleoabundans* ~44%, *Nitzschia sp.* ~46%, *P. tricornutum* ~25%, *Schizochytrium sp.* ~63%, *T. sueica* ~19%, *Porphyridium cruentum* 9.5%, *Oedogonium* and *Cladophora* yielded simultaneously 26% and 20% bio-oil. Unprecedentedly microalgae *Laminaria saccharinait* produce 79% bio-oil yield after the hydrothermal liquefaction (Sharma and Sharma, 2017; Frac et al., 2010; Hidalgo et al., 2013; Harun et al., 2010; Satyanarayana et al., 2011; Kirrolia et al., 2013). *Nannochloropsis* sp., *S. platensis*, *Chlorococcum* sp., *C. vulgaris* microalgal species recovers 25%, 77.9%, 81.7%, 12.5 % bio-oil due to various oil extraction methods such as SC-CO_2, soxhlet, ionic liquids methods, etc. (Pragya et al., 2013). N-hexane, α-pinene, para-cymene, terpene are also used as solvents for microalgal bio-oil extraction (Tanzi et al., 2012).

4.2.2 ALGAE-BASED HIGH ESTEEM APPENDED COMMODITIES

Currently microalgal regimes represent for generating high esteem appended molecules due to their large amount of biodiversity, ability to represent reservoir of novel value-added molecules and products in all over the world. Out of 30,000 known species few are susceptible to generate different value-added products like pigments, proteins, vitamins, high amount of PUFA (Poly unsaturated fatty acid), etc. (Sharma and Sharma, 2017, Chew et al., 2017). It can adapt very well to any environment in different conditions for survive and produce different types of secondary metabolites which cannot be found in another organism. The three types of microalgae such as *Dunaliella species, Chlorella species* and *Spirulina species* mostly used to produce high value-added components (Priyadarshani and Rath, 2012; Caporgno and Mathys, 2018). *D. salina, H. pluvialis, Odontella aurita, P. cruentum, Porphyridium sp, I. galbana, P. tricornutum, L. majuscule,* and *Muriellopsis sp*; these microalgal species are hugely used in industry for different value-added compounds production (Liu et al., 2016; Gong et al., 2011). Poly unsaturated fatty acid (PUFA) such as DHA (Docosahexaenoic acid), EPA (Eicosapentaenoic acid) has naturally derived from *Schizochytrium* sp., *Ulkenia* sp., *I. galbana, C. pyrenoidosa, C. ellipsoidea* and *Crypthecodinium* sp. microalgal species. *Phaeodactylum tricornutum, Monodus subterraneus,* and *Nannochloropsis sp.,* contains (2.2–3.9% of DW), (3% of DW), and (2.8–4.3%), EPA (eicosapentaenoic acid), respectively. The universal commercial impact is ~700 million US$ per year.1.8 billion US$ in the year of 2019 has been expected global market value of algal-carotenoids (Bhalamurugan et al., 2018; Kumara et al., 2019). Lutein reaches US $309 million by 2018 market value (Andrade et al., 2018). Phycobiliproteins are used in various sectors as a high value-added molecule. It has been derived from *Arthrospira platensis* species having various therapeutics nutraceutical applications (Manirafasha et al., 2016). *Isochrysis galbana* have the highest production yield of fatty acid 10 % on dry matter (Sánchez et al., 2016). *Arthrospira* species contains highest proteins and high value-added molecules which are used in food industry (Varfolomeeva and Wassermanb, 2011). *Lyngbya, Symploca, and Oscillatoria* like different marine organism are used for building of amino acid through the combinatorial enzymatic activities including polyketide synthases (PKSs), non-ribosomal peptide synthetases (NRPSs) (Gangl et al., 2015). The unicellular green algae *H. pluvialis* provides high value-added red pigments up to 5% on dry weight basis (Leu and Boussiba, 2014).

4.2.2.1 PIGMENTS

Pigments are colorful chemical constituents which participate in photosynthesis and ATP generations as a light energy absorbers. It absorbs certain wavelengths of visible light (Markou and Nerantzis, 2013). Microalgal pigments are one of the highest potential products in the market due to their high market value, excessive availability, and excellent health promoting properties. Mainly microalgal species contains three major pigments. Those are carotenoids, (provide orange, yellow color), phycobilin (red or blue natural colorant), and chlorophyll (green colorant). Different microalgal species provides various types of pigments and give high value for various products. ~500 out of 16,000 Green microalgal species provide chlorophyll-α, chlorophyll-β, β-carotene, prasinoxanthin, siphonaxanthin, and astaxanthin like pigments. >200 out of 100,000 brown microalgae contain Chlorophyll-α, Chlorophyll-c, β-carotene, fucoxanthin, diadinoxanthin pigments. Moreover, ~12–23 out of 200 cryptomonads species contains chlorophyll-α, chlorophyll-c, carotenoids and phycobiliproteins. ~10 out of 2000 Blue-green microalgae contains chlorophyll-α, xanthophyll and phycobiliproteins. About 40/900 and 130/220 Euglenoids, Dinoflagellates provides chlorophyll-α, chlorophyll-β, diadinoxanthin, neoxanthin, and β carotene, peridinin (Begum et al., 2016). These biopigments contain provitamin A (E160a) and vitamin-E (E306, E307, and E308) which have been used for pharmaceuticals, veterinary, medical purposes (e.g., anti-oxidants, anti-inflammatory, anticancer, antiaging, anti-tumor, etc.), as well as cosmetic, food, and aquafeed industry (Koller et al., 2014; Lafarga, 2020). *Chlorella vulgaris* is one of the potential microalgal species which provides different types of pigments in different culture condition (β-Carotene: 7–12,000 $\mu g.gm^{-1}$ dw), (Astaxanthin: 550,000 $\mu g.gm^{-1}$ dw), (Cantaxanthin: 362,000 $\mu g\ g^{-1}$ dw), (Lutein: 52–3830 $\mu g.gm^{-1}$ dw), (Chlorophyll-α: 250–9630 $\mu g.gm^{-1}$ dw), (Chlorophyll-β: 72–5770 $\mu g\ g^{-1}$ dw), (Pheophytin-a: 2310–5640 $\mu g\ g^{-1}$ dw), (Violoxanthin: 10–37 $\mu g.gm^{-1}$ dw) (Safi et al., 2014). Carotenoids are additional bio-pigments having potential involvements in photosynthesis. Carotenoids shows 5-carbon basic unit of isoprene which polymerized enzymatically form 40 carbon structures those are highly conjugated. 400 known microalgal species contains carotenoid (i.e., xanthophylls, β carotene, astaxanthin, lutein, lycopene and zeaxanthin) pigments. Carotenoid have highly nutritional value (especially β-carotene contains provitamin A) that's why these are used in feed additives as a natural colorant. It is also used in pharmaceutical, cosmetics industries for largely attributed to their antioxidant, anti-inflammatory, anticancer properties (Chu, 2012). Brown-orange microalgal species like *Dunaliella salina, Dunaliella bardawil, Botryococcus braunii* contains yellow color ß-Carotene pigments. Which provides multivitamin, provitamin-A, antioxidant, food additives, i.e., E160a, colorants of egg yolk, reddish color in salmon and trout fishes (Koller et al., 2014; Guedes et al., 2011; Priyadarshani and Rath, 2012). *Dunaliella salina* provides β-carotene upto 14% of its dry weight. *P. tricornutum* produces 2.3 ±0.4 (mg/g DCW), 1.0 ± 0.3 (mg/g DCW), β-Carotene, 8.0 ± 1.6 (mg/g DCW), 12.2 ± 1.1 (mg/g DCW), fucoxanthin under different wavelength of light intensities (Wichuk et al., 2014). Moreover, *C. closterium, E. magnus, E. polyphem, E. vischeri, P. tricornutum, V. helvetica, V. punctata, V. stellate* microalgae provide sufficient amount of fucoxanthin (Gateau et al., 2017). *Dunaliella bardawil* have rich in β-carotene which inhibits the oxidation of lower density lipoproteins (LDLs) and increase the plasma triglycerides, cholesterols and high-density lipoproteins (HDLs) levels in mouse and humans. Also this microalgal sp. contains phytoene and phytofluene for their carotenoids pigments production protect from UV rays oxidative damage and prevent premature aging and others disorders. *Haematococcus pluviali* provides keto carotenoids (e.g., astaxanthin), produce 4–5% dry weight astaxanthin which are mainly used for natural food colorants, feed additives, in acquaculture, poultry also the supplementary feed of salmon fish, trout fish and shrimp. Recently astaxanthins have been used for pharmaceuticals sectors also due to it contains antioxidant properties and it has shown great ability to prevent various heart diseases, chronic diabetes, and inflammatory diseases (Chu, 2012, Liu et al., 2016, Bilal et al., 2017; Shariati and Hadi, 2011). *Chlorococcum, Scenedesmus, Chlorella zofingiensis,* microalgal species are promising producer of highest amount of astaxanthin (Liu et al., 2014; Odjadjare et al., 2017). *H. pluvialis* is another major micro algal reservoir of astaxanthin; 3.8 to 7.6 mg per day astaxanthin can take as a dietary supplement (Shah et al., 2016). Xanthophylls, lutein, zeaxanthin are applied for chicken skin color improvement (Guedes et al., 2011). Even zeaxanthin is enriched in corn karnels, gulmohar plants, orange fruits, berries, and marigold flowers. Those are utilized for different medicinal purposes. *Arthrospira sp., B. braunii, C. acidophila, D. salina, M. aeruginosa, N. excentricum, Nannochloropsis oculata,* and *Scenedesmus almeriensis* produce zeaxanthin pigments of around 0.34 $mg.gm^{-1}$ with an average commercial impact 10 US$ per 1,000 mg (Bhalamurugan et al., 2018; Gateau et al., 2017). *Chlorella ellipsoidea,* and *Chlorella vulgaris* provide luteins 2–4 mg/g DW, which prevents the molecular degeneration, colon cancer (Guedes et al., 2011). Hence, it bears antioxidant properties and decrease approximately 60 chronic diseases (Gong and Bassi, 2016). *S. almeriensis* provides 4.77 $mg.L^{-1}.d^{-1}$, with the market value 2.5 US$ per 1000 mg of lutein (Bhalamurugan et al., 2018). Violaxanthin and canthaxanthin are the members of xanthophylls in recent studies shows that the most Violaxanthin pigment producing microalgal species are *Dunaliella tertiolecta, Chlorella ellipsodea,* and *B. braunii.* It has anti-inflammatory properties, strong antiproliferation properties against human mammary cancer cells (Koller et al., 2014; Sathasivam and ki, 2018). Canthaxanthin pigment is used in food additives which has been derived using *Nannochloropsis sp.* especially *N. salina, N. oculata,* and *N. gladitana,* etc. It is mainly feed in poultry for their skin and egg yolk coloration, in aquaculture salmonid, and sausage colorants. In Canada, canthaxanthin are commercialized for tanning pills (Koller et al., 2014). Now a days Canthaxanthin widely using as food colorants in various nations like Austria, Portugal, United States, Israel, and Spain (Raposo et al., 2013). *C. pyrenoidosa* is most potential microalgae which contains lutein. The highest extraction yield is 1240.1 (μg/g) for USCCE-ultrasound-enhanced subcritical CO_2 extraction (Poojary et al., 2016). *Dunaliella salina* is one of the novel microalgal species which contains various types of pigments (0.66% w/w lutein), (12% w/w *β*-Carotene), 13.8%w/w All-trans-*β*-carotene), 1.1%w/w All-trans-zeaxanthin) (Morais et al., 2015). Phycobiliproteins are one of the major accessories photosynthetic pigments which are namely differing for their spectral properties (Like phycocyanin, allophycocyanin and phycoerythrin) derived from various microalgal species (e.g., *Spirulina platensis, Arthrospira sp*., and *Porphyridium sp*.). The major applications of Phycocyanin, Phycoerythrin pigments as natural colorants (beverage, ice creams, and sweet); cosmetics; immunofluorescence technical applications; antibody labeling agents, receptor and other biological molecule (Chu, 2012; Koller et al., 2014). *Spirulina platensis* produce phycocyanin ~0.11 to 12% dry biomass at different light intensity. Huge Antioxidant present in phycocyanin, which is more effective than trolox (vitamin E analog) and vitamin C (Chu, 2012, Borowitzka, 2013). *Arthrospira platensis* does rich in phycobilliproteins. Respectively contains C-phycocyanin (17.5% DW), allophycocyanins (3.8% DW), and phycoerythrin (1.2% DW). *Porphyridium cruentum* contains allo-phycocyanins (~5%), R-phycocyanin (~11%), B-phycoerythrin (~42%) (Pignolet et al., 2013). *Arthrospira platensis* has been used in animal-based feed industries as natural pigment food color, candies, jellies, dairy products also (Hua et al., 2018). Carotenoid pigments contents (%) some microalgal species are *H. pluvialis* 3–7%, *C. vulgaris* 12.5% TC, *C. vulgaris* 55.5% TC, *C. zofingiensis* ~0.7%, *C. striolata Var. multistriata* ~0.15%, *D. salina* ~3–13%, *C. zofingiensis* ~0.9%, *C. striolata Var. multistriata* ~0.7%, *S. platensis* ~70–80% TC, *C. pyrenoidosa* ~0.2–0.4%, *B. braunii* ~0.16%, *B. braunii* ~75% TC, *C. vulgaris* ~45% TC, *P. tricornutum* ~1.65%, *Isochrysis aff. galbana* ~1.8%, *C. closterium* ~0.5%, *O. aurita* ~2.2%, *C. striolata Var. multistriata* ~4.7%, *C. zofingiensis* ~25% TC, *C. vulgaris* ~36% TC, *B. braunii* ~0.17%, *Nannochloropsis* sp. ~0.1%, *Scenedesmus sps* ~0.69%, *Chlorococcum* sp. ~0.25% (Ambati et al., 2019). Lycopin is one if the useful provitamin derives from carotenoids. It has shown different biological activities. Used as an antioxidant, reduce cancer cells growth, prevent oxidative DNA damage, etc. various pharmaceutical sectors (Santhosh et al., 2016). More than 70 companies are producing *chlorella sp.* Taiwan *Chlorella* Manufacturing Company has stood highest *chlorella* producer in across the world (400,000 tons/year biomass produced) (Morais et al., 2015). Chlorophyll is a green color pigments involve in photosynthesis.

Chlorella sp. is one of the best sources to provide chlorophyll pigments. Chlorophyll a, chlorophyll b different types of pigments are present. *Chlorella* contains 7% of the biomass chlorophyll which are quantity wise higher than *Spirullina sp.* (Khanra et al., 2018; Harun et al., 2010). It has huge nutritional value, also use as food colorant for dietary supplement. Chlorophyllin derive from chlorophylls which used to protect against colon cancer. Chlorophyllin can also be utilized for personal hygiene purposes (e.g., deodorants, pastilles, commercialized products for bad breath, etc.) considering higher deodorant efficacy (Koller et al., 2014). On the other hand, chlorophyllin provides afatoxin-B1, heterocyclic amines, polycyclic aromatic hydrocarbons, which has shown antimutagenic effects (Matos, 2017). *A. falcatus, C. reinhardtii, C. vulgaris, M. dybowskii, P. lutheri,* and *S. dimorphus* microalgae do contain cellular chlorophyll (Ferreiral et al., 2017; Miazek et al., 2015). Chlorophyll-α, chlorophyll-β, pheophytin-α, pheophytin-β, pheophorbide-α, and pheophorbide-β are effective for some topical treatment like ulcers, hemostatic eczema, etc. antioxidants, anti-inflammatory properties present in chlorophyll pigments. Which have different approaches in pharmaceutical industries (Ferreiral et al., 2017). when *Coelastrum cf. pseudomicroporum Korshikov* microalgal species cultivated in salt stress condition it gives huge carotenoid pigments (1.73–91.2 $pg.cell^{-1}$) (De Souza et al., 2018). In Fed-batch culture mode heterotrophic condition, *Chlorella protothecoides* produces higher lutein than other conditions ~22.7 $mg.L^{-1}.d^{-1}$ and *C. zofingiensis* produces ~1.4 $mg.L^{-1}.d^{-1}$ astaxanthin in batch culture under heterotrophic condition (Sun et al., 2015).

4.2.2.2 COSMETICS

Microalgae contain bioactive compounds involve for ameliorating treatment of skin associated functions (Bhalamurugan et al., 2018). Each specific microalga contains different types of pigments which are also used as a colorant for various cosmetics. Microalgae are utilized as biomolecules, antioxidant agents, and water absorbing biomolecules. It is applied for skin and hair care value added biomolecules. Microalgae have potentiality to protect the skin and hair from harmful UV rays (Priyadarshani and Rath, 2012). Recently microalgae are receiving huge attention for it is effectively used as various skin disorders like aging, tanning, antioxidant, etc. (Bhalamurugan et al., 2018). Natural pigments also have ability to protect against cutaneous inflammation, skin cancer, and melanoma disorders (Sathasivam and Ki,

2018). Some microalgal sp. mainly (e.g., *Arthrospira, Chlorella, spirullina*) are implicated as therapeutic biocomedities for taking care of skin and hair care through the manufacturing of lotions, creams, shampoo, conditioner, etc. (Bhalamurugan et al., 2018; De Souza et al., 2018). Some classical microalgal sp. have potential application in cosmetics industries including *C. crispus, M. stellatus, A. nodosum, A. esculenta, S. platensis, N. oculata, C. vulgaris, D. salina,* and *Euglena gracilis* (Matos, 2017; Priyadarshani and Rath, 2012). Microalgae are polyphyletic and biochemically diverse microorganism. Which are contain chlorophyll-α, it is capable for oxygenic photosynthesis that are observed UV radiation. Some microalgae produce organic metabolite like sporopollenin (*C. terrestre, C. microporum, E. coelastroides, Scenedesmus sp., S. chlorelloidea, S. rubescens, S. spongiosa, D. salina,* and *C. fusca*), scytonemin (*Chlorogloeopsis sp., Calothrix sp., Scytonema sp., Rivularia sp.,* and *N. commune Lyngbya cf. aestuarii Chroococcidiopsis sp., N. punctiforme*), and mycosporine like amino acids (*A. spiralis, C. minutissima, C.sorokiniana, D. tertiolecta, S. chlorelloidea, Isochrysis sp., P. gyrans, C. criophilum, T. tumida, P. pseudodenticulata, S. microtrias, T. weissflogii,* and *A. catenella*) to defend against ultra violet (UV) irradiation while permitting visible light spectrum involves in photosynthesis (Priyadarshani and Rath, 2012; Sharma and Sharma, 2017). Some marine microalgae contain phenolic compounds, which bottle up generation of matrix metal lo-proteinase. It prevents premature aging of skin. Some micro algae derived exopolysaccharides compounds are used for cosmetics production. Like Fucoidans, carrageenans, ulvans acquired from brown, red and green algae, these are also used as suspending agents, hair conditioner, wound-healing agents (Bhalamurugan et al., 2018). Moreover, astaxanthin based compounds are outstanding antioxidants which contain elevated amount of antioxidant, vitamin E and C these bioactive molecules help to prevent the protein and lipid in human lymphocyte which encourages the oxidative power of super oxide dismutase and catalase enzymatic efficiencies. It also used for suppressed the skin high per pigmentation; inhibit the melanin function, improvement the texture of all skin type layers (Sathasivam and Ki, 2018). *Spirulina* whitening facial mask, *Spirulina* firming algae mask, *Spirulina* facial scrub, etc. Cosmetics products contains herbal extracts and proteins which are improves the skin's immunity, moisture balance, removes the deadly skin surface tissue cells from skins as cleaning agent and vitalize the tissues associated with skin surfaces (Sharma and Sharma, 2017).

4.2.3 PHARMACEUTICALS AND THERAPEUTIC APPLICATIONS WITH SPECIAL EMPHASIS ON ANTIVIRAL, ANTI-OXIDANT, ANTI-MICROBIAL, AND ANTICANCER ACTIVITY OF MICROALGAL COMPOUNDS

Microalgae are the fertile reservoirs of bioactive molecules. Since last 50–60 years, the large amount of microalgal species has taken a reliable place in therapeutic and pharmaceutical sectors. Due to its natural pigments or colorants the aforementioned high value-added vitamins, nutrition, high antioxidant, antitumor, anticancer, anti-inflammatory, antiaging capacities (Sharma and Sharma, 2017; Borowitzka, 2013). About 300 microalgal species are toxin producer. Those toxins (e.g., cytotoxicity, neurotoxicity, hepatoxicity, etc.) can fight against much human, domestic, wildlife health hazards. The wide ranges of phycotoxins are used in health sectors for their therapeutic activities such as antifungal, antitumor, antibiotic, neurotoxic activities, immunosuppressant, etc. (Sigamani et al., 2016). A broad spectrum of *Arthrospira* sp., *Chlorella sp.*, *D. salina*, and *H. pluvialis* microalgal species have been used in pharmaceutical sectors for various purposes. *Spirullina* is a multicellular filamentous blue green microalgae contains chlorophyll a, carotenoid pigments, rich in vitamins, mineral unsaturated fatty acids, essential amino acids, proteins (~46–71%), carbohydrate (~8–16%), lipids (~4–9%) of its dry weight, etc., *spirulina* provides (2.330×10^3 IU/kg) provitamin A, (~140 mg per 100 gm algal biomass) b-carotene, (~100 mg per 100 gm algal biomass) vitamin-E, (~4.0 mg per 100 gm algal biomass) Riboflavin B2, (~0.8 mg per 100 gm algal biomass) Vitamin-B6, (~0.32 mg per 100 gm algal biomass) Vitamin-B12, (~0.005 mg per 100 gm algal biomass) biotin, (~2.2 mg per 100 gm algal biomass) vitamin-K, etc. essential compounds which are used for various therapeutic purposes like (mal nutrition, anti-aging, anti-oxidant, anti-inflammatory, etc.). *Chlorella* species are single cell microalgae which contains various compounds (e.g., (55,500IU/kg) Provitamin-A, (~1.5 mg per 100 gm algal biomass) Thiamin-B1, (~4.8 mg per 100 gm algal biomass) Riboflavin-B2, (~125.9 mg per 100 gm algal biomass) Vitamin B1, (~191.6 mg per 100 gm algal biomass) Biotin, etc. which are used in various therapeutic purposes. *Dunaliella salina* provides ~49–57% protein, ~6–8% lipids, ~4–32% carbohydrate, of its dry weight. *H. pluvialis* is an unicellular freshwater Chlorophyta group of microalga. It provides astaxanthin (up to ~2–3% on dry algal biomass). Those are used as strong antioxidant therapeutic agents (Mobin and Alam, 2017). *Chlorella vulgaris* and *Spirulina* sp. produce high quantity of omega 3 fatty acid. Daily uptake of omega 3 fatty acids can prevent cardiovascular damage, hypertensions, myocardial infarctions, thrombosis cardiac arrhythmias, psoriasis, ulcerative colitis, asthma like various diseases. Omega 3 fatty acid also contains eicosatetraenoic acid (EPA) and docosahexaenoic acid (DHA) having beneficial impact on pregnant women for developments of fetal brain, give positive effects of nervous systems (Vega et al., 2012). However, some microalgae produces elevated quality based biological therapeutic agents towards pharmaceutical industries (e.g., *Chlorella sp.* contains Lutein, β carotene, α carotene, α tocopherol, *C. cohnii* contains docosahexaenoic acid, *H. pluvalis* contains carotenoids, astaxanthin, lutein, *N. gaditana* contains eicosatetraenoic acid, *S. almeriensis* contains lutein, β carotene, *C. reinhardtii* contains glycerol) (Bhalamurugan et al., 2018). *I. galbana, N. gaditana, Nannochloropsis sp., P. tricornutum, Attheya ussurensis sp. nov., Diacronema vlkianum, Fragilaria pinnata, Glaucocystis nostochinearum, Karenia brevis, Karenia papilionacea, Micromonas pusilla, Nematochrysopsis sp., Nitzschia closterium, Porphyridium cruentum, Pyramimonas cordata, Schizochytrium aggregatum, Takayama tasmanica,* and *Thalassi-onema nitzschioides* like various microalgae contain different types of phytosterols (e.g., 23-methylcholesta-5,22-E-dien-3β-ol, 23-methyl-5α-cholest-22-E-en-3β-ol, lathosterol, ergosterol, stigmasterol, 24-Ethylcholesta-5,7,22-trienol, stigmasta-7,24-(241)-dien-3β-ol, 27-Nor-(24R)-4α-methyl-5α-ergosta-8(14),22-dien-3β-ol brevesterol (24S)-4α-Methyl-5α-ergosta-8(14),22-dien-3β, its 27-Nor derivative, etc.) 7 to 34 g per kg or (0.7%–3.4%). It reduces the risk of any cardiovascular disorders (Luo et al., 2015). *Chlorella pyrenoidosa* and *Arthrospira platensis*contains produce 13.2–25.8 $mg.gm^{-1}$ dry algal biomass and (17–43.2 $mg.gm^{-1}$ dry algal biomass phenolic compounds. Phenolic compounds are used as antioxidant agents; prevent damaging the cells against UV radiation, etc. (Barkia et al., 2019). *C. vulgaris, C. ellipsiodea, A. platensis,* and *N. oculata* produce ACE-inhibitory peptides used for various treatments like hypertension, heart diseases, body inflammation, etc. (Camacho et al., 2019). *Chlorella sp* produces b-1,3-glucan, which are an important bioactive compound used for various therapeutic purposes tumor immunotherapy, bone marrow chemo therapy, etc. (Odjadjare et al., 2017). *Nostoc spp.* contains some essential fatty acids, poly unsaturated fatty acids, Vitamin B complex, ascorbic acid, α-tocopherol, etc. compounds which are helps to decrease cancer risk, prevents molecular degeneration. *Dunaliella spp.* Contains high amount of enzymes and vitamins which are used as antimicrobial agents (Bhalamurugan et al., 2018). *Pyrophyridium* sp., *Synechocystis* sp., *P. ceylanicum, Limnothrix* sp., and *S. lividus* microalgal clades have biosynthesized phycobilliproteins. It is used for physiological, pharmaceutical, medicine purposes (Manirafasha et al., 2016). *Spirulina* contains a unique polysaccharide which used as an anti-carcinogenic agents. It Is also have anti-inflammatory, immunomodulating, antiviral, antibacterial activities (Hoseini et al., 2013). Different microalgae contain different bioactive compounds which has huge health benefits also. *Spirulina Chlorella vulagris, Haematococcus, Chlorella pyrenoidosa, Chlorella ellipsoidea, Porphyridium, Nostac flegelliforme, M. griffithii, M. salina, M. subterraneus, M. arcuatum, M. contortum, M. dybowskii, M. neglectum, M. terrestre,* and *M. tortile* contain proteins, peptides, and amino acids, phycocyanobilin, hexadecatetraenoic acid, glycoprotein, polysaccharides, photosynthetic pigments β-carotene, astaxanthin (Up to 7% DW; 75% TC), phycocyanobilin, cantaxanthin (4.75% DW), fucoxanthin, lutein (45% TC) have been used for various health purposes likely minimization of brain damage, enhancement of anti-inflammatory activity, amelioration of antioxidant activity, reduction in allergic symptoms, protective effects against cellular damage, hepatoprotective, protective activities against neuro-degenerative diseases, minimization of possibilities of gastric ulcers, cancer, antimicrobial activity, immunostimulating activity, reduction in anti-inflammatory, and immunological responses, possible increment as dietary supplements, etc. (Hamed, 2016; Sathasivam et al., 2019; Sathasivam and Ki, 2018). Microalgae *odontella aurita* has been used as a healthy nutritional feed stock (Posten and Chen, 2016). *Arthrospira platensis, S. limacinum, C. cohnii, P. lutheri, P. cruentum, Mortieriella alpina, P. incise, Nannochloropsis oculata, Phaeodactylum tricornutum, Monodus subterraneus,* and *Isochrysis galbana* microalgal species contain large amounts of Omega 3 α linolenic acid (ALA), eicosatetraenoic acid (EPA), omega 6 γ linolenic acid (GLA), arachidonic acid (ARA), docosahexaenoic acid (DHA) which are used as different therapeutic purposes like human brain and eye development at fetus, anti-inflammatory, muscle anabolic formulations, good for cardiovascular health, etc. (Matos, 2017). Microalgae *Kirchneriella sp, Ecklonia radiata* provide high amount of fatty acids, carbohydrates, photosynthetic pigments which are beneficial for various health purposes (e.g., antioxidant activities, prevent against atherosclerosis, cancer, many heart diseases, etc.) (De Souza et al., 2018). Few phytoplankton species contains huge amount of sterols. *Chaetoceros sp.* and *Skeletonema sp.* produce ~27.7 and 2.0 μg sterols.gm^{-1} dry algal biomass. *Thalassiosira* sp. and Pavlov asp. do provide higher amount of sterols (e.g., brassicasterol, campesterol, stigmasterol, etc.) these sterols compounds provides some health benefits (e.g., hypocholesterolemic) (Raposo et al., 2013). Fucoxanthin is a natural pigment derived from many microalgal species

(*E. polyphem, E. vischeri, P. tricornutum, V. helvetica,* etc.) which inhibit the proliferation of human blood cells and formation of tube in human umbilical vein endothelial cells (Khan et al., 2018; Gateau et al., 2017). *Chlamydomonas reinhardtii* has been used as proteins deficiency therapeutic agents. It produces large scale of recombinant proteins and 2–20% total soluble protein (TSP) which is industrially beneficial (Rasala and Mayfield, 2011). *Botryococcus braunii Kützing* microalgae are being used as potential therapeutic agents. It contains 15–35% dry weight of triterpenes (i.e., botryococcene, squalene) which have great antitumoral activities. *I. galbana* and *P. tricornutum* provides ~22.6 and ~29.8% per total fatty acids of EPA which interact in various pharmaceutical purposes for their antioxidant, anti-inflammatory, anti-cancerous activity (Sánchez et al., 2016). *Aphanizomenon flosaquae* have positive impacts human health issues due to its bioactive properties (Varfolomeeva and Wassermanb, 2011). Studies have been shows that *Chlorella vulgaris* is a potential producer of natural bioactive compounds. It releases silver-based nanoparticles (AgNPs) (Ebrahiminezhad et al., 2016). *Dunaliella salina, A. platensis, B. braunii, C. sp., H. pluvialis, N. wimmeri, S. oocystiformis, C. closterium, E. magnus, E. polyphem, Vischeria helvetica, V. punctata, V. stellate,* and *Chlorella protothecoides*, microalgal species contains various types of vitamins, carotenoids (e.g., β carotene, astaxanthin, canthaxanthin, fucoxanthin, lutein, and zeaxanthin) which has been applied as health promoting biologically active compounds (Gateau et al., 2017). *Dunaliella sp.* and *Chlorella sp.* show positive impacts against pathogenic microorganisms in human (*B. subtilis, S. aureus, Streptococcus pyogenes, S. epidermidis,* and *Micrococcus luteus*) (Falaise et al., 2016). *Spirullina sp*., *Pseudanabaena tenuis* are potential microalgal species. It has anti-inflammatory, antioxidant, anticancerous ability used as pharmaceutical agents (Ghosh et al., 2015). In various stress conditions *Chlorella zofingiensis* produces huge amounts of astaxanthin which are used in pharmaceutical sectors in various purposes (Skjånes et al., 2013). Marine microalgae *C. closterium, N. salinarum, P. tricornutum, H. ostrearia, S. costatum,* and *Amphora* sp. provide various types of polysaccharides such as sPS, sEPS, EPS, PS, etc. which covers a huge arena in pharmaceutical industry (De Jesus Raposo et al., 2014). *Arthrospira platensis, Arthrospira maxima,* and *Porphyridium cruentum* contain high amount of proteins (50–60% of dry biomass), carbohydrates (40–57 %), unsaturated fatty acids which are used as nutritional feed supplements to combat malnutrition (Buonoa et al., 2014).

About 1000 microalgal species were screened for those various defensive activities against different health hazards. *Spirulina platensis, Porphyridium* contains polysaccharides which have been shown antiviral activities on human cytomegalo viruses, herpes simplex, measles viruses and fight for different HIV-1, HSV-2 pathogenecities. A novel marine microalgal species *Poteriochromonas malhamensis* shows various enzyme activities like protein tyrosine kinase, protein kinase C. which are destroy the human leukemic cells and also its shown many capabilities against preventing the apoptosis during the growth of cancer cells, affects cell signaling by signaling enzymes activation due to their biologically active compounds. *Nostoc ATCC 53789 provides* cryptophycin which have huge anticancer activity. Scytonemin microalgal compounds isolated from *Stigonema* inhibit the protein serine which is used as a drug for anti-inflammatory, antiproliferative activities (Sharma and Sharma, 2017; Sigamani et al., 2016; Forján et al., 2014). *C. protothecoides, C. zofingiensis, C. vulgaris,* and *C. minutissima* contain many biologically active compounds (e.g., lutein ~4.60 % W/W, astaxanthin ~1.50 % W/W, phenolic ~0.20% W/W, terpenoids ~0.09% W/W, alkaloids 2.45% W/W, phytol 2.70% W/W, phenol ~1.81% W/W) which are used as antioxidant, antitumor, anticoagulant, antibacterial, antihiperlipidemia agents. *Dunaliella salina* contains β-carotene 12% W/W, trans β carotene ~13.8%W/W, trans-zeaxanthin ~1.1%W/W, trans lutein ~0.66%W/W, sterols ~1.3%W/W. those are used in therapeutic sectors for their antioxidant, anti-inflammatory analgesic, hepatoprotective, antiedemal, and bronchodilatory properties. Some extracted microalgal biomass such as *spirullina, Porphyridium cruentum,* and *Synechococcus* provides C-phycocyanin, carotenoids like various pigments which are contains high antioxidant properties, pigmentation ability (Gateau et al., 2017; Morais et al., 2015). *Spirulina sp., Chlorella sp.,* and *Dunaliella sp.,* has been shown some antimicrobial abilities due to their rich amount of fatty acids, carbohydrates, acrylic acids, terpenoids, sterols, acetogenins, phenols α- and β-ionone, β-cyclocitral, neophytadiene, etc. Mainly inflammation causes high swelling, heat, redness, pain. Those microalgae has potentiality to prevent Anti-Inflammation due to its anti-inflammatory compounds (Morais et al., 2015). *Bigelowiella* sp., *Chattonella antique, Chattonella marina, Chattonella subsalsa, Chrysowaernella, Cyanophora paradoxa, Diacronema vlkianum, Fragilaria pinnata, Glaucocystis nostochinearum, Isochrysis galbana, Karenia brevis, Micromonas aff, Pusilla sp., Navicula incerta, Nematochrysopsis sp., Nitzschia closterium, Karenia umbella, Pycnococcus provasolii, Takayama helix, Tetraselmis chui, Thalassi-onema nitzschioides, Rhizosolenia setigera,* and *Polarella glacialis* are the different microalgae that contain different types of phytosterols (e.g., Ergosterol, 7-Dehydroporiferasterol, Ergosterol peroxide, 7-Oxocholesterol, 4α-Methyl sterols, Dinosterols, Dehydrodinosterol 4α,24-Dimethyl-cholestan-3β-ol 4α,24-Dimethylcholest-5-en-3β-ol Cholesta-5,7-dien-3β-ol, 27-Nor-(24R)-4α-methyl-5α-ergosta-8(14),22-dien-3β-ol Brevesterol, Gymnodinosterol (24R)-4α-Methyl-5α-ergosta-8(14), 24R)-4α-Methyl-5α-ergosta-8(14),22-dien-3β-ol Gymnodinosterol etc.). Some studies have been shows that highest phytosterol biomass contains by *P. lutheri* (5.1% dry weight). Mainly phytosterol contains tetracyclic cyclopenta α phenanthrene structure (ring A, B, C and D) and an aliphatic side chain (R) at C17 of ring D which have been used for various health benefits for their anti-inflammatory, antihypercholesterolemic, antioxidant, anticancer, anti-diabetic properties (Camacho et al., 2019). *A. platensis, Porphyridium sp.,* and *Synechococcus sp* enrich with higher phicobilliproteins contents. Most of microalgal species accumulate phycobiliproteins under various environmental stress conditions. These compounds contain non carcinogenic, non-toxic, antioxidant, anti-inflammatory, immunomodulating properties implicate to treat various therapeutic prospective likely some acute viral diseases, heart diseases, etc. (Manirafasha et al., 2016). *Haematococcus sp.* produces astaxanthin and provitamin-A which have been used as antioxidant, anti-inflammatory, anti-cancerous, antiaging agents. Some toxin producing microalgae (e.g., *Nitzschia pungens, Alexandrium lusitanicum,* etc.) have shown different bioactive characteristics likely antifungal, anticanceruous, antimicrobial activities (Raposo et al., 2013). *Undaria pinnatifda, C. closterium, P. tricornutum* contains various types of natural bioactive compounds which have shown as antioxidant, Anti-angiogenic, anticancerous activities. *Nostoc sp.* contains cryptophycin which has been used as human colorectal cancer treatment. *Oscillatoria boryana* extract which has been used for breast cancer treatment. *Microcystis sp.* also has been used as anti-cancerous agent (Khan et al., 2018). *Undaria pinnatifida* contains high amount of neoxanthin and fucoxanthin which are potential biomolecules for reducing the risk of prostate cancer, human colorectal carcinoma (HCT116), umbilical cord fibroblasts in human (HUC-Fm), mice melanoma, etc. (HH and GS, 2013; Galasso et al., 2017). *Porphyridium sp., Gyrodinium impudicum, Chlorella stigmatophora, Haematococcus pluvialis,* and *Chlorococcum sp.* are most essential micro algal regime which are rich in astaxanthin, sulfated polysaccharides, p-KG03, hydrosoluble components, and sterols, etc. bioactive compounds. These biologically active compounds inhibit the intracellular reactive oxygen species accumulation upon regular cellular function. Decrease in H_2O_2 activates the transcriptional messenger nuclear factor NF-κB; partial blocking of adhesion to endothelial cells, used as Immunostimulatory, anti-inflammatory compounds (Guedes et al., 2011). *Phormidium autumnale* contains sterols compounds such as stigmasterol

(~455.3 μg.gm^{-1}), βsitosterol (~279 μg.gm^{-1}) which are used as an antioxidant agents (Oliveira et al., 2020). *Parachlorella kessleri* and *Galdieria sulfuraria* contain ~0.28 mg.gm^{-1} and ~0.4 mg.gm^{-1} astaxanthin and lutein respectively which have shown anticancer, antioxidant properties. *Nitzschia cf. carinospeciosa* contains ~5.5 mg.gm^{-1}, *Cyclotella meneghiniana* contains ~2.3 mg.g^{-1}, and *Mallomonas sp. SBV13* contains ~26.6 mg.gm^{-1} of fucoxanthin which has anti-obesity anti-oxidant properties. Fucaxanthin have also potential for anti-angiogenic activities in human umbilical vein endothelial cell lines (HUVECs) (Sathasivam and Ki, 2018). *Alexandrium minutum* (Am2-M), *A. tamarense WE* (At2-M), *G. australes* (Ga-C, Ga-M), *P. hoffmannianum 1031* (Ph3-M), and *P. reticulatum (Pr3-M)* have shown the highest anticancerous activities during the clinical trials (Vera et al., 2018).

4.2.4 MICROALGAL-DERIVED COMPOUNDS FOR HUMAN USAGE

4.2.4.1 BIO-FERTILIZER

Since many decades, microalgae are one of the anomalous components which are used as a biofertilizer. It is used instead of chemical fertilizer. Because in many cases it is a naïve component than chemical fertilizer, also it is cheap and contains natural compounds. Various microalgae contain different types of nutrients which used for plant growth, seed germination, soil conditioner in various agricultural yields (Sharma and Sharma, 2017; Singh and Ahluwalia, 2013). In some coastal areas after extracted lipid and carbohydrate the left over microalgal biomass contains so many nutrients. Those are used as a potential biofertilizer. It has ability to fix the atmospheric nitrogen. Biofertilizers play a crutial role to growth in plants and enhance soil fertility. Some algae contain vitamin-B12 like *Cylindrospermum* sp., *T. tenuis, Nostocmuscorum, Hapalosiphon fontinalis. Nostoc,* and *Hapalosiphon sp.* contain auxin like Indole-3-acetic acidindole-3-propionic acid or 3-methyl indole which helps to promote the growth in plants and binding soil agents and also aggregation (Sharma and Sharma, 2017; Priyadarshani and Rath, 2012). *Trichodesmium* microalgal species fix atmospheric nitrogen (Barra et al., 2014). Most essential nutrients for agricultural residue is nitrogen, phosphorus, potassium. Some microalgal sp. contains majorly these nutrients. And those are extensively used as a good fertilizer. The NPK ratio helps the plant growths, improve the root systems. *N. salina* enhances sequestration of CO_2 in soil ecosystem. *C. vulgaris* and *L. majuscula* have been used as sustainable biofertilizers. Moreover, ~30–60% of nitrogen (N), phosphorus

(P), magnesium (M)g, calcium (Ca), sulfur (S) and ~15–25% of manganese (Mn) & iron (Fe) are available in *Auxenochlorella protothecoides* microalgal biomass. To this end, it's also used as a potential biofertilizers in agricultural applications (Maurya et al., 2016). *Chlorella KR-1* strain has shown higher efficiency of CO_2 fixation up to ranges from 10–50% CO_2 (v/v). *Spirulina sp.* shows inversely proportional CO_2 concentration during the growth rate, biofixation rate (Cheah et al., 2014). *S. incrassatulus* and *S. dimorphus* show elevating carbon sequestration proportion than other microalgal species. Some potential microalgal species and their carbon fixation rates are *Scenedesmus incrassatulu* (1.50 $g.L^{-1}.d^{-1}$), *Scenedesmus dimorphu* (1.27 $g.L^{-1}.d^{-1}$), *Spirulina platensis*,(~0.92 $g.L^{-1}.d^{-1}$) *Chlorella* sp. (~0.7 $g.L^{-1}.d^{-1}$) *Scenedesmus sp.* (~0.61 $g.L^{-1}.d^{-1}$), *Scenedesmus obliquus* (~0.55 $g.L^{-1}.d^{-1}$), *C. vulgaris* (~0.43 $g.L^{-1}.d^{-1}$), *Dunaliella* sp. (~0.31 $g.L^{-1}.d^{-1}$), *Dunaliella tertiolecta* (~0.27 $g.L^{-1}.d^{-1}$), *Chlorella vulgaris* (0.25 $g.L^{-1}.d^{-1}$), *Chlorella kessleri* (~0.16 $g.L^{-1}.d^{-1}$), *Haematococcus pluvialis* (~0.14 $g.L^{-1}.d^{-1}$), etc. (Patel et al., 2016). Dry biomass of *C. vulgaris* increase the pigment content like chlorophyll a, chlorophyll b, carotenoids in different seedlings (e.g., *Lactuca sativa*) (Bhalamurugan et al., 2018). Microalgae *Haematoccus pluvialis,* and *Nannochloropsis salina* are one of the novel species for fuel production but they also consume nitrogen during the production of nitrogen-based bio-fertilizer 0.37 and 0.29 kg per kg oil, respectively (Lam and Lee, 2012). Microalgal lipid polysaccharide layers provide moisture to the soil. *Acutodesmus dimorphus* dry biomass use for tomato plant growth and seed germination. Since, the microalgal fertilizer are heavily used in agricultural yields, so it should be more concern about how much it toxic or nontoxic for soil and how much it reliable for environmental and soil health to became worldwide safe products for agricultural yields (Koutra, 2018).

4.2.4.2 BIOFIBERS IN PAPER INDUSTRIES

Sulphur enriched algal polysaccharides provide massive structural constancy in broad spectrum of algal species. Some cellulose containing microalgal species exist as potential feedstock which can be used for paper production. The current study shows that an algal mixture was taken from municipal waste water plant. Ten percent pulp is mixed in it, after that the result is that it is increasing the texture of paper without decreasing brightness and it is being done with 45% lower material cost (Sharma and Sharma, 2017).

4.2.5 MICROALGAE AS FOOD AND DIETARY SUPPLEMENTS

The terminology "Functional Food" first concept has been developed in 1980s in Japanese continent (Borowitzka, 2013). Various microalgae have been implies as supplements of main food resources inclusive animal feed additives due to high nutritional value, high contents of vitamins, amino-acids, and omega 3 fatty acids, etc. (Bhalamurugan et al., 2018). There are three types of animals uptakes dried microalgal biomass as dietary supplements. Weaning pigs diets consist of dried microalgal biomass of ~7.2 and ~15% (w/w) algae meal which replace soybean meal (48% crude protein) (Maurya et al., 2016). *Spirulina* sp. is one of the emergent microalgal regimes (Suganya et al., 2016) which produces worldwide (3000 t/year) as dietary supplements from many years (Chu, 2012; Priyadarshani and Rath, 2012; Khanra et al., 2018). It contains high nutrition such as essential protein, γ linolenic acid, vitamin and mineral. Microalgae also protect health against diabetes, hypertension, arthritis, anemia, cardiovascular diseases and cancer like many diseases. *Spirulina* is also known for its antioxidant compounds like phycocyanin and vitamin E (Chu, 2012). 7 gm Spirulina dry biomass contains ~4gm protein residues, ~1gm fat residues having omega 3 and omega 6 fatty acid, vitamin B1, vitamin B2 and vitamin B3 (Koyande et al., 2019). *Nostoc flagelliforme* contains high amount of pigments (echinenone, myxoxanthophyll, allophycocyanin, phycocyanin and chlorophyll), 19 types of amino acids (~ 8 amino acids are used for human nutrition which making up 35.8–38.0% of the total amino acids). This micro algal species has also been used in the past 400 years in China to treat different types of diseases like diarrhea, hypertension, and hepatitis. *chlorella* is one of the huge cultured microalgae used as processed nutritional food, pill, and medicine purpose. It contains high amount of proteins (~51–58% protein content per unit algal dry biomass), carotenoid and huge amount of vitamin which helps for balance the blood lipid amount (Chu, 2012). Micro algal biomass is applied to support the dietary aid for leprosy patients to increase their body weight, energy and health. Several types of food products are developed from *Chlorella ellipsoidea* likely soups, green tea, noodles, bread and rolls, cookies, ice cream, etc. (Sathasivam et al., 2019). Hainan Simai Pharmacy Company, China; Earthrise Nutritional Factory, California, USA; Cyanotech Corporation Hawaii, USA; and Myanmar Spirulina factory, Myanmar are some producers which are produce those preservative and functional foods from *spirulina sp.* (Priyadarshani and Rath, 2012). *Isochrysis galbana* biomass contains omega 3 fatty acids which are used for making biscuits as a processed food (Borowitzka, 2013). *Chlorella vulgaris* has been used as a supplement as fish foods. Some well-known producers like Taiwan Chlorella Manufacturing Company, Taiwan; and Klötze, Germany are produced *Chlorella*-based products having world production per year is 2000 tons. It's also used as a noodles, tablets, powders, nectar, fruits and vegetable preservative. One of the most important species is *Dunaliella salina* (Suganya et al., 2016) which are grown in all over the world (1200 t/year) (e.g., Cognis Nutrition and Health – Australia) for huge sources of photosynthetic pigments and beta carotene, use as a vitamin C supplement (De Souza et al., 2018). Several micro algal species has been utilized as a food suppliements including *Anabaena cylindrica* (enriched with 43–56% protein, 25–30% carbohydrates, 4–7% lipid on a dry biomass), *Botryococcus braunii* (enriched with 40% protein, 2% carbohydrates, 33% lipid), *Chlamydomonas rheinhardii* (enriched of 48% protein, 17% carbohydrates, 21% lipid), *C. pyrenoidosa* (57% protein, 26% carbohydrates, 2% lipid per dry biomass), *Dunaliella bioculata* (bears 49% protein, 4% carbohydrates, 8% lipid per dry biomass), *Dunaliella tertiolecta* (29% protein, 14% carbohydrates, 11% lipid per dry biomass), *Euglena gracilis* (contains 39–61% protein, 14–18% Carbohydrates, 14–20% Lipid on a dry matter basis), *Porphyridium cruentum* (28–39% protein, 40–57% carbohydrates, 9–14% Lipid per dry biomass), *Prymnesium parvum* (28–45%protein, 25–33% carbohydrates, 22–39% lipid per dry biomass), *Scenedesmus dimorphus* (8–18% protein, 21–52% carbohydrates, 16–40% lipid per dry biomass), *S. obliquus* (contains 50–56% protein, 10–17% carbohydrates, 12–14% lipid per dry biomass), *Scenedesmus quadricauda* (47% protein, 10% Carbohydrates, 1.9 % lipid per dry biomass), *Spirogyra sp,* (contains 6–20% protein, 33–64% carbohydrates, 11–21% lipid per dry biomass), *Spirulina maxima* (60–71% protein, 13–16% carbohydrates, 6–7% lipid per dry biomass), *Synechoccus sp.* (contains 63% protein, 15% carbohydrates, 11% lipid per dry biomass), *Tetraselmis maculate* (contain 52% protein, 15% carbohydrates, 3% lipid per dry biomass) (Priyadarshani and Rath, 2012; Barkia et al., 2019; Pignolet et al., 2013; Suganya et al., 2016). Eicosapentaenoic acid (EPA) derived from *Nannochloropsis sp.,* and *Phaeodactylum tricornutum* play a significant role as nutrient suppliment. However, *Nannochloropsis sp.,* and *Phaeodactylum tricornutum* have been enriched with omega 3 fatty acids. Docosahexaenoic acid (DHA) is one of the components which are used as dietary food suppliment, derived from various microalgal sp. such as *C. cohnii, Schizochytrium,* and/or *P. lutheri*. It contain PUFA (poly unsaturated fatty acid) which are essential for infantile brain, eye, human fetus and breast milk development. Arachidonic acid (ARA) contains four-fold unsaturated ω 6 fatty acid, γ Linolenic acid (GLA) contains a ω 6 unsaturated fatty acid which are derived from different microalgal species and used as a nutritional food bevarages (Koller et al., 2014; Camacho et al., 2019; Shah et al., 2018; Bhalamurugan et al., 2018; Goldberg et al., 2011; Varfolomeeva and Wassermanb, 2011). Various microalgal sp. like *Attheya ussurensis sp. nov., Bigelowiella, Chattonella antique, Chattonella marina, Chattonella subsalsa, Chlorella vulgaris, Chrysoderma* sp., *Chrysomeris, Crypthecodinium cohnii, Diacronema vlkianum, Dunaliella salina, Dunaliella tertiolecta, Glaucocystis nostochinearum, Karenia brevis, Karenia mikimotoi, Micromonas affpusilla Micromonas pusilla, Pycnococcus provasolii, Phaeodactylum tricornutum, Schizochytrium aggregatum,* and *Tetraselmis suecica*, are contains different types of phytosterol like 24-Ethylcholest-5-en-3β-ol, crinosterol, stigmasterol, isofucosterol, ergosterol, 7-Dehydroporiferasterol, Ergosterol peroxide, 7-Oxocholesterol, 4α-Methyl sterols, Dinosterols, Dehydrodinosterol 4α,24-Dimethyl-cholestan-3β-ol 4α,24-Dimethylcholest-5-en-3β-ol Cholesta-5,7-dien-3β-ol, 24-Oxocholesterol acetate, ergost-5-en-3β-ol, cholest-5-en-24–1,3-(acetyloxy)-,3β-ol, 27-Nor-(24R)-4α-methyl-5α-ergosta-8(14),22-dien-3β-ol Brevesterol (24S)-4α-Methyl-5α-ergosta-8(14), 22-dien-3β, its 27-Nor derivative, 24-Methycholesta-5,24(28)-dien-3β-ol 24-Methylcholesta-5-en-3β-ol 28-Isofucosterol, etc. these are naturally present in these micraoalgal sp. phytosterol contents 7 to 34 g per kg (~0.7%–3.4%) microalgae oil extracts (i.e., *I. galbana, N. gaditana, N. sp.* and *P. tricornutum*). 19–57 thousand L oil per acre, varies from different microalgal sp. which are better than vegetable oil. Those are heavily riched in vitamin-A, vitamin-B1, vitamin-B2, vitamin-B6, vitamin-B12 and vitamin-K, folic acid, niacin, minerals (calcium, phosphorous, iron, iodine, magnesium, zinc, selenium, copper, potassium, manganese and sodium) and various antioxidant compounds (i.e., carotenoid, xanthophyll and chlorophyll), etc. because of these reasons those are highly used for human consumptions (Luo et al., 2015; Bhalamurugan et al., 2018; Koutra et al., 2018; Garc et al., 2017). Different sources like fish, dairy products, plant animals proteins produce ACE (angiotensin-I converting enzyme) which is the bioactive peptides. Their antihypertensive properties inhibit the enzyme activity and help to regulate the mammalian blood pressure. Now a days some microalgal sp. (*C. vulgaris, C. ellipsiodea, A. platensis,* and *N. oculata*) are applied for producing ACE like peptides which are used as algal proteins for nutritional foods. Since last 70 years microalgal single cell protein (almost 75%) was produced for food supplements (Barkia et al., 2019; Lopes et al., 2019; Lee and Gonzalez-Marino, 2010; De and Ghosh, 2018). *S. platensis* is a richest protein supplement which contain proteins approximately 460–630 $gm.kg^{-1}$ dry biomasses. The huge amount of *s. platensis* microalgal species used for

SCP (single cell protein) production all over the world. These are also used as well as a meat meal supplements (Lupatini et al., 2017; De and Ghosh, 2018). Besides *Spirulina sp.* also contains several types of micronutrients, minerals (i.e., iron ~0.58–1.8 $g.kg^{-1}$, calcium ~1.3–14 $g.kg^{-1}$, phosphorus ~6.7–9.0 $g.kg^{-1}$ and potassium ~6.4–15.4 $g.kg^{-1}$ algal dry biomass), vitamins (pseudovitamin B12, provitamin A, B1 (thiamine), B2 (riboflavin) and B3 (niacin), etc. which are also provides nutritional supplements in vegetarians diet (Hoseini et al., 2013). 1.0% of *S. platensis* has been added in a biscuits which shows better results (86.9% digestibility) compared with other cookies (Vaz et al., 2016). *Dunaliella bardawil* contains highest amino acid Glu (12.7g/100 protein) (Christaki et al., 2011). *Arthrospira platensis* contains highest dietary fibers having 42.82±1.20% w/w per algal dry biomass (Molino et al., 2018). In daily basis different types of pigments have been taken by human for their antioxidants activities which are derived from different types of microalgae *Dunaliella salina, D. bardawi* contain β-Carotene the daily basis intake recommendation is 2–7 mg, *Haematococcus pluvialis, Botryococcus braunii* contain Astaxanthin which intake capacity is 6–12 mg/day, *Arthrospira, Spirulina* contains phycobilins (phycocyanin, phycoerythrin), the daily basis intake recommended 200–400 mg/day (Matos, 2017; Milledge, 2011). *Arthrospira sp,* contains exopolyssacharides which stimulate the growth of probiotic bacteria in yogurt, fermented milk. *Chlorella spp.* Simunteniously incorporated into yogurt, cheese like food products to enhance the nutritional quality (Caporgno and Mathys, 2018). Microalgal *Arthrospira platensis* has been used as a dietary supplement for ruminants (<20%), poultry (<21%), rabbits (<15%), *Schizochytrium sp.* have been applied in poultry (7.4%), *Laminaria digitata* as a supplement of pig feed (up to 10% daily basis) for inhibit their growth and nutritional value which provides better quality meat for human consumption (Madeira et al., 2017; Harun et al., 2010; Christaki et al., 2011). *Chlorella* sp. has also been recognized as safe food ingredient as generally recognized as safe (GRAS) status by Food and Drug Administration, USA. It has been shown different intact roles in health sectors and cure many diseases (e.g., increase immune system, hypoglycemic effects, prevent tumor and cancer cells, lowering blood pressure, etc.) (Matos et al., 2016; Liu and Chen, 2016).

4.2.5.1 MICROALGAE AS POTENTIAL AQUACULTURE FEED

Currently microalgae are used as a high quality aquaculture feed for aquafeed industry. Microalgae have great potential of proteins, amino acid, lipid, vitamins, carotenoid, and energy in feed. Some studies show that generally microalgae have been enriched with ~30–40 % (w/w) protein, ~10–20 % (w/w) lipids and ~5–15 % (w/w) carbohydrates at late exponential growth phase (Shah et al., 2018; Guedes and Malcata, 2012; Ejike et al., 2016). These are mainly used as food for some aquatic species (i.e., larvae and juvenile shell and finfish and zoo plankton also). Most versatile microalgal species has been recognized as aquaculture feeds including *Chlorella* sp., *Tetraselmis* sp., *Isochrysis* sp., *Pavlova* sp., *Phaeodactylum* sp., *Chaetoceros* sp., *Nannochloropsis* sp., *Skeletonema* sp., and *Thalassiosira sp.* These species rich in highter content of protein aminoacids, which helps to survive of aquatic species. Beside this, *Chaetoceros, Thalassiosira, Tetraselmis, Isochrysis*, and *Nannochloropsis* those micro algae are also used for larval feed. These algal communities are applied in aquatic species. Indirectly feeding means the algae providing through artemia, rotifers, and daphnia which are fed to the target larval organisms. Various manufacturing units have been producing different types of aquaculture feed by mixing of *chlorella* sp. and *spirulina sp.* because of these two microalgal species contain high amount of proteins. *Hypneacervicornis* and *Cryptonemia crenulata* are also used in aquaculture for rich in protein in shrimp diets (Priyadarshani and Rath, 2012). *Spirulina* sp. contains different types of natural pigments which provides color in prawns, salmonid fish, ornamental fish, and other high-value fish (Chen et al., 2015). *C. vulguris* have protective effects through reducing oxidative stress against heavy metal contamination, also some toxic compounds like (lead, cadmium, and naphthalene). It contains high amount of carotenoids which use as antioxidant agents also which useful for aquatic species feed (Safi et al., 2014). More than 40 microalgal sp. are cultivated for using aquaculture. *Haematococcus pluvialis sp.* contains carotenoid which is enriched with canthaxanthin and astaxanthin which acts as coloring agent of salmonids flesh (Forján et al., 2014). Current studies have shown that some microalgal biomass have great application in aquaculture feed. Pigments of microalgae *Arthrospira platensis* helps to improve the fish color in Red tilapia (*Oreochromis* sp.). *Dunaliella salina sp.* enhances the immunological antioxidant rates and increases the survival rate in *Shrimp Penaeus monodon* aquaculture. Moreover, *Nannochloropsis* sp., *Isochrysis* sp., *Nannochloropsis gaditana, Tetraselmis chuii, and Phaeodactylum tricornutum, Arthrospira* sp., *Nanofrustulum* sp., *Tetraselmis* sp., *Haematococcus pluvialis, Arthrospira maxima, Tetraselmis suecica, A. platensis, Arthrospira* sp., and *Isochrysis* sp. have increased the growth rate, white blood cells, RBC counts, hemoglobin content, and albumin levels of fish. Amelioration of pigmentation helps to gain weight in different aquatic sp. like *Gilthead seabream (Sparus aurata), Golden barb (Puntius gelius), Atlantic salmon (Salmo salar), common carp (Cyprinus carpio), Pacific white shrimp (Litopenaeus vannamei), Pacific white shrimp (L. vannamei), Red tilapia fingerling (Oreochromis sp.), Ornamental carp (C. carpio), Nile tilapia (Oreochromis niloticus),* and *Parrot fish (Oplegnathus fasciatus)*, etc. Aforementioned microalgal sp. works without hampering any growth rates, no negative effects on nutrient utilization, digestion as well as survival of all aquatic species (Shah et al., 2018; Kaparapu, 2018). Mixed microalgal consortium approaches have been used as feed suppliments because of its availability as higher balanced nutritional value. Major advantages of usage of microalgae are rich in nutritional value, higher special availability, lesser amount of toxicity, easily digestive cell wall, etc. (Sirakov et al., 2015). In the contrary, proteins, vitamins, and poly unsaturated fatty acids are also enriched in novel microalgal species which have great impacts on aquafeed. The lack of PUFA omega-3 causes various types of negative effects of fruitfulness, fecundation, and hatching protentials even decreases egg quality in different fishes (Forján et al., 2014). Many different methods are using now a days for improving those compounds. Like Manipulation light intensity, temperature and nutrient allows for higher lipid contents, optimization in various ways for overall yield and productivity (Sirakov et al., 2015). Sometimes microalgae have been used as a food improvement quality compound. It effects against some pathogenic bacteria in aquaculture by its antibacterial activities. It has been confabbed entire protective measures against *V. campbellii* and *V. proteolyticus* and *Dunaliella tertiolecta* aquatic species (Falaise et al., 2016). Microalgae provides many phyto nutrient besides this it also provides PUFA, DHA, EPA, AA (arachidonate) which are also essential for growth and metamorphosis for aquatic ecobiological system. DHA, EPA, AA ratios are very important at the absolute levels. Some microalgal communities contain higher quantities of EPA (~7–34%). Some prymnesiophytes (i.e., *Pavlova spp.* and *Isochrysis sp.*) and cryptomonads are rich in DHA (~0.2–11%). *Nannochloropsis spp.* contains higher amount of AA (up to 4%). And when the correlational statistics lies amongst microalgal regimes and their corresponding nutritional assesses as long as PUFA are presents in adequate proportion then those are produce better growths of *juvenile oysters, larval scallops sp.* As well as it provides high dietary proteins for those aquatic species (Guedes and Malcata, 2012; Chauton et al., 2015). In last few decades over hundreds of microalgal regimes have been used for aquafeed but still less than twenty species have reached importance in aquaculture industry (Sharma and Sharma, 2017).

4.2.5.2 MICROALGAE AS GELIFY AND THICKENING AGENTS

Sodium and/or Calcium salts of alginic acid are the most potential phycocolloid having major commercial impact as thickening agents in various food industries. Alginate contains α L guluronate and β D-mannuronate (Khanra et al., 2018). It has been derived from brown algal species such as *Lessonia* sp., *Macrocystis* sp., *Laminaria* sp., *A. nodosum,* and *Durvillaea* sp., microalgal species (Drexler and Yeh, 2014). Alginate has geling and chileting properties. It has a high importance in textile industry for sizing cotton yarn. Due to its chileting properties it is capable of making highly viscous solutions for these reasons it is used as food and pharmaceutical sector (Sharma and Sharma, 2017). In food industry alginate are regularly used for its moisture retention properties (Khanra et al., 2018). Agar is the mixture of polysaccharides compounds. Those are found in some marine microalagal species. Agar derived from red algal regimes, i.e., *G. gracilis, G. dura, Pterocladia sp.* (Khanra et al., 2018; Drexler et al., 2014) which are also used as a food additives like sweet dishes, candy chocolates, fruit juice, jam along with industrial usage in sizing of papers, printing in textiles, application in molecular biology used as geling parts of agar molecules (Sharma and Sharma, 2017). *Gracilaria and Gelidium* are the most important algal species which are extensively used in industry for agar producing. In 2019, 12,000 tons/year agar has been produced from diffrerent algal species. Carrageenan is other potential algae derived natural polysaccharides having huge impact in health management. USA is the largest Carrageenan producing country. Carrageenan has been produced more than 30 different types for food applications like dairy, meet, desserts, etc. It's derived from *Gigartina* sp., *Chrondrus* sp. and *Euchema* sp which belong to the clade rodophyta (Khanra et al., 2018). Carrageenans are water soluble compounds; it is used for empulsify and stebilizing agents in various food industries (e.g., jellies, jams, deserts, meat products). Besides, carrageenans act as antitumor, antiviral, anticoagulant agents (Sharma and Sharma, 2017).

4.2.6 MICROALGAE AS ENZYME BIOFACTORIES

Diverse ranges of industrially viable enzymes (cellulases, amylases, galactosidases, proteases, lipases, phytases, laccases, antioxidant enzymes etc.) have been derived from diverse algal regimes (Brasila et al., 2017). The protein proportion in the microalgal regimes also increases due to the presence of different enzymes. Enzymes help to improve the protein extraction yields (Sari et al., 2013). Microalgal cellular walls are stiff and need unique parameter for lysis. While 52.8% gene sets are presents in *H, sapiens, M. musculus* and *D. rerio*, their only16.3% orthologous gene sets are found in various microalgal species like *C. reinhardtii, T. pseudonana* and *C. merolae*. Now a day some microalgal genes are uncovering for industrial enzyme production. Approximately 14,500 annotated microalgal genes are present in gene bank. Cellulose, hemicellulose, pectinase, Amylolytic enzymes, Galactosidases, these types of enzymes are capable to break down the microalgal cell wall, carbohydrate polymers in algae. *Euglena* species produce paramylon for carbon storage due to their primary energy production, and then they metabolize β 1,3 glucan hydrolase enzymes to aggregate carbohydrate moeties. *Anabena* strains are capable to produce hydrolytic enzymes. Amylase and β-amylase are ascertained in *D. parva.* Starch phosphorylase has been identified in *D tertiolecta.* Intracellular α galactosidase enzymetic catalytic efficiency has been found in golden brown algae *Poterioochromonas malhamensis*. InEight to nine species of cyanophycea and chlorophycea families was shown β-D-galactosidase enzyme acitivity. (i.e., *Cosmarium* sp., and *Acutodesmus obliquus*). *Arthospira platensis, A. variabilis* have potential for protease production. *Nannochloropsis oceanica, Nannochloropsis oceanica*are capable to lipase production. *Shewanella xiamenensis, Tethacystis aeria* have been shown laccase enzymatic efficacies upon substrate, i.e., 2′-azino-bis-3-ethylbenzthiazoline-6-sulfonic acid (ABTS). *Porphyridium cruentum* have been shown Superoxide dismutase activities. Which are also known as metalloenzyme, which transform super oxide radicals into molecular O_2 and H_2O_2, and used for various therapeutic applications (Brasila et al., 2017; Raposo et al., 2013). *Porphyridium cruentum* contains arachidonic acid which is used as nutritional feed in infants (Matos et al., 2017). A study have been shown that recombinant microalgal strain *Chlorella minutissima UTEX2341* was shown Xylanase enzyme acivity and capable to produce oil (Liu et al., 2017). A marine microalgae *I. galbana* is one of the great producers of carbonic anhydrase. It's a kind of metalloenzyme which is present in the blood cells (Raposo et al., 2013). *Isochrysis galbana* microalge produces carbonic anhydrase which plays a crucial role to convert CO_2 in to carbonic acid (Khan et al., 2018).

4.2.7 MICROALGAE USED AS BIOREACTORS FOR BIOPLASTIC PRODUCTION

Bioplastic and/or bioplastic precursor biosynthesis from microalgae has been emerging as most promising and economically feasible technology. *Spirulina* sp. and *Synechocystis* sp. have biosynthesized poly β hydroxybutyrate (PHB) as precursor of biodegradable plastic. *Stichiciccis sp., Chlorella sp., U. lactuca, Anabaena sp* and *Scenedesmus sp.,* strains have been implied as high value-added polymer producers. These microbial communities are used to generate diverse ranges of bioplastics and/or bioplastic precursors of hybrid plastic, cellulose based plastic, poly-lactic acid and biopolyethylene, etc. There are several bioprocess industries including Sun Petro, Dow Chemical Pvt. Ltd, Cereplast Pvt. Ltd has already taken a bold initiative to manufacture bioplastics from algae derived precursors. Moreover, Biotechnology Institute in Soley has already be started synthesizing biodegradable plastic from microalgal species *Spirulina (Arthrospira platensis*) (Khanra et al., 2018). Microalgae such as diatoms (e.g., *Phaeodactylum tricornutum*) are rich in lipids and silicate which are also helpful for PHB biosynthesis. After the usage of specific codon and large scale of synthesis *P. tricornutum* the high level of PHB upto 10.6 % of algal dry weigh have been observed (Hempel et al., 2011).

4.3 FUTURE OUTLOOK

Since very ancient days microalgal biomass have been considered as most value-added biomolecule producer in biorefinery approaches. Approximately 30,000 microalgal species have already been identified to have its promosing bioprocessing ability to synthesize bio active products and value-added by-products. Current extensive literature survey on microalgal biocommodities has inevitably shownd that microalgae have huge enomorous potentiality (as viable or resting phase) towards biomass production which can be the primary reservoir of various ranges of high value-added products. The Aeneid thoughtfulness of microalgae derived biologically compounds are relevant for industrial aspects. Current efforts are driving attention by various industries and business purposes to producing microalgal biomass derived value-added products. Microalgal biomass has been considered economically feasible and commercialization forvaluable resources. Microalgae are not only potential energy reservoir but microalgal biomass plays a precious role in therapeutic, aquaculture, biopolymers, enzyme derivation, cosmetic industries, bioactive compounds, high value-added pigments production purposes. However, several technological and economic challanges restrict the micro algal production platform through the utilization of waste biomass. To this end, extentive usage of bioprocess engineering, and extentive research need to be implimented for vast generation of microalgal species which contributes towards potential range of various industrial large-scale productions. Thus, microalgae-based scale up for value added biomolecule generations is demandable for market sustainability in near future.

ACKNOWLEDGMENT

Author(s) are thankful to the authority of Head, Division of Microbiology, ICAR-IARI, New Delhi; JIS University Kolkata; and Principal, Vishnu Institute of Pharmaceutical Education and Research, Narsapur, Medak, Telangana for constant support and encouragements.

KEYWORDS

- **biorefinery approach**
- **carbon dioxide sequestration**
- **high value-added biomolecules**
- **microalgae**
- **microalgal biomass**
- **microbial cell factory**
- **waste biomass utilization**

REFERENCES

Abdelaziz, A. E. M., Ghosh, D., & Hallenbeck, P. C., (2014). Characterization of growth and lipid production by *Chlorella* sp. PCH90, a microalga native to Quebec. *Bioresource Technology, 156*, 20–28.

Aguirre, A. M., Bassi, A., Saxena, P., (2013). Engineering challenges in biodiesel production from microalgae. *Critical Reviews in Biotechnology, 33*, 293–308.

Ambati, R. R., Gogisetty, D., Aswathanarayana, R. G., Ravi, S., Bikkina, P. N., Lei, B. L., & Yuepeng, S., (2019). Industrial potential of carotenoid pigments from microalgae: Current trends and future prospects. *Critical Reviews in Food Science and Nutrition, 59*, 1880–1902.

Andrade, L. M., Andrade, C. J., Dias, M., Nascimento, C. A., & Mendes, M. A., (2018). Chlorella and spirulina microalgae as sources of functional foods, nutraceuticals, and food supplements; an overview. *MOJ Food Processing & Technology, 6*, 45–58.

Arenas, E. G., Palacio, M. C. R., Juantorena, A. U., Fernando, S. E. L., & Sebastian, P. J., (2016). Microalgae as a potential source for biodiesel production: techniques, methods, and other challenges. *International Journal of Energy Research., 41*, 761–789.

Bahadar, A., & Khan, M. B., (2013). Progress in energy from microalgae: A review. *Renewable and Sustainable Energy Reviews, 27*, 128–148.

Barkia, I., Saari, N., & Manning, S. R., (2019). Microalgae for high-value products towards human health and nutrition. *Marine Drugs, 17*, 304.

Barra, L., Chandrasekaran, R., Corato, F., & Brunet, C., (2014). The challenge of ecophysiological biodiversity for biotechnological applications of marine microalgae. *Mar. Drugs, 12*, 1641–1675.

Begum, H., Yusoff, F. M. D., Banerjee, S., Khatoon, H., & Shariff, M., (2016). Availability and utilization of pigments from microalgae. *Critical Reviews in Food Science and Nutrition, 56*, 2209–2222.

Bhalamurugan, G. L., Valerie, O., & Mark, L., (2018). Valuable bioproducts obtained from microalgal biomass and their commercial applications: A review. *Environ. Eng. Res., 23*, 229–241.

Bilal, M., Rasheed, T., Ahmed, I., & Iqbal, H. M. N., (2017). High-value compounds from microalgae with industrial exploitability: A review. *Frontiers in Bioscience, 9*, 319–342.

Borowitzka, M. A., (2013). High-value products from microalgae—their development and commercialization. *J. Appl. Phycol., 25*, 743–756.

Brasila, B. D. S. A. F., Siqueiraa, F. G. D., Saluma, T. F. C., Zanetteb, C. M., & Spierb, M. R., (2017). Microalgae and cyanobacteria as enzyme biofactories. *Algal Research, 25*, 76–89.

Buonoa, S., Langellottia, A. L., Martelloa, A., Rinnaa, F., & Fogliano, V., (2014). Functional Ingredients from Microalgae. *Food & Function, 5*, 1669–1685.

Bux, F., (2013). *Potential Biotechnological Applications of Microalgae: Biodiesel and Value Added Products, 1*, 260. CRC Press.

Camacho, F., Macedo, A., & Malcata, F., (2019). Potential industrial applications and commercialization of microalgae in the functional food and feed industries: A short review. *Marine Drugs, 17*, 312.

Caporgno, M. P., & Mathys, A., (2018). Trends in microalgae incorporation into innovative food products with potential health benefits. *Front. Nutr., 5*, 58.

Chandra, T. S., Deepak, R. S., Kumar, M. M., Mukherji, S., Chauhan, V. S., Sarada, R., & Mudliar, S. N., (2016). Evaluation of indigenous fresh water microalga *Scenedesmus obtusus* for feed and fuel applications: Effect of Carbon dioxide, light and nutrient sources on growth and biochemical characteristics. *Bioresource Technology, 207*, 430–439.

Chauton, M. S., Reitan, K. I., Norsker, N. H., Tveterås, R., & Kleivdal, H. T., (2015). A techno-economic analysis of industrial production of marine microalgae as a source of EPA and DHA-rich raw material for aquafeed: Research challenges and possibilities. *Aquaculture, 436*, 95–103.

Cheah, W. Y., Show, P. L., Chang, J. S., Ling, T. C., & Juan, J. C., (2014). Biosequestration of atmospheric CO_2 and flue gas-containing CO_2 by microalgae. *Bioresource Technology, 184*, 190–201.

Chen, J., Wang, Y., Benemann, J. R., Zhang, X., Hu, H., & Qin, S., (2015). Microalgal industry in China: Challenges and prospects. *J. Appl. Phycol., 28*, 715–725.

Chew, K. W., Yap, J. Y., Show, P. L., Suan, N. H., Juan, J. C., Tau, C. L. T. C., Lee, D. J., & Chang, J. S., (2017). Microalgae biorefinery: High value products perspectives. *Bioresource Technology, 229*, 53–62.

Choi, H. J., (2016). Dairy wastewater treatment using microalgae for potential biodiesel application. *Environ. Eng. Res., 21*, 393–400.

Christaki, E., Paneri, P. F., & BONOS, E., (2011). Microalgae: A novel ingredient in nutrition. *International Journal of Food Sciences and Nutrition., 62*, 794–799.

Chu, W. L., (2012). Biotechnological applications of microalgae. *IeJSME., 6*, S24–S37.

Chunga, Y. S., Leeb, J. W., & Chung, C. H., (2017). Molecular challenges in microalgae towards cost-effective production of quality biodiesel. *Renewable and Sustainable Energy Reviews, 74*, 139–144.

Costa, J. A. V., & Morais, M. G. D., (2011). The role of biochemical engineering in the production of biofuels from microalgae. *Bioresource Technology, 102*, 2–9.

De Jesus, R. M. F., De Morais A. M. M. B., & De Morais, R. M. S. C., (2014). *Bioactivity and Applications of Polysaccharides from Marine Microalgae, 1*, 1683–1727. Springer International Publishing: Cham, Switzerland.

De, K., & Ghosh, D., (2018). Microbial single cell protein generation: A comparative existing state of art. *Biotechnology and Nature, 1*, 94–98.

Doan, Q. C., Moheimani, N. R., Mastrangelo, A. J., & Lewis, D. M., (2012). Microalgal biomass for bioethanol fermentation: Implications for hypersaline systems with an industrial focus. *Biomass and Bioenergy, 46*, 79–88.

Drexler, I. L. C., & Yeh, D. H., (2014). Membrane applications for microalgae cultivation and harvesting: A review. *Rev. Environ. Sci. Biotechnol., 13*, 487–504.

Ebrahiminezhad, A., Bagheri, M., Taghizadeh, S. M., Berenjian, A., & Ghasemi, Y., (2016). Biomimetic synthesis of silver nanoparticles using microalgal secretory carbohydrates as a novel anticancer and antimicrobial. *Advances in Natural Sciences: Nanoscience and Nanotechnology, 7*, 015018.

Eduardo, D. F. S. C., & Alberto, B., (2016). Bioethanol from microalgae and cyanobacteria: A review and technological outlook. *Process Biochemistry, 51*, 1833–1842.

Ejike, C. E. C. C., Collins, S. A., Balasuriya, N., Swanson, A. K., Mason, B., & Udenigwe, C. C., (2016). Prospects of microalgae proteins in producing peptide-based functional foods for promoting cardiovascular health. *Trends in Food Science & Technology, 59*, 30–36.

Enamalaa, M. K., Enamalab, S., Chavalic, M., Donepudid, J., Yadavallie, R., Kolapallia, B., Aradhyulaf, T. V., et al., (2018). Production of biofuels from microalgae – a review on cultivation, harvesting, lipid extraction, and numerous applications of microalgae. *Renewable and Sustainable Energy Reviews, 94*, 49–68.

Falaise, C., François, C., Travers, M. A., Morga, B., Haure, J., Tremblay, R., Turcotte, F., et al., (2016). Antimicrobial compounds from eukaryotic microalgae against human pathogens and diseases in aquaculture. *Mar. Drugs, 14*, 159.

Farieda, M., Samera, M., Abdelsalamb, E., Yousefc, R. S., Attiab, Y. A., & Ali, A. S., (2017). Biodiesel production from microalgae: Processes, technologies and recent advancements. *Renewable and Sustainable Energy Reviews, 79*, 893–913.

Ferreiral, V. D. S., & Anna, C. S., (2017). Impact of culture conditions on the chlorophyll content of microalgae for biotechnological applications. *World J. Microbiol. Biotechnol., 33*, 20.

Forján, E., Navarro, F., Cuaresma, M., Vaquero, I., Domínguez, M. C. R., Gojkovic, Z., Vázquez, M., et al., (2014). Microalgae: Fast-growth sustainable green factories. *Critical Reviews in Environmental Science and Technology, 45*, 1705–1755.

Frac, M., Tys, S. J., & Tys, J., (2010). Microalgae for biofuels production and environmental applications: A review. *African Journal of Biotechnology, 9*, 9227–9236.

Galasso, C., Corinaldesi, C., & Sansone, C., (2017). Carotenoids from Marine Organisms: Biological Functions and Industrial Applications. *Antioxidants, 6*, 96.

Gangl, D., Zedler, J. A. Z., Rajakumar, P. D., Martinez, E. M. R., Riseley, A., Włodarczyk, A., Purton, S., et al., (2015). Biotechnological exploitation of microalgae. *Journal of Experimental Botany, 66*, 6975–6990.

Garc, J. L., Vicente, M. D., & Galan, B., (2017). Microalgae, old sustainable food and fashion nutraceuticals. *Microbial Biotechnology, 10*, 1017–1024.

Garciaa, L. M., Adjalléb, K., Barnabéb, S., & Raghavana, G. S. V., (2017). Microalgae biomass production for a biorefinery system: Recent advances and the way towards sustainability. *Renewable and Sustainable Energy Reviews, 76*, 493–506.

Gateau, H., Solymosi, K., Marchand, J., & Schoef, B., (2017). Carotenoids of microalgae used in food industry and medicine. *Mini. Rev. Med. Chem., 17*, 1140–1172.

Ghosh, D., & Hallenbeck, P. C., (2015). Cyanobacteria and photosynthetic bacteria: Metabolic engineering of hydrogen production. *Advances in Enzymatic Conversion of Biomass to Biofuels, 1*, 112–124.

Ghosh, D., & Hallenbeck, P. C., (2015). Metabolic engineering of hydrogen production by green algae. *Advances in Enzymatic Conversion of Biomass to Biofuel, 1*, 96–110.

Ghosh, T., Paliwal, C., Maurya, R. K., & Mishra, S., (2015). Microalgal rainbow colors for nutraceutical and pharmaceutical applications. *Plant Biology and Biotechnology, 69*, 85–91.

Goh, B. H. H., Ong, H. C., Cheah, M. Y., Chen, W. H., Yu, K. L., & Mahlia, T. M. I., (2019). Sustainability of direct biodiesel synthesis from microalgae biomass: A critical review. *Renewable and Sustainable Energy Reviews, 107*, 59–74.

Goldberg, I. K., Iskandarov, U., & Cohen, Z., (2011). LC-PUFA from photosynthetic microalgae: Occurrence, biosynthesis, and prospects in biotechnology. *Appl. Microbiol. Biotechnol., 91*, 905–915.

Gong, M., & Bassi, A., (2016). Carotenoids from microalgae: A review of recent developments. *Biotechnology Advances, 34*, 1396–1412.

Gong, Y., Hu, H., Gao, Y., Xu, X., & Gao, H., (2011). Microalgae as platforms for production of recombinant proteins and valuable compounds: Progress and prospects. *J. Ind. Microbiol. Biotechnol., 38*, 1879–1890.

Gonzalez, J. R., Olivieri, G., Vree, J. D., Bosma, R., Willems, P., Reith, H., Eppink, M., et al., (2016). Towards industrial products from microalgae. *Energy & Environmental Science, 9*, 3036–3043.

Gouveia, L., (2011). Microalgae as a feedstock for biofuels. *Springer Briefs in Microbiology, 1*, 1–69.

Guedes, A. C., & Malcata, F. X., (2012). Nutritional value and uses of microalgae in aquaculture. *Aquaculture, 10*, 59–78.

Guedes, A. C., Amaro, H. M., & Malcata, F. X., (2011). Microalgae as Sources of Carotenoids. *Marine Drugs, 9*, 625–644.

Guedes, A. C., Amaro, H. M., & Malcata, F. X., (2011). Microalgae as sources of high added-value compounds—A brief review of recent work. *Biotechnol. Prog, 27*, 597–613.

Guldhe, A., Kumari, S., Ramanna, L., Ramsundar, P., Singh, P., Rawat, I., & Bux, F., (2017). Prospects, recent advancements and challenges of different wastewater streams for microalgal cultivation. *Journal of Environmental Management, 203*, 299–315.

Hamed, I., (2016). The evolution and versatility of microalgal biotechnology: A review. *Comprehensive Reviews in Food Science and Food Safety, 15*, 1104–1123.

Han, S. F., Jin, W. B., Tu1, R. J., Wu, W. M., (2015). Biofuel production from microalgae as feedstock: Current status and potential. *Crit. Rev. Biotechnol., 35*, 255–268.

Harun, R., Manjinder, S. M., Forde, G. M., & Danquah, M. K., (2010). Bioprocess engineering of microalgae to produce a variety of consumer products. *Renewable and Sustainable Energy Reviews, 14*, 1037–1047.

Hempel, F., Bozarth, A. S., Lindenkamp, N., Klingl, A., Zauner, S., Linne, U., Steinbüchel, A., & Maier, U. G., (2011). Microalgae as bioreactors for bioplastic production. *Microbial Cell Factories, 10*, 81.

HH, A. E. B., & GS, E. B., (2013). Healthy benefit of microalgal bioactive substances. *Journal of Aquatic Science, 1*, 11–23.

Hidalgo, P., Toro, C., Ciudad, G., & Navia, R., (2013). Advances in direct transesterification of microalgal biomass for biodiesel production. *Rev. Environ. Sci. Biotechnol., 12*, 179–199.

Hoseini, S. M., Darani, K. K., & Mozafari, M. R., (2013). Nutritional and medical applications of spirulina microalgae. *Mini-Reviews in Medicinal Chemistry, 13*, 1231–1237.

Hua, J., Nagarajan, D., Zhanga, Q., Chang, J. S., & Lee, D. J., (2018). Heterotrophic cultivation of microalgae for pigment production: A review. *Biotechnology Advances., 36*, 54–67.

Kaparapu, J., (2018). Application of microalgae in aquaculture. *48*, 21–26 Phykos.

Kapoore, R. V., Butler, T. O., Pandhal, J., & Vaidyanathan, S., (2018). Microwave-assisted extraction for microalgae: From biofuels to biorefinery. *Biology, 7*, 18.

Katiyar, R., Gurjar, B. R., Biswas, S., Pruthi, V., Nalin, K. N., & Kumar, P., (2017). Microalgae: An emerging source of energy based bio-products and a solution for environmental issues. *Renewable and Sustainable Energy Reviews, 72*, 1083–1093.

Khan, M. I., Shin, J. H., & Kim, J. D., (2018). The promising future of microalgae: Current status, challenges, and optimization of a sustainable and renewable industry for biofuels, feed, and other products. *Microb. Cell Fact., 17*, 36.

Khanra, S., Mondal, M., Halder, G., Tiwari, O. N., Gayen, K., & Bhowmick, T. K., (2018). Downstream processing of microalgae for pigments, protein and carbohydrate in industrial application: A review. *Food and Bioproducts Processing, 110*, 60–84.

Kiran, B., Kumar, R., & Deshmukh, D., (2014). Perspectives of microalgal biofuels as a renewable source of energy. *Energy Conversion and Management, 88*, 1228–1244.

Kirrolia, A., Bishnoi, N. R., & Singh, R., (2013). Microalgae as a boon for sustainable energy production and its future research & development aspects. *Renewable and Sustainable Energy Reviews, 20*, 642–656.

Klinthong, W., Yang, Y. H., Huang, C. H., & Tan, C. S., (2015). A review: Microalgae and their applications in CO_2 capture and renewable energy. *Aerosol and Air Quality Research, 15*, 712–742.

Koller, M., Muhr, A., & Brauneg, G., (2014). Microalgae as versatile cellular factories for valued products. *Algal Research, 6*, 52–63.

Koller, M., Salerno, A., Tuffner, P., Koinigg, M., Böchzelt, H., Schober, S., Pieber, S., et al., (2012). Characteristics and potential of micro algal cultivation strategies: A review. *Journal of Cleaner Production, 37*, 377–388.

Koopmans, M. V., Wijffels, R. H., Barbosa, M. J., & Eppink, M. H. M., (2013). Biorefinery of microalgae for food and fuel. *Bioresource Technology, 135*, 142–149.

Koutra, E., Economou, C. N., Tsafrakidou, P., & Kornaros, M., (2018). Bio-based products from microalgae cultivated in digestates. *Trends in Biotechnology, 36*, 819–833.

Koyande, A. K., Chew, K. W., Rambabu, K., Tao, Y., Chu, D. T., & Show, P. L., (2019). Microalgae: A potential alternative to health supplementation for humans. *Food Science and Human Wellness, 8*, 16–24.

Kumar, A., Ergas, S., Yuan, X., Sahu, A., Zhang, Q., Dewulf, J., Malcata, F. X., & Langenhove, H. V., (2010). Enhanced CO_2 fixation and biofuel production via microalgae: Recent developments and future directions. *Trends in Biotechnology, 28*, 371–380.

Kumar, S., Hablot, E., Moscoso, J. L. G., Obeid, W., Hatcher, P. G., DuQuette, B. M., Graiver, D., et al., (2014). Polyurethanes preparation using proteins obtained from microalgae. *J. Mater. Sci., 49*, 7824–7833.

Kumara, B. R., Deviramb, G., Mathimanic, T., Ducd, P. A., & Pugazhendhi, A., (2019). Microalgae as rich source of polyunsaturated fatty acids. *Biocatalysis and Agricultural Biotechnology, 17*, 583–588.

Lafarga, T., (2020). Cultured microalgae and compounds derived thereof for food applications: Strain selection, cultivation, drying, and processing strategies. *Food Reviews International, 36*, 559–583.

Lam, M. K., & Lee, K. T., (2012). Microalgae biofuels: A critical review of issues, problems and the way forward. *Biotechnology Advances, 30*, 673–690.

Lee, T. L. C., & Gonzalez-Marino, G. E., (2010). Microalgae for "healthy" foods—Possibilities and challenges. *Comprehensive Reviews in Food Science and Food Safety, 9*, 655–675.

Leu, S., & Boussiba, S., (2014). Advances in the production of high-value products by microalgae. *Industrial Biotechnology, 10*, 169–183.

Liu, C., Hao, Y., Jiang, J., & Liu, W., (2017). Valorization of untreated rice bran towards bio-focculant using a lignocellulose-degrading strain and its use in microalgal biomass harvest. *Biotechnology for Biofuels, 10*, 90.

Liu, J., & Chen, F., (2016). Biology and industrial applications of chlorella: Advances and prospects. *Adv. Biochem. Eng. Biotechnol., 153*, 1–35.

Liu, J., Sun, Z., Gerken, H., Liu, Z., Jiang, Y., & Chen, F., (2014). *Chlorella zofingiensis* as an alternative microalgal producer of astaxanthin: Biology and industrial potential. *Mar. Drugs, 12*, 3487–3515.

Liu, L., Pohnert, G., & Wei, D., (2016). Extracellular metabolites from industrial microalgae and their biotechnological potential. *Marine Drugs, 14*, 191.

Lopes, E. J., Maroneze, M. M., Depra, M. C., Sartori, R. B., Dias, R. R., & Zepka, L. Q., (2019). Bioactive food compounds from microalgae: An innovative framework on industrial biorefineries. *Current Opinion in Food Science, 25*, 1–7.

Luo, X., Su, P., & Zhang, W., (2015). Advances in microalgae-derived phytosterols for functional food and pharmaceutical applications. *Marine Drugs, 13*, 4231–4254.

Lupatini, A. L., Colla, L. M., Canana, C., & Colla, E., (2017). Potential application of microalgae *Spirulina platensis* as a protein source. *J. Sci. Food Agric., 97*, 724–732.

Madeira, M. S., Cardoso, C., Lopes, P. A., Coelho, D., Afonso, C., Bandarra, N. M., & Prates, J. A. M., (2017). Microalgae as feed ingredients for livestock production and meat quality: A review. *Livestock Science, 205*, 111–121.

Maeda, Y., Yoshino, T., Matsunaga, T., Matsumoto, M., & Tanaka, T., (2018). Marine microalgae for production of biofuels and chemicals. *Current Opinion in Biotechnology, 50*, 111–120.

Manirafasha, E., Ndikubwimana, T., Zeng, X., Lu, Y., & Jing, K., (2016). Phycobiliprotein: Potential microalgae derived pharmaceutical and biological reagent. *Biochemical Engineering Journal, 109*, 282–296.

Markou, G., & Nerantzis, E., (2013). Microalgae for high-value compounds and biofuels production: A review with focus on cultivation under stress conditions. *Biotechnology Advances, 31*, 1532–1542.

Matos, A. P., (2017). The impact of microalgae in food science and technology. *J. Am. Oil Chem. Soc., 94*, 1333–1350.

Matos, A. P., Feller, R., Moecke, E. H. S., Oliveira, J. V. D., Junior, A. F., Derner, R. B., & Sant'Anna, E. S., (2016). Chemical characterization of six microalgae with potential utility for food application. *J. Am. Oil Chem. Soc., 93*, 963–972.

Matos, J., Cardoso, C., Bandarra, N. M., & Afonso, C., (2017). Microalgae as a healthy ingredient for functional food: A review. *Food Funct., 8*, 2672–2685.

Maurya, R., Paliwal, C., Ghosh, T., Pancha, I., Chokshi, K., Mitra, M., Ghosh, A., & Mishra, S., (2016). Applications of de-oiled microalgal biomass towards development of sustainable biorefinery. *Bioresource Technology, 214*, 787–796.

McGinn, P. J., Dickinson, K. E., Bhatti, S., Frigon, J. C., Guiot, S. R., O'Leary, S. J. B., (2011). Integration of microalgae cultivation with industrial waste remediation for biofuel and bioenergy production: Opportunities and limitations. *Photosynth Res., 109*, 231–247.

Medipally, S. R., Yusoff, F. Md., Banerjee, S., & Sharif, M., (2015). Microalgae as sustainable renewable energy feedstock for biofuel production. *BioMed Research International, 3*, 519513.

Miazek, K., Iwanek, W., Remacle, C., Richel, A., & Goffin, D., (2015). Effect of metals, metalloids and metallic nanoparticles on microalgae growth and industrial product biosynthesis: A Review. *Int. J. Mol. Sci., 16*, 23929–23969.

Milledge, J. J., (2011). Commercial application of microalgae other than as biofuels: A brief review. *Rev Environ Sci Biotechnol., 10*, 31–41.

Mobin, S., & Alam, F., (2017). Some promising microalgal species for commercial applications: A review. *Energy Procedia, 110*, 510–517.

Molino, A., Iovine, A., Casella, P., Mehariya, S., Chianese, S., Cerbone, A., Rimauro, J., & Dino, M. D., (2018). Microalgae characterization for consolidated and new application in human food, animal feed and nutraceuticals. *Int. J. Environ. Res., 15*, 2436.

Morais, M. G. D., Vaz, B. D. S., Morais, E. G. D., & Costa, J. A. V., (2015). Biologically active metabolites synthesized by microalgae. *BioMed Research International, 2015*, 15.

Odjadjare, E. C., Mutanda, T., & Olaniran, A. O., (2017). Potential biotechnological application of microalgae: A critical review. *Critical Reviews in Biotechnology, 37*, 37–52.

Oliveira, D. T. D., Costa, A. A. F. D., Costa, F. F., Filho, G. N. D. R., & Nascimento, L. A. S. D. N., (2020). Advances in the biotechnological potential of Brazilian marine microalgae and cyanobacteria. *Molecules, 25*, 2908.

Patel, A., Gami, B., Patel, P., & Patel, B., (2016). Microalgae: Antiquity to era of integrated technology. *Renewable and Sustainable Energy Reviews, 71*, 535–547.

Phwana, C. K., Ong, H. C., Chen, W. H., Linga, T. C., Nge, E. P., & Show, P. L., (2018). Overview: Comparison of pretreatment technologies and fermentation processes of bioethanol from microalgae. *Energy Conversion and Management, 173*, 81–94.

Picardo, M. C., Medeiros, J. L. D., Monteiro, J. G. M., Chaloub, R. M., Giordano, M., & Arau´jo, O. D. Q. F., (2013). A methodology for screening of microalgae as a decision making tool for energy and green chemical process applications. *Clean Technologies and Environmental Policy, 15*, 275–291.

Pignolet, O., Jubeau, S., Garcia, C. V., & Michaud, P., (2013). Highly valuable microalgae: Biochemical and topological aspects. *Journal of Industrial Microbiology and Biotechnology, 40*, 781–796.

Poojary, M. M., Barba, F. J., Aliakbarian, B., Donsì, F., Pataro, G., Dias, D. A., & Juliano, P., (2016). Innovative alternative technologies to extract carotenoids from microalgae and seaweeds. *Marine Drugs, 14*, 214.

Posten, C., & Chen, S. F., (2016). Advances in biochemical engineering/biotechnology. *Microalgae Biotechnology* (Vol. 153, p. 188). Springer International Publishing.

Pragya, N., Pandey, K. K., & Sahoo, P. K., (2013). A review on harvesting, oil extraction and biofuels production technologies from microalgae. *Renewable and Sustainable Energy Reviews, 24*, 159–171.

Priyadarshani, I., & Rath, B., (2012). Commercial and industrial applications of micro algae: A review. *Journal of Algal Biomass Utilization, 3*, 89–100.

Raheem, A., Prinsen, P., Vuppaladadiyam, A. K., Zhao, M., & Luque, R., (2018). A review on sustainable microalgae based biofuel and bioenergy production: Recent developments. *Journal of Cleaner Production, 181*, 42e59.

Raposo, M. F. D. J., Morais, R. M. S. C. D., & Morais, A. M. M. B. D., (2013). Health applications of bioactive compounds from marine microalgae. *Life Sciences, 93*, 479–486.

Rasala, B. A., & Mayfield, S. P., (2011). The microalga *Chlamydomonas reinhardtii* as a platform for the production of human protein therapeutics. *Bioengineered Bugs, 2*, 50–54.

Ratha, S. K., & Prasanna, R., (2012). Bioprospecting microalgae as potential sources of "green energy"—Challenges and perspectives (Review). *Applied Biochemistry and Microbiology, 48*, 109–125.

Rodrigues, C. S., Villela, H. D. M., Martins, A. P., Marques, L. G., Colepicolo, P., & Tonon, A. P., (2015). Microalgae for economic applications: Advantages and perspectives for bioethanol. *Journal of Experimental Botany, 66*, 4097–4108.

Safi, C., Zebib, B., Merah, O., Pontalier, P. Y., & Garcia, C. V., (2014). Morphology, composition, production, processing and applications of *Chlorella vulgaris*: A review. *Renewable and Sustainable Energy Reviews, 35*, 265–278.

Sánchez, N. F. S., Blanco, R. V., Carlos, B. H., Ariño, A. T., Mendoza, P. C. G., & Coronado, R. S., (2016). Lipids rich in ω-3 polyunsaturated fatty acids from microalgae. *Appl. Microbiol. Biotechnol., 100*, 8667–8684.

Sani, Y. M., Daud, W. M. A. W., & Aziz, A. R. A., (2013). Solid acid-catalyzed biodiesel production from microalgal oil—The dual advantage. *Journal of Environmental Chemical Engineering, 1*, 113–121.

Santhosh, S., Dhandapani, R., & Hemalatha, N., (2016). Bioactive compounds from Microalgae and its different applications- a review. *Advances in Applied Science Research, 7*, 153–158.

Sari, Y. W., Bruins, M. E., & Sanders, J. P. M., (2013). Enzyme assisted protein extraction from rapeseed, soybean, and microalgae meals. *Industrial Crops and Products, 43*, 78–83.

Sathasivam, R., & Ki, J. S., (2018). A review of the biological activities of microalgal carotenoids and their potential use in healthcare and cosmetic industries. *Mar. Drugs, 16*, 26.

Sathasivam, R., Radhakrishnan, R., Hashem, A., & Allah, E. F. A., (2019). Microalgae metabolites: A rich source for food and medicine. *Saudi Journal of Biological Sciences, 26*, 709–722.

Satyanarayana, K. G., Mariano, A. B., & Vargas, J. V. C., (2011). A review on microalgae, a versatile source for sustainable energy and materials. *International Journal of Energy Research, 35*, 291–311.

Shah, M. R. M. D., Lutzu, G. A., Alam, A. M. D., Sarker, P., Chowdhury, M. A. K., Parsaeimehr, A., Liang, Y., Daroch, M., (2018). Microalgae in aquafeeds for a sustainable aquaculture industry. *J. Appl. Phycol., 30*, 197–213.

Shah, Md. M. R., Liang, Y., Cheng, J. J., & Daroch, M., (2016). Astaxanthin-producing green microalga *Haematococcus pluvialis*: From single cell to high value commercial products. *Front. Plant Sci., 7*, 531.

Shariati, M., & Hadi, M. R., (2011). Microalgal biotechnology and bioenergy in *Dunaliella. Molecular and Environmental Bioengineering, 1*, 483–505.

Sharma, P., & Sharma, N., (2017). Industrial and biotechnological applications of algae: A review. *Journal of Advances in Plant Biology, 1*, 01–25.

Show, P. L., Tang, M. S. Y., Nagarajan, D., Ling, T. C., Ooi, C. W., & Jo-Shu, C. J. S., (2017). A Holistic Approach to Managing Microalgae for Biofuel Applications. *Int. J. Mol. Sci., 18*, 215.

Shubaa, E. S., & Kifle, D., (2018). Microalgae to biofuels: 'Promising' alternative and renewable energy, review. *Renewable and Sustainable Energy Reviews, 81*, 743–755.

Sigamani, S., Ramamurthy, D., & Natarajan, H., (2016). A review on potential biotechnological applications of microalgae. *Journal of Applied Pharmaceutical Science, 6*, 179–184.

Singh, B., Guldhe, A., Rawat, I., & Bux, F., (2014). Towards a sustainable approach for development of biodiesel from plant and microalgae. *Renewable and Sustainable Energy Reviews, 29*, 216–245.

Singh, U. B., & Ahluwalia, A. S., (2013). Microalgae: A promising tool for carbon sequestration. *Mitig. Adapt. Strateg. Glob. Change, 18*, 73–95.

Sirakov, I., Velichkova, K., Stoyanova, S., & Staykov, Y., (2015). The importance of microalgae for aquaculture industry: Review. *International Journal of Fisheries and Aquatic Studies, 2*, 81–84.

Skjånes, K., Rebours, C., & Lindblad, P., (2013). Potential for green microalgae to produce hydrogen, pharmaceuticals and other high value products in a combined process. *Critical Reviews in Biotechnology, 33*, 172–215.

Souza, M. P. D., Hoeltz, M., Gressler, P. D., Benitez, L. B., & Schneider, R. C. S., (2018). Potential of microalgal bioproducts: General perspectives and main challenges. *Waste and Biomass Valorization, 10*, 2139–2156.

Sudhakar, K., & Premalatha, M., (2015). Characterization of micro algal biomass through FTIR/TGA /CHN analysis: Application to *Scenedesmus sp. Energy Sources, Part A: Recovery, Utilization, and Environmental Effects, 37*. 2330–2337.

Suganya, T., Varman, M., Masjuki, H. H., & Renganathan, S., (2016). Macroalgae and microalgae as a potential source for commercial applications along with biofuels production: A biorefinery approach. *Renewable and Sustainable Energy Reviews, 55*, 909–941.

Sun, Z., Li, T., Zhou, Z. G., & Jiang, Y., (2015). Microalgae as a source of lutein: Chemistry, biosynthesis, and carotenogenesis. *Adv. Biochem. Eng. Biotechnol., 153*, 37–58.

Taleb, A., Pruvost, J., Legrand, J., Marec, H., Gouic, B. L., Mirabella, B., Legeret, B., et al., (2015). Development and validation of a screening procedure of microalgae for biodiesel production: Application to the genus of marine microalgae *Nannochloropsis. Bioresource Technology, 177*, 224- 232.

Tanzi, C. D., Vian, M. A., Ginies, C., Elmaataoui, M., & Chemat, F., (2012). Terpenes as green solvents for extraction of oil from microalgae. *Molecules, 17*, 8196–8205.

Tiwari, O. N., Bhunia, B., Muthuraj, M., Bandyopadhyay, T. K., Ghosh, D., et al., (2020). Optimization of process parameters on lipid biosynthesis for sustainable biodiesel production and evaluation of its fuel characteristics. *Fuel, 269*, 117471.

Uduman, N., Qi, Y., Danquah, M. K., Forde, G. M., & Hoadley, A., (2010). Dewatering of microalgal cultures: A major bottleneck to algae-based fuels. *J. Renewable Sustainable Energy, 2*, 012701.

Usher, P. K., Ross, A. B., Valero, M. A. C., Tomlin, A. S., & Gale, W. F., (2014). An overview of the potential environmental impacts of large-scale microalgae cultivation. *Biofuels, 5*, 331–349.

Varfolomeev, S. D., & Wassermanb, L. A., (2011). Microalgae as source of biofuel, food, fodder, and medicines. *Applied Biochemistry and Microbiology, 47*, 789–806.

Vaz, B. S., Moreira, J. B., Morais, M. G., & Costa, J. A. V., (2016). Microalgae as a new source of bioactive compounds in food supplements. *Current Opinion in Food Science, 7*, 73–77.

Vega, T. C. A., Lim, D. K. Y., Timmins, M., Vernen, F., Li, Y., & Schenk, P. M., (2012). Microalgal biofactories: A promising approach towards sustainable omega-3 fatty acid production. *Microbial Cell Factories, 11*, 96.

Venkata, M. S., Nikhil, G. N., Chiranjeevi, P., Reddy, C. N., Rohit, M. V., Kumar, A. N., & Sarkar, O., (2016). Waste biorefinery models towards sustainable circular bioeconomy: Critical review and future perspectives. *Bioresource Technology, 215*, 2–12.

Ventura, S. P. M., Nobre, B. P., Ertekin, F., Hayes, M., Vaquero, M. G., Vieira, F., Kocjj, M., et al., (2017). Extraction of value-added compounds from microalgae. *Microalgae-Based Biofuels and Bioproducts, 1*, 461–483.

Vera, C. R. D., Crespín, G. D., Daranas, A. H., Looga, S. M., Lillsunde, K. E., Tammela, P., Perälä, M., et al., (2018). Marine microalgae: Promising source for new bioactive compounds. *Mar. Drugs, 16*, 317.

Wichuk, K., Brynjólfsson, S., & Fu, W., (2014). Biotechnological production of value-added carotenoids from microalgae. *Bioengineered, 5*, 204–208.

Wijffels, R. H., & Barbosa, M. J., (2010). An outlook on microalgal biofuels. *Science, 329*, 796.

Wijffels, R. H., Barbosa, M. J., & Eppink, M. H. M., (2010). Microalgae for the production of bulk chemicals and biofuels. *Biofuels, Bioprod. Bioref., 4*, 287–295.

Wolkers, H., Barbosa, M., Kleinegris, D. M. M., Bosma, R., & Wijffels, R. H., (2011). Microalgae: The green gold of the future?: Large-scale sustainable cultivation of microalgae for the production of bulk commodities. *Wageningen: Wageningen UR – Food & Biobased Research, 34*.

Wu, L. F., Chen, P. C., Huang, A. P., & Lee, C. M., (2012). The feasibility of biodiesel production by microalgae using industrial wastewater. *Bioresource Technology, 113*, 14–18.

Zhao, C., Brück, T., & Lercher, J. A., (2013). Catalytic deoxygenation of microalgae oil to green hydrocarbons. *Green Chemistry, 15*, 1720–1739.

Zhu, L. D., & Hiltunen, E., (2016). Application of livestock waste compost to cultivate microalgae for bioproducts production: A feasible framework. *Renewable and Sustainable Energy Reviews, 54*, 1285–1290.

CHAPTER 5

Biological Approaches for Advanced Fuels Development from Biological/Plant Residues or Wastes

RAJESH K. SRIVASTAVA

Department of Biotechnology, GIT, GITAM (Deemed to be University, Visakhapatnam, 530045, Andhra Pradesh, India,
E-mail: rajeshksrivastava73@yahoo.co.in

ABSTRACT

In the present scenario, the whole world is facing energy scarcity and utilization from current stocks of non-renewable resources. It causes a high rate of environmental degradation by influencing its biotic and non-biological components. So, we need to develop fuels from biological waste matter available in huge quantities and at low prices. This waste utilization is a critical challenge for society to make or produce sustainable power from biological application processes (for conversion into monomers and fuels). This approach could help in the reduction of pollution and also enhance energy security in the world. In this regard, advanced biofuel and biogas development could be alternative options to non-renewable fuel, creating adverse effects to biological or non-biological components in the environment. Biofuel usage for our daily needs can maintain healthy environments with more energy security for the current and future generations. The sun, geothermal, wind energy, or biomass energy can provide massive amounts of sustainable or renewable energy. These come from various available resources. These energies can help maintain the green environment by mitigating global issues and maintaining good health for all the components of the environment.

Biotechnology for Waste Biomass Utilization. Prakash Kumar Sarangi & Latika Bhatia (Eds.)

5.1 INTRODUCTION

From the ancient to the current period, most developmental works were done by consuming traditional, conventional, or non-renewable fuel energy such as petroleum oils, gases, or coals. This energy consumption in the world has created many problems such as their stock shortage in the future, emission of toxic gases or byproducts, enhanced global environmental pollution, or other issues such as climatic changes and greenhouse gas production (Yang and Chen, 2012). So, our society is searching for alternative fuel options for daily needs energy consumption in our fast and busy lives. In this regard, advanced biofuel or biogas development can play critical roles in providing additional or non-conventional fuel energy to the world. And it can keep our environment healthy with greenery also (Atabani et al., 2012).

Biofuel is utilized for cooking and heating tasks, and in developing countries, including Asia and Africa, people are working to generate and utilization biofuel for daily needs energy. In the 21st century, biofuel could be found as one of the critical fuel energy with a reduction in the particulate matter or toxic gases and also reduced carbon dioxides emission in our environment (Mofijur et al., 2012). Plant biomasses as waste matter are reported to this advanced biofuel biosynthesis via practical biological approaches for hydrolysis and its conversion into biofuels or biogases. It can reduce fossil fuel energy use with the mitigation of greenhouse gases (GHGs) in atmospheres. Further, this approach can help reduce pollution and improve waste management tasks (Tan et al., 2009; Mofijur et al., 2012). Now, advanced research tasks are involved in the primary justification of bio-mediated agents in fuel development via providing the environmental benefits and displacement of conventional or non-renewable fuel utilization for our energy needs in daily life. In the current period, the most efficient bioproduction of ethanol, butanol, propanol, and methanol is reported from biological matter utilization and plant products or structures found in complex, simple, or both forms (Tan et al., 2009).

Some dedicated crops report for these biofuel productions, but it can create food security issues for the world once food matter or products are used as natural carbon sources. So we need to develop effective, efficient technologies for the utilization of waste plant residues or issues present in abundant quantities (Mofijur et al., 2012). But it created an air pollution problem, once burnt by a farmer in their croplands. So, advanced biofuel development can be started via non-food crops and non-oils crops/plant biomasses and reported waste materials for society. Some examples are Jatropha species, reported for biofuel development as raw carbon substrates and dedicated for ethanol production, without affecting the food supply and demands (Jangu and Meena, 2014).

Many advanced research works and technological development in recent years reported improving the design of procedures and becoming applicable for biofuel generation from waste plant biomasses without affecting human food consumption. So we can minimize the negative environmental consequence via the utilization of biofuel energy for different energy needs tasks. Government and non-governmental recommended policies can also promote advanced biofuel development by developing habits or adaption at local people levels via utilization of agricultural wastes (Atabani et al., 2012; Jangu and Meena, 2014).

Further, we can use this fuel energy for business-oriented tasks but it needs technological development to energy crop producers with more economical ways or less negative environmental impact. Biofuel also links with social aspects such as food security, food prices, or resources (Atabani et al., 2012; Jangu and Meena, 2014). So in this chapter, the author will emphasize biofuel development with other forms of renewable energy that can benefit energy security for the world. It will also discuss plant biomass sources from different crops considered as waste plant biomass. Further discussion will focus on microbial systems for a few advanced biofuels and biogases development.

5.2 PLANT BIOMASS AND OTHER PARTS AS WASTE MATTER

Biomass is a source of organic matter, coming from plant or animal sources, crop residues, waste wood, or waste tree parts included as plant biomasses. Still, burning biomass releases chemical energy as heat and this can be applied for electric power generation by steam turbine utilization. Lower plant or microbial biomass generated point is a clean and renewable fuel. It is a green fuel and is an alternative to coal or petroleum oil (fossil fuel), which is now utilized for electricity generation (Chinnici et al., 2015).

Biofuel development can provide energy security and reduce adverse effects on our environment. It can also drive forces worldwide for clean energy development. This biofuel development is only possible via the utilization of the potential of agro-industry stimulation, and it is shown in Figure 5.1. But its fuel energy can create other problems such as automotive engine compatibility in long-term operation. Other advanced biofuel uses can minimize the food shortage issue globally and utilize raw materials such as non-oil plants, seeds, or crops or non-food sources (lignocelluloses) (Ou et al., 2009).

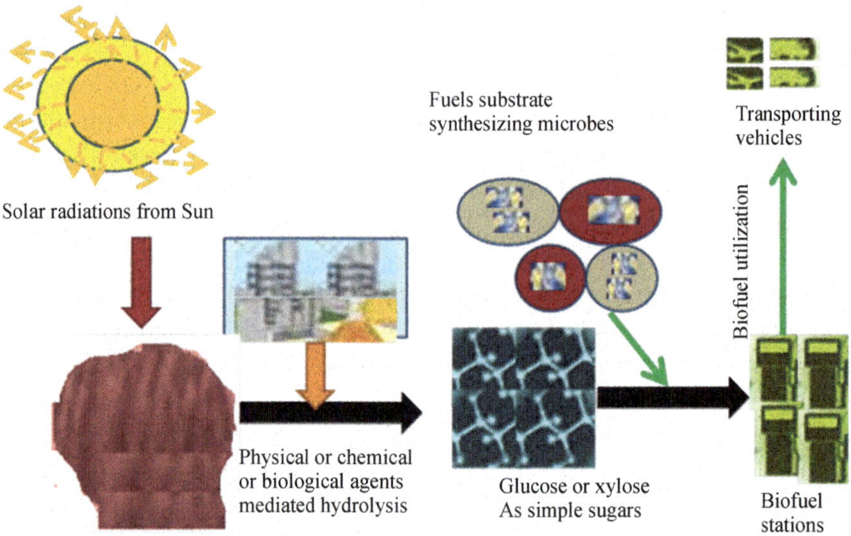

FIGURE 5.1 Solar radiation captured plant waste biomasses, utilized via its proper hydrolysis that was used for biofuels development for energy needs.

5.2.1 PLANT BIOMASSES SOURCE AND ITS ANALYSIS FOR FUELS

Recent and applied science has helped to characterize the many forms of plant biomass that come from the forest regions (forest plant biomasses produced a higher quantity of carbon emission than fossil fuel). Forest biomass utilization for fuel energy can have negative consequences on biodiversity. But some other forms of biomass energy can serve as low carbon-containing energy options under correct circumstances. In this regard, sawdust and wood chips as waste from sawmills can exhibit the quick capacity to decompose and release the carbon in the atmosphere with low carbon energy sources (Chinnici et al., 2015; Avcıoğlu et al., 2019).

Further, the grassland communities on the Qinghai-Tibetan Plateau (QTP) grassland generates due to environmental and anthropogenic activities. A comparison of plants' diversity, plants' biomass, or total plant-covered areas across different locations discusses. The different types of grasslands show on the QIP location. And various kinds of relationships between diversity of plant indices profiles such as Shannon-Weiner or Sampson Evenness profiles with plant cover, plant biomasses, or plants covered area with soil elements profiles are discussed using correlation analysis (Avcıoğlu et al., 2019). Plant biomass with soil fertility is compared with the covering of plants or biomass of plants relation, which is done via a simple linear regression model. There are reports on weak positive relationships between the Evenness index profile and plant biomasses and a positive relationship. These reports between plants cover area and biomasses of plants (Fayiah et al., 2019).

Using local wood biomass is reported for bioenergy and earlier, plant biomass of Alpine area was also used. North Italy found the desirable option for recycling wood residues that supported heat or electric power generation. Further, local managers' support and policymakers' contributions are committed to planning and implementing renewable energy strategies by considering the circular economy pattern (Kumari and Singh, 2018). Integration of environmental accountancy with other assessment methods reports via exploring the economic profitability or social desirability for wood-based bioenergy. This biofuel synthesis shows in mountain or hilly areas that show low population density and high forest-covered sites (Kumari and Singh, 2018; Buonocore et al., 2019).

A vast quantity of burning wood can result in deforestation and natural habitat reduction by converting from renewable to non-renewable energy. In the world, non-renewable energy stocks are found in an unequal manner. These have created a big issue for getting fuel energy and availability in the future. Further, its fuel energy-burning harms the environment (Wang et al., 2017).

5.3 RENEWABLE FORMS OF ENERGY

In our environment, all energy sources do not make renewable energy, and biomass or large hydroelectric dams can create difficulty in tradeoff via consideration of the impact on wildlife, 5.3climatic change, or other issues. Different types of renewable energy sources using emerging technologies can promote their generation as clean energy sources via natural resources or processes via constantly replenishing (Wang et al., 2017). And the sunlight and wind sources report via providing shine our light and blowing benefits respectively. These sources can be available depending on time and weather conditions. Renewable energy generates huge quantities using new and advanced technology to harness natural power and utilizes heating, transportation, lighting, or other tasks (Nassar et al., 2019).

The renewable energy system contrasts with non-renewable sources (i.e., fossil fuels), and there is a strong linkage between them and economic growth. Many investigating the development of renewable energy sources and hydropower, bioenergy, or solar energy have shown their impacts on the rural household economy area (Nassar et al., 2019). In China, during 2003–2017, the employment of panel data is shown about this linkage. In the last few years, renewable energy plant infrastructure has two ways of fixed-effect models for revealing the relationship between energy investment and household income or consumption profiles. Further, the Granger causality test panels are used to verify the impact of renewable energy and economic growth profiles. It found that bioenergy, solar energy, or hydropower are needed to improve China's rural household economies (Huang et al., 2020).

5.3.1 FUEL ENERGY FROM WASTES PLANT MATTERS

5.3.1.1 BIOFUELS WITH COMPETENT TO INTERNAL ENGINE

Further problems can report for severe corrosion, carbon deposition, and the wearing of engine parts for fuel supply system components that can come from biodiesel uses as energy needs. The advantages and disadvantages of biodiesel discuss biodiesel engine's dedication to providing the ultimate solution for biodiesel commercialization tasks. In this regard, Brazil has boosted its bioethanol generation tasks with its marketing by introducing flexible-fuel vehicles (FFV) for transportation purposes (Sgroi et al., 2005).

This approach has dedicated the engine for ethanol and gasoline fuel energy. Further, a similar process can help in bringing a breakthrough in the commercialization tasks of biofuel. We need to focus on the biofuel scenario in the present by assessing biofuels' policies and standards for fuels worldwide. Different biofuel processing technology has to be discussed with some guidelines on the dedication of biofuel engines. In contemporary engine designs, minor modification affecting cost or price is essential by utilizing advanced research works and technology development (Sgroi et al., 2005; López et al., 2009).

5.3.1.2 ENZYMATIC-MEDIATED BIOMASSES HYDROLYSIS

Next, xylanases are discussed as the industrial enzyme and show more demand in the paper or pulp industry, food, and animal feed industries at a commercial scale. These enzymes now report for their application via biological agents-mediated conversion of lignocelluloses wastes. It helps in conversion into valuable products or fermentable sugar forms. This enzyme reports as a hydrolase enzyme. These enzymes demonstrate the capability to catalyze the β-1,4 linkage or bonds in heteropolysaccharides xylan and form xylose monomers (Buonocore et al., 2019).

Fungi are mainly used to synthesize the xylanase enzyme. The other utility of this sugar reports for fermentation by other microbial cells that quickly converted into biofuel. It reported that fungi use is reported for low-cost saccharification processes in practical ways for agricultural wastes. Xylose is the most energy-efficient source for biofuel production (Javed et al., 2019; Yan et al., 2020).

5.3.1.3 APPROACHES FOR BIOFUELS SYNTHESIS

Further, kinetic equations report analysis of the compensation effect show for complete algal biomasses' pyrolysis process. In this process, two reaction zones were utilized and analyzed by carrying out the individual reaction. The first zone was used for activation energies and evaluated within the conversion zone of 0.05 to 0.95. These processes have shown the different activation energies (~125 to 147 KJ/mole for lignocellulosic biomass and 113 to 138 KJ/mol for zone^{-1} algal biomasses) (Yan et al., 2020).

Next, the pre-exponential factor calculates for these biomasses. The compensation effect reported between activation energy and pre-exponential factors—regression analysis conducted with the kinetic model for plant biomass. Master plots (Zα) used the nucleation model that find dominant at a lower or higher temperature for plant biomass. Further, the pre-exponential factor ultimately finds in zone-2 of algal biomasses. And lignocelluloses biomass degradation or hydrolysis sees in Figure 5.2 (Vasudev et al., 2019).

Aquatic plant biomass also contains a high quantity of water contents or moisture holding similar lignocellulosic structures. This plant biomass utilizes anaerobic digestion (AD) techniques. It has shown a high amount of methane recovery per volume in simultaneous ways due to a high biodegradation biomass rate. For this aquatic plant, there are two novel stages: serial of wet-AD (~W-AD) and solid-state AD (~SS-AD) and proposed for utilization for biogases generation. These processes were found with quick digestion of labile cytoplasm and reported from the wet-AD reactor quickly. But slow digestion in SS-AD says for lignocellulosic fractions by taking long retention time (Kumar et al., 2020).

Further, this AD system's serial approach and methane recovery were discussed with the chemical oxygen demand (COD) value balance from aquatic plant biomasses and examined in a semi-continuous operation mode. We discuss the *Elodea muttallii* aquatic plant which is found to grow in the basin of the south of Lake Biwa, Japan, which can be utilized as a substrate for AD plants. This single-stage AD operation uses different hydraulic retention times (HRT) compared with methane recoveries, and it has shown in the reduction of COD value 47.6% to 33.1% in shortening HRT value (15 days) (Iweh et al., 2020).

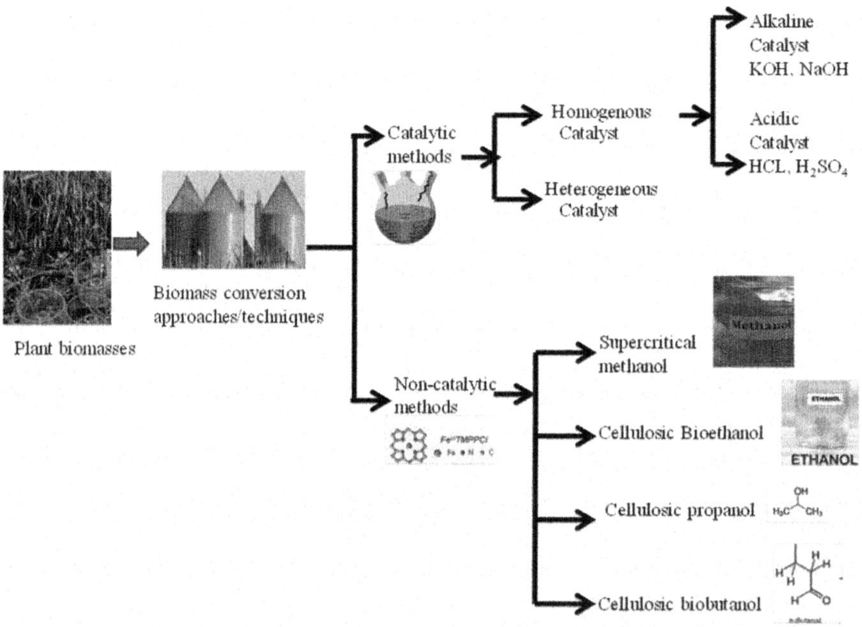

FIGURE 5.2 Various approaches of pretreatment for hydrolysis of plant biomasses and helped in release of monomers sugars utilization biofuel developments.

Many reports on dual-chamber microbial electrolysis cells (MEC) have been discussed. With the concentric cylinder, find for hydrogen synthesis from lignocellulosic material. This chamber fabricates with suitable fuel synthesis. Three different lignocellulosic materials utilize by using simultaneous saccharification and fermentation (SSF). It reported the maximum hydrogen (H2) production rate (HPR- 2.46 m.moles/L/D) with a total energy recovery efficiency (11.3%) and leading hydrogen volumetric yield (18.7 L/kg) from these hydrolyzed substrates. Further, the concentration of reducing sugars and organic acid, pH, or currents in the MEC system during hydrogen monitor (Zhang et al., 2019). Bioethanol and biodiesel are discussed below as suitable microbial systems and raw matters, but n-butane and biomethane synthesis are not discussed in more detail.

5.3.1.4 BIOETHANOL AND BIODIESEL

Many research works are done on integrated biomasses conversion or transformation tools and utilized for algal biomasses. For checking these efficiencies, first algal biomass collected from rivers bodies, including Song river, Dehradun, via keeping the laboratory conditions. Later, it is processed under the same laboratory parameters. For this work completion, around 0.65 dry weight per meter square microalgal biomasses from two microalgal species were harvested from this freshwater river (Kumar et al., 2018). Later, lipid extraction was performed by applying the Soxhlet extraction methods approach or tools, and this process used reagents such as chloroform and methanol (2:1). And 18.6% of lipids are found from the extraction of microalgal biomasses. The blending of algal biodiesel found with butanol (A5B25D70), and petroleum diesel (A5B30D60) find by using Inline blending methods on a volume basis. Next, oils extracted from microalgal biomasses are hydrolyzed to release fermentable sugars used for ethanol production by microbial fermentation and found around 61% (Kumar et al., 2018; Moncada et al., 2014).

There are many reports on developing new technologies that can apply and show the increased quantity of biofuels synthesis, and it can compare without the need for food source utilization. In this regard, microalgal biomasses cultivation does in recent years, and this biomasses discussed the concept of biorefinery development and industrial sustainability substrates for bioethanol synthesis. Further, the cultivation approaches discussed the accumulation of nearly 50% carbohydrates in their biomasses (dry cell weight). It also showed the importance of water with a nutrient recycling nature (de Farias Silva and Bertucco, 2019) and is shown in Figure 5.3.

For microalgal biomasses, the saccharification approach discussed the application of hydrolyzing enzymes or acids (in optimum concentration) via hydrothermal liquefaction methods or tools and discussed flash hydrolysis. After the saccharification processes, the main monosaccharides in microalgal biomasses discusses as glucose (Markou et al., 2013). These algal biomasses showed a high hydrolysis rate that was utilized in the fermentation process (with 80% utilization efficiency). Anaerobic digestion approaches treated the vinasses and recycled the carbon dioxide from ethanol production fermentation processes with increased sustainability (Kim et al., 2014).

Next, alternative techniques or tools were discussed for bioethanol extraction or converting in concentrated form from fermentation broth to utilize this fuel in transportation. Many advantages from microalgal biomasses utilization discuss for ethanol or biodiesel production via comparing to higher classes' plant biomasses (reference plant). But its cultivation cost is still high and that needs to be reduced (Kim et al., 2014; de Farias Silva and Bertucco, 2019).

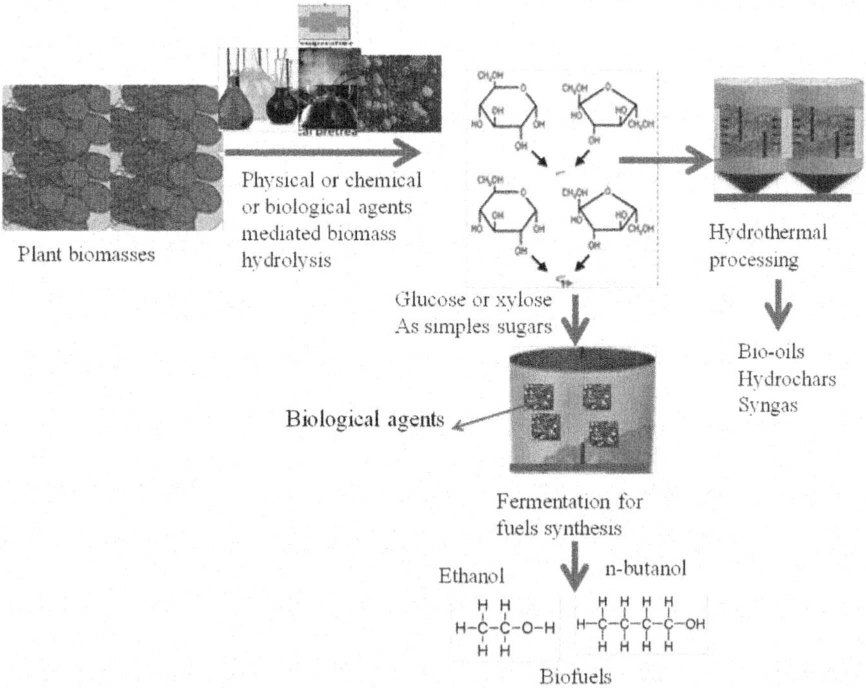

FIGURE 5.3 Biofuel synthesis from plant biomasses via utilization of different conversion approaches for fuel development.

Bioethanol synthesis from algal biomasses or feedstock is discussed with increasing interest in advanced research works in recent years. It is different from fossil fuel, and it is involved in bioethanol synthesis but it found still extensive as well as less processing time. The current study was focused on crucial green alga (i.e., *Chlorella valgaris* or *C. valgaris*) species and found in bioethanol development tasks (Varaprasad et al., 2020). Further, *C. valgaris* discussed its growth initiation using the universal Tris-acetate phosphate (TAP) medium containing various concentrations and combinations of vitamins like thiamin, biotin, or cobalamin (B1, B7, and B12). These combinations of compositions enhance the algal biomasses used in ethanol synthesis (Ruangsomboon et al., 2018). The optimal concentration of vitamins such as CV2 media (TAP with 0.4 $g.L^{-1}$ of B1, 0.002 $g.L^{-1}$ of B7, or

0.007 g.L^{-1} of B12) showed the higher algal biomasses synthesis containing lipid contents or quantity.

Further, all the algal biomasses or feedstock used for ethanol production utilize an efficient microbial System (*Saccharomyces cerevisiae*) that utilized light and dark fermentation (Liu et al., 2018). Next, from these algal biomasses or feedstock from CV2 medium, a high ethanol concentration was achieved in 48 h in dark fermentation and compared with other feedstocks. Also, light fermentation produced ethanol yield at different time intervals. All these results on bioethanol production from these algal biomasses are compared with grabbing investors' attention in the bioenergy sector. Next, bioethanol production is reported at the commercial level or scale from these algal feedstocks or algal biomasses wastes and seen in Figure 5.4 (Kim et al., 2014; Varaprasad et al., 2020). Some examples of other renewable energy are discussed below.

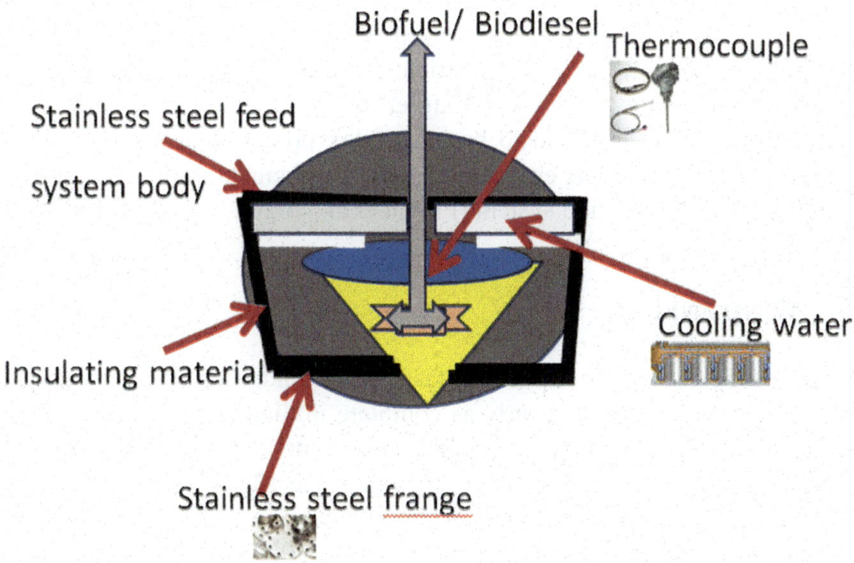

FIGURE 5.4 Biofuel injector used for biodiesel operation from utilization of algal plants

5.4 SOLAR RADIATION ENERGY

The distribution of solar systems at the local level for homes or businesses can generate electricity. It is achieved by keeping roof-top panels' arrangement or using community project usages to power the whole neighborhood area. Next, solar farms can also create the thousand homes' ability via a mirror system to concentrate the sunlight energy across the acres of the solar cells system (Pravin et al., 2020). Further, floating solar systems or float-voltaic system is found as a practical application of wastewater nutrients utilization facilities or bodies of water for the non-sensitive area's ecology. 1% of U.S. electricity generation is reported from solar energy supply and now nearly the third position of all new generating capacities from the solar system in 2017. It is said that it is second only to natural gas. The solar energy system is a clean energy source that does not produce any pollutants or greenhouse gases (Hughes and Meckling, 2017) and sees in Table 5.1 with renewable energy forms.

Here is information on new solar energy installation that has provided the most significant or huge boost in renewable energy, i.e., half of all the new forms. Next, wind energy source power accounted for hydropower's third capacity (15%)—newly installed solar energy capacity involved for electricity-producing technology from a few decades to the current time. As we know, our fertile soils are capable of profitable growth of plants or crops. And that can provide the foods, nutrients, and other essential components to biotic components in nature (Wang et al., 2020). The burning of plant biomasses or wood is found to be a complex process when compared to coal. And wood renewable energy resources are obtained from a sustainably managed forest production system (Pravin et al., 2020).

5.5 WIND ENERGY

Wind energy has come from old-fashioned windmills and turbines or skyscrapers (with turbines nearly as complete in diameter). It gained more attention around the world, and wind energy turns turbine blades which feed an electric generator for the production of electric power. Wind energy is nearly 6% of power registered in the U.S. and generates cheap energy sources in many parts of the country (Li et al., 2020). In the U.S., the top windpower-producing states are Califonia, Texas, Oklahoma, or Lowa. The turbine then utilizes high-speed wind such as a hilltops system or open plains or offshore area in open water system (Tang 2018; Li et al., 2020).

5.6 TIDAL ENERGY

Tidal energy is another form of hydro-energy, and it is utilized twice daily. Tidal current is used to drive the turbine generator. Tidal flow is not similar

TABLE 5.1 Renewable Resources Including Biomasses Fuel for Sustainable Energy for World Need with More Energy Security and Clean Fuel Energy

Renewable energy	**Energy capacity from resource processes**	**References**
1. Optimized performance of solar-steam generation shown for solar energy	The new strategy uses is found for an evaporation rate (2.94 $kg.m^{-2}h^{-2}$) that reported for energy efficiency of solar–steam generation. Further using of design configuration of photothermal material reported for warm and cold evaporation surfaces in the solar system	Wang et al., (2020)
2. Development of a new model estimated the technical potential for net-zero energy buildings with different locations, climates or building types or vintages ding types of vintages	The potential for building-integrated solar energy generation is reported in all worlds. Further deep efficiency scenarios reported with the allowance of more types of building with the meeting of zero energy levels by 2050. Global and region levels potentials, estimated for building-integrated solar energy generation system.	Petrichenko et al., (2019)
3. An integrated solar-tri-generation system (ISTGS) has used linear Fresnel reflector (LFR) technology	This has produced 110 MWe of electricity from steam turbines, 4.5 x 10^4 m^3/day of fresh water, and 2.3 x 10^3 kg/s of chilled water Optimized configuration of integration of 83.6 hectares of LFR solar field from the tri-generation plant (130 MWe), gave Levelized electricity cost ~ 6.37 US₵/kWh with 96.40 k-ton reductions in annual CO_2 emission.	Dabwan et al., (2020)
4. Wind power generation probabilistic model is used to generate and allocate sustainable electrical energy	Accurate wind rate forecasts reported the minimal necessity of auxiliary energy balancing systems and reserving power to incorporate wind energy with minimal instability of wind energy generation. These wind power forecasts can address the complexities of balancing supply and demand in a wind power system	Banik et al., (2020)
5. The relationship is reported between financial mechanisms and wind capacity. And wind power capacity was reported across fifteen countries and 10 U.S. states from 2005 to 2017.	Increased capacity for wind power is reported from 17GW (in 2000) to 513 GW (in 2017) in those countries or states and shows 88% of total wind energy generation capacity. A further contribution of individual wind energy capacity to overall electricity generation is also found an increase from 0.15–24.2 (2005) and 1.2 to 38.5% (in 2017)	Gupta et al., (2019)

TABLE 5.1 *(Continued)*

Renewable energy	Energy capacity from resource processes	References
6. The load-perturbation control system is reported on hour shaping timescales and total systems of capital costs. And it is found for renewable energy with optimal relative installed capacities of wind, solar or inter-stream tidal generation.	This has influenced the approximate 20% more than the cost of renewable energy generation without storage or output shaping. Further timescales of days or weeks favored the higher capacity of installed wind, solar, or tidal energy generation. And energy storage for smooth and shape follow to electrical load perturbation in Nova Scotia Canada.	Pearre and Swan, (2020)
7. Fully holistic treatment of tidal barrages is recognized the energy generation in environmental, societal and economic opportunities	This fulfills the societal needs of the people and also the necessity for compensatory habitat. An organic architectural design can harness the energy from vast tidal resources and can help to combat the twin challenges of climatic change or energy security. This can apply to the North West Hydro Resource Model.	Petley et al., (2019)
8. Electric power generation is achieved from the partially depleted natural gas reservoir that is suitable for carbon dioxides-based geothermal energy extraction	Extracting heat produced from natural gas is reported that utilized the supercritical CO_2 as working fluid and it has exhibited dual tasks of enhanced gas recovery (EGR) as well as extracting geothermal energy (CO_2-Plume Geothermal~CPG). It is found from deep natural gas reservoirs for electric power generation.	Ezekiel et al., (2020)
9. Small-sized biomass gasification system reported for power generation (<200kW)	This power generation is discussed with improvement on the implementation of small-sized biomass gasification system or process that helped in gas fuels for power generation	Situmorang et al., (2020)
10. Role of woody biomasses reported for reduction of fossil GHGs emission in future.	Woody biomass for power or heat energy has an estimated 5–14% in 2040. It can reduce the cost of power or heat energy. 1Twh amount of woody bioheat can reduce fossil GHGs emissions by 10–17 $MTCO_2$.	Jåstad et al., (2020)
11. Cellulosic ethanol production from optimum dilute acid pretreatment condition, sugar yields from the straw of the 30 wheat varieties	33.7–36.3% glucan (proxy for cellulose), 16.8–19.5% xylan (proxy for hemicellulose), and 18.4–20.6% lignin reported from dilute acid pretreatment. It showed total sugar recovery in vary levels between 0.239 and 0.401 g g−1 dry matter of raw biomass across wheat straw varieties.	Fitria et al., (2019)

TABLE 5.1 *(Continued)*

Renewable energy	Energy capacity from resource processes	References
12. Wheat straw (WS) pretreated by ammonium sulfite (AS) showed for efficient acetone-butanol-ethanol (ABE) production by *Clostridium acetobutylicum* ATCC 824	Higher ABE titer and yield achieved in simultaneous saccharification and fermentation (SSF) that used different level of AS pretreated WS (6%, 7.5%, 9% and 10.5%, w/v) and SHF is found for increased concentration ABE (butanol) titer from 6% to 10.5%, or 12.28 (7.52) to 17.75 (11.25) g/L. ABE (butanol) yield reduced from 161 (98) to 133 (84) g/kg raw WS	Qi et al., (2019)

to other hydro-energy sources (that is not constant). This energy is highly predictable with compensating for the periods for tide current is the low situation (Petley et al., 2019). The tidal energy approach can harm wildlife via tidal barrages, like a dam, located in an ocean bay or lagoon location. Similar to tidal power, wave power can be found on barrier-like structures or ocean floor-anchored devices on or just below the water surface (Pearre and Swan, 2020).

5.7 GEOTHERMAL ENERGY

Geothermal energy harnesses natural heat below the Earth's surface, which is used to heat homes or to generate electricity. In oceans, tidal and waveform energy reports as a developmental phase or stage. Ocean area can be controlled by the moon's gravity that harnesses or captures its capacity as an interactive option (Longa et al., 2020). Geothermal energy is not shown to be much crucial in the U.K. compared to Iceland. In Iceland, geothermal heat is easily or freely available. Now more essential resources are biomass energy that converts solid fuel and is made from plant material with easy conversion into electricity (Ezekiel et al., 2020; Longa et al., 2020).

Nowadays, geothermal energy which is harnessed from hot springs is used for humankind's needs. In this energy source, the Earth's core area is considered hot as the sun's surface due to the slow decay of radioactive particles or elements in the rocks at the center of the Earth's planet. In this energy source, drilling a deep well brings the boiling underground water to the Earth's surface as hydrothermal energy or resources. Then it is pumped via a turbine device for the generation of electric power (Pearre and Swan, 2020). Geothermal energy plants have shown low emission for pumping the steam water due to the reservoir system's back. This source of energy can increase the risks of earthquakes in that area due to geological hot spots. Geothermal system installation is expensive at the initial stage, but it has shown fewer maintenance issues and longer life than air (Alirahmi et al., 2019).

5.8 CONCLUSION

This chapter focuses on biomasses-developed fuel sources, and these fuels are reported as fewer carbon emitters for our environment and maintain energy security for the world. It shows in non-renewable fuel sources that can create its storage or stock availability and toxic particle formation and gases in our environment. This chapter emphasizes waste matters in our climate if burned, and then produces poisonous gases or particles. So these wastes reported with bioagent or microbial cells mediated fuel development. This fuel development can maintain the good health of biotics and non-biotics components via keeping greenery for earth planets. Next, bioethanol and biodiesel discuss with algal biomasses utilization that has shown more chance for commercial-scale fuels. This chapter also discussed the solar, wind, or geothermal energy source that can help us get renewable energy. So we can conclude with the development of plant-based advanced fuels development.

KEYWORDS

- **advanced fuel**
- **bioenergy**
- **biological processes**
- **plant residue**
- **renewable energy**
- **waste matters**

REFERENCES

Alirahmi, S. M., Dabbagh, S. R., Ahmadi, P., & Wongwises, S., (2020). Multi-objective design optimization of a multi-generation energy system based on geothermal and solar energy. *Ener. Conver. Manag., 205*, 112426.

Atabani, A. E., Silitonga, A. S., Badruddin, I. A., Mahlia, T. M. I., Masjuki, H. H., & Mekhilef, S., (2012). A comprehensive review on biodiesel as an alternative energy resource and its characteristics. *Renew. Sustain. Ener. Rev., 16,* 2070–2093.

Avcıoğlu, A. O., Dayıoğlu, M. A., & Türker, U., (2019). Assessment of the energy potential of agricultural biomass residues in Turkey. *Renew.Ener., 138*, 610–619.

Banik, R., Das, P., Ray, S., & Biswas, A., (2020). Wind power generation probabilistic modeling using ensemble learning techniques. *Materials Today: Proceedings, 26*, 2157–2162.

Buonocore, E., Paletto, A., Russo, G. F., & Franzese, P. P., (2019). Indicators of environmental performance to assess wood-based bioenergy production: A case study in Northern Italy. *J. Clean. Product., 221*, 242–248.

Chinnici, G., Amico, M. D., Rizzo, M., & Pecorino, B., (2015). Analysis of biomass availability for energy use in sicily. *Renew. Sustain. Ener. Rev., 52*, 1025–1030.

Dabwan, Y. N., Pei, G., Gao, G., Feng, J., & Li, J., (2020). A novel integrated solar tri-generation system for cooling, freshwater and electricity production purpose: Energy, economic and environmental performance analysis. *Solar Ener., 198*, 139–150.

De Farias, S. C. E., & Bertucco, A., (2019). Bioethanol from Microalgal biomass: A promising approach in biorefinery/ *Braz. Arch. Biol. Technol., 62*, e19160816.

Ezekiel, J., Ebigbo, A., Adams, B. M., & Saar, M. O., (2020). Combining natural gas recovery and CO2-based geothermal energy extraction for electric power generation. *Appl. Ener., 269*, 115012.

Fayiah, M., Dong, S., Li, Y., Xu, Y., Gao, X., Li, S., Shen, H., et al., (2019). The relationships between plant diversity, plant cover, plant biomass and soil fertility vary with grassland type on Qinghai-Tibetan plateau. *Agricult. Ecosys. Environ., 286*, 106659.

Fitria, Ruan, H., Fransen, S. C., Carter, A. H., Tao, H., & Yang, B., (2019). Selecting winter wheat straw for cellulosic ethanol production in the Pacific Northwest, U.S.A. *Biomas. Bioener., 123*, 59–69.

Gupta, D., Das, A., & Garg, A., (2019). Financial support vis-à-vis share of wind generation: Is there an inflection point?. *Ener.,181*, 1064–1074.

Huang, J., Li, W., Guo, L., Hu, X., & Hall, J. W., (2020). Renewable energy and household economy in rural China. *Renew. Ener., 155*, 669–676.

Hughes, L., & Meckling, J., (2017). The politics of renewable energy trade: The US-China solar dispute. *Ener. Polic., 105*, 256–262.

Iweh, N. S., Koyam, M., Akizuki, S., Ban, S., & Tod, T., (2020). Novel wet-solid states serial anaerobic digestion process for enhancing methane recovery of aquatic plant biomass. *Sci. Total Environ., 730*, 138993.

Jangu, S., & Meena, B. L. A. L., (2014). Biofuel development in India: Prospects and problems. *J. Biofuel, 5*(2), 96–11.

Jåstad, E. O., Bolkesjø, T. F., Trømborg, E., & Rørstad, P. K., (2020). The role of woody biomass for reduction of fossil GHG emissions in the future North European energy sector. *Appl. Ener., 274*, 115360.

Javed, U., Ansari, A., Aman, A., & Qade, S. A. U., (2019). Fermentation and saccharification of agro-industrial wastes: A cost-effective approach for dual use of plant biomass wastes for xylose production. *Biocataly. Agricul. Biotechnol., 21*, 101341.

Kim, K. H., Choi, I. S., Kim, H. M., Wi, S. G., & Bae, H., (2014). Bioethanol production from the nutrient stress-induced microalga *Chlorella vulgaris* by enzymatic hydrolysis and immobilized yeast fermentation. *Bioresour. Technol., 153*, 47–54.

Kumar, V., Kumar, P., Kumar, P., & Singh, J., (2020). Anaerobic digestion of *Azolla pinnata* biomass grown in integrated industrial effluent for enhanced biogas production and COD reduction: Optimization and kinetics studies. *Environ. Technol. Innovat., 17*, 100627.

Kumar, V., Nanda, M., Joshi, H. C., Singh, A., Sharma, S., & Verma, M., (2018). Production of biodiesel and bioethanol using algal biomass harvested from fresh water river. *Renew. Ener., 116*, 606–612.

Kumari, D., & Singh, R., (2018). Pretreatment of lignocellulosic wastes for biofuel production: A critical review. *Renew. Sustain. Ener. Rev., 90*, 877–891.

Li, Y., Huang, X., Tee, K. F., Li, Q., & Wu, X. P., (2020). Comparative study of onshore and offshore wind characteristics and wind energy potentials: A case study for southeast coastal region of China. *Sustain. Ener. Technol. Assessm., 39*, 100711.

Liu, H., Sun, J., Chang, J. S., & Shukla, P., (2018). Engineering microbes for direct fermentation of cellulose to bioethanol. *Crit. Rev. Biotechnol., 10*, 1–17.

Longa, F. D., Nogueira, L. P., Limberger, J., Van, W. J. D., & Der, Z. B., (2020). Scenarios for geothermal energy deployment in Europe. *Ener., 206*, 118060.
López, J. M., Gómez, Á., Aparicio, F., & Sánchez, F. J., (2009). Comparison of GHG emissions from diesel, biodiesel and natural gas refuse trucks of the city of Madrid. *Appl. Ener., 86*, 610–615.
Markou, G., Angelidaki, I., Nerantzis, E., & Georgakakis, D., (2013). Bioethanol production by carbohydrate-enriched biomass of *Antrospira (Spirulina) platensis*. *Energies, 6*, 3937–3950.
Mofijur, M., Masjuki, H. H., Kalam, M. A., Hazrat, M. A., Liaquat, A. M., Shahabuddin, M., & Varman, M., (2012). Prospects of biodiesel from *Jatropha* in Malaysia. *Renew. Sustain. Ener. Rev., 16*(7), 5007–5020.
Moncada, J., Tamayo, J. A., & Cardona, C. A., (2014). Integrating first, second, and third generation biorefineries: Incorporating microalgae into the sugarcane biorefinery. *Chem. Eng. Sci., 118*, 126–140.
Nassar, I. A., Hossam, K., & Abdella, M. M., (2019). Economic and environmental benefits of increasing the renewable energy sources in the power system. *Ener. Rep., 5*, 1082–1088.
Ou, X., Zhang, X., Chang, S., & Guo, Q., (2009). Energy consumption and GHG emissions of six biofuel pathways by LCA in (the) People's Republic of China. *Appl. Ener., 86*, S197–S208.
Pearre, N., & Swan, L., (2020). Combining wind, solar, and in-stream tidal electricity generation with energy storage using a load-perturbation control strategy. *Ener., 203*, 117898.
Petley, S., Starr, D., Parish, L., Underwood, Z., & Aggidis, G. A., (2019). Opportunities for tidal range projects beyond energy generation: Using Mersey barrage as a case study. *Front. Architect Resear., 8*(4), 620–633.
Petrichenko, K., Ürge-Vorsatz, D., & Cabeza, L. F., (2019). Modeling global and regional potentials for building-integrated solar energy generation. *Energ. Build., 198*, 329–339.
Pravin, P. S., Misra, S., Bhartiya, S., & Gudi, R. D., (2020). A reactive scheduling and control framework for integration of renewable energy sources with a reformer-based fuel cell system and an energy storage device. *J. Proc. Cont., 87*, 147–165.
Qi, G., Huang, D., Wang, J., Shen, Y., & Gao, X., (2019). Enhanced butanol production from ammonium sulfite pretreated wheat straw by separate hydrolysis and fermentation and simultaneous saccharification and fermentation. *Sustain. Ener. Technolog. Assessm., 36*, 100549.
Ruangsomboon, S., Sornchai, P., & Prachom, N., (2018). Enhanced hydrocarbon production and improved biodiesel qualities of *Botryococcus braunii* KMITL 5 by vitamins thiamine, biotin and cobalamin supplementation. *Algal Res., 29*, 159–169.
Sgroi, M., Bollito, G., Saracco, G., & Specchia, S., (2005). BIOFEAT: Biodiesel fuel processor for a vehicle fuel cell auxiliary power unit: Study of the feed system. *J. Power. Sour., 149*, 8–14.
Situmorang, Y. A., Zhao, Z., Yoshida, A., Abudula, A., & Guan, G., (2020). Small-scale biomass gasification systems for power generation (<200 kW class): A review. *Renew. Sustain. Ener. Rev., 117*, 109486.
Tan, K. T., Lee, K. T., Mohamed, A. R., & Bhatia, S., (2009). Palm oil: Addressing issues and towards sustainable development. *Renew. Sustain. Ener. Rev., 13*(2), 420–427.
Tang, T., (2018). Explaining technological change in the US winds industry: Energy policies, technological learning, and collaboration. *Ener. Polic., 120*, 197–212.
Varaprasad, D., Ragasudha, N., Paramesh, K., Shankarm, P. C., Parveen, S. N., & Chandrasekhar, T., (2020). Production of bioethanol from green alga *Chlorella Vulgaris*:

An important approach to utilize algal feedstock or waste. In: Ghosh, S., Sen, R., Chanakya, H., & Pariatamby, A., (eds.), *Bioresource Utilization and Bioprocess* (pp. 57–65). Springer, Singapore.

Vasudev, V., Ku, X., & Lin, J., (2019). Kinetic study and pyrolysis characteristics of algal and lignocellulosic biomasses. *Bioresour. Technol., 288*, 121496.

Wang Y., Wu X., Shao B., Yang X., Owens G., & Xu, H., (2020). Boosting solar steam generation by structure enhanced energy management. *Sci. Bullet., 65*(16), 1380–1388.

Wang, C., Chang, Y., Zhang, L., Pang, M., & Hao, Y., (2017). A life-cycle comparison of the energy, environmental and economic impacts of coal versus wood pellets for generating heat in China. *Ener., 120*, 374–384.

Wang, Y., Fan, Y., Wang, D., Liu, Y., Qiu, Z., & Liu, J., (2020). Optimization of the areas of solar collectors and photovoltaic panels in liquid desiccant air-conditioning systems using solar energy in isolated low-latitude islands. *Ener., 198,* 117324.

Yan, P., Xiao, C., Xu, L., Yu, G., Li, A., Piao, S., & He, N., (2020). Biomass energy in China's terrestrial ecosystems: Insights into the nation's sustainable energy supply. *Renewable and Sustainable Ener. Rev., 127*, 109857.

Yang, Q., & Chen, G. Q., (2012). Nonrenewable energy cost of corn-ethanol in China. *Ener. Polic., 41*, 340–347.

Zhang, L., Wang, Y. Z., Zhao, T., & Xu, T., (2019). Hydrogen production from simultaneous saccharification and fermentation of lignocellulosic materials in a dual-chamber microbial electrolysis cell. *Internat. J. Hydrog. Ener., 44*(57), 30024–30030.

CHAPTER 6

Water Hyacinth (*Eichhornia crassipes*): Novel Applications and Products Through Biotechnological Interventions

HARIT JHA and NEHA NAMDEO

Department of Biotechnology, Guru Ghasidas Vishwavidyalaya, Bilaspur, Chhattisgarh, India, E-mail: harit74@yahoo.co.in (Harit Jha)

ABSTRACT

Water hyacinth well-known as *Eichhornia crassipes* is a marine plant that is indigenous to the Amazon basin and is referred to as one of the most deleterious aquatic wild plants in the world. It is one of the fastest-growing plants that reproduce both asexually and sexually through runners or stolons and through seeds, respectively. There are several problems caused by water hyacinth, i.e., blockage of water bodies, providing a breeding ground for mosquitoes, hinders water transportation, loss of biodiversity, prevents the penetration of sunlight, less opportunity for fishing, harmful effects on human health, etc. Water hyacinth biomass is easily available and can be used for several biotechnological remediations like wastewater treatment (phytoremediation), biogas production, use as compost, bioethanol production, etc. Several phytochemicals like alkaloids, flavonoids, terpenoids, tannin, phenolic compounds, sterols, etc. have also been reported in water hyacinth. It has many applications like paper manufacturing, fiberboard production, basket, footwear and handicraft manufacturing, etc. In this chapter, water hyacinth biomass utilization is discussed for various bioremediation processes as well as the harmful effects and various applications of water hyacinth have also been elaborated.

Biotechnology for Waste Biomass Utilization. Prakash Kumar Sarangi & Latika Bhatia (Eds.)

6.1 INTRODUCTION

Water hyacinth (WH; *Eichhornia crassipes*) is a perennial marine flowering plant that is indigenous to South America located in the Amazon basin (Dölle et al., 2020). This is a freshwater hydrophyte mainly found in tropical and subtropical areas of the globe (Solms et al., 2016). It is known as "Florida devil" in South Africa, "Blue devil" or "Bengal terror" in India, "Water terror" in South-Western Nigeria, and "German weed" in Bangladesh (Das et al., 2015).

It is a plant species that mostly depends on moderate to high temperatures and is therefore found in subtropical and tropical regions. The major limiting factor for the growth of WH is salinity, therefore they cannot be found in the coastal areas (Ting et al., 2018). This weed is a marine macrophyte and monocotyledon which belongs to the family Pontederiaceae. *E. crassipes* is freely suspended with an eye-catching bunch of flowers and swollen green leaves (Figure 6.1a). The high degree of invasion is due to its adaptation to a severe environment and rapid growth rate (Breeding, 2015). This flora may be very useful as it is a lukewarm water weed that has underwater roots and floating leaves like developing macrophytes (Lalitha et al., 2012).

It has wide, smooth, broad, and ellipsoidal leaves and, it may grow on the top of the water up to 3 feet (1 m) tall. The leaves are 10–20 cm (4–8 inches) transversely on a stem and its floating via afloat bulb-like nodules at its base on top of the water level (Magar et al., 2018). The leaves and stems contain air-filled sacs, which help them to float in water (Bhattacharya and Kumar, 2010). They have elongated, squishy, and swollen stalks (Figure 6.1b). The roots of "beautiful blue devil" WH is freely hanging, feathery, and is lavender-black. An upright stalk supports only a barb of around 8–15 flowers, generally violet to pink color with 6 petals. It is known as one of the fastest-growing plants and reproduces both sexually by seeds and, asexually by runners or stolons, which ultimately form daughter plants, and its growth is promoted in dry conditions, and they usually germinate in six months.

Thousands of seeds can be produced by a particular plant every year and they can stay feasible under the water in adverse environmental circumstances (low light intensity and temperature) for more than 28 years and its doubling time is just two weeks (Magar et al., 2018; Ting et al., 2018).

6.2 DEVELOPMENT AND COMPOSITION OF WATER HYACINTH

The optimum growth of WH can be seen in water full of nutrients, and favorable pH value between 6.5 to 8.5, temperature from 28°C to 30°C and, salinity < 2%, 20 mg/L N, 3 mg/LP, and 53 mg/L K (Dersseh et al., 2019). The growth of WH can be described in two ways either by studying the density of plants in units of moist plant matter per unit of surface area or by studying the water surface percentage covered in a specific time phase. Its growth also depends on many biological factors. In standard conditions, the free-floating can wrap up the surface of the water at relatively low plant density (10 kg/m^2 wet weight) and can attain the highest density (50 kg/m^2 wet weight) before its growth ceases (Lalitha et al., 2012).

(a)

(b)

FIGURE 6.1 (a) Water body covered by water hyacinth (*Eichhornia crassipes*). (b) Water Hyacinth (*Eichhornia crassipes)* plant.

In appropriate environmental conditions, its growth rate can be up to 220 kg/ha/day (Bayrakci and Koçar, 2014) and its annual growth rate is around 40–140 tons /dry matter/hectare (Santos et al., 2018). *E. crassipes* are highly invasive and also can form thick and impenetrable suspended mats on the water surface (Dölle et al., 2020).

It was being used as an attractive species to beautify the water bodies. It has three floral morphs known as "tristylous," and they are named because of the size of their pistil, i.e., long, medium, and short. It is assumed to be the most assiduous plant on earth as well as the nastiest aquatic weed which causes severe environmental problems. It has a highly toxic character and has been listed among 100 of the world's worst insidious extraterrestrial species. It has radical-scavenging activity and has many compounds like terpenoids, phenolic acids, vitamins, lignin, alkaloids, sterols, stilbenes, and many metabolites with significant antioxidant activity (Tyagi and Agarwal, 2017). The chemical composition of WH includes lignin, cellulose, hemi-celluloses ash, and nitrogen (Table 6.1). It has organic matter–75.8%, ash–24.2%, and nitrogen–1.5%. The composition of ash is 28.7% K_2O, 21.0% Cl, 12.8% CaO, 7.0% P_2O_5, and 1.8% Na_2O (Rezania et al., 2015). The chemical composition of WH may vary depending on geographical location (Lalitha et al., 2012).

TABLE 6.1 Composition of Water Hyacinth

Components	Composition (in %)
Nitrogen	03
Ash	20
Lignin	10
Cellulose	25
Hemicellulose	35

6.3 PHYTOCHEMICAL PROPERTIES

The plant chemicals which are non-nutritive but have protective or disease preventive properties are known as phytochemicals (Lata and Dubey, 2010). They are important as synthetic medicines because they are the only source of medicine in some regions (Lalitha et al., 2012). These are produced by plants to protect themselves, but now many of them are also being used to protect humans against diseases.

6.3.1 PHYTOCHEMICALS IN WATER HYACINTH

6.3.1.1 FLAVONOID

The presence of phytochemical flavonoids in water hyacinth makes it a good resource for antioxidants. They are found in the aqueous and petroleum extracts (Lalitha et al., 2012). The existence of apigenin, tricin, gossypetin, chrysoeiol and azaeleatin, have been reported in the rhizome and shoot of the weed (Nyananyo et al., 2007). Quercetin, isovitexin, kaempferol, and orientin was also detected in rhizome and shoot (Lata and Dubey, 2010)

6.3.1.2 ALKALOID

Phytochemical analysis of WH shows the presence of alkaloids (Goud et al., 2009). Qualitative estimation by thin-layer chromatography showed the existence of tomatine and cytosine both in the rhizome and shoot, whereas thebaine, quinine, and codeine in the shoot area and nicotine in the rhizome (Lata and Dubey, 2010). The presence of alkaloids was also reported in the solvent extracts of leaf of WH (Kayathri et al., 2015).

6.3.1.3 TANNIN

In the methanol and aqueous extracts of WH, tannins are present. The dead and decaying matters as well as plant debris also release tannin-like compounds (Lalitha et al., 2012). Tannins were also found in the acetone and methanol leaf extracts of WH (Aravind et al., 2013). Braymer's test was performed for checking the presence of tannin and in the ethanol, ethyl acetate, and chloroform extract of WH, the tannin was detected (Tyagi and Agarwal, 2017).

6.3.1.4 PHENOLS

Phenolic compounds are a bulky and miscellaneous set of molecules, including diverse types of aromatic secondary metabolites are found in plants. Phenols were found in methanol extract, aqueous extract, and ethyl acetate extract of the leaves of water hyacinth (Jayanthi et al., 2011; Thamaraiselvi and Jayanthi, 2012). The phenolic compounds reported in the shoot extract include catechol, pyrogallol, 2-methylresorcinol, 4-methylresorcinol, genetisic, salicylic, syringic, and vanillic acids. The thin-layer chromatography of ethanolic extract of rhizomes demonstrated the presence of genetisic, salicylic acids, catechol, resorcinol, and 4-methylresorcinol (Jayanthi et al., 2011).

6.3.1.5 TERPENOIDS

In various extracts of WH, terpenoids were detected. GC-MS of ethanol extract detected 3,7,11,15-tetramethyl-2-hexadecen-1-ol and phytol (Muthunarayanan et al., 2011). Growth regulating substances, i.e., indole and gibberellins were also found in the roots of WH (Sircar and Ray, 1961).

6.3.1.6 GLYCOSIDES

Cardiac glycosides are reported in the aqueous and chloroform extracts of the shoot of WH. The roots, leaf stalks, and flowers are reported to have phospholipids, phosphatidylcholine, phosphatidylglycerol, and phosphatidylethanolamine respectively. The major fatty acids found in the roots, leaf stalks, leaves, and flowers of WH include palmitic, linoleic, palmitic, linolenic, and linoleic acid respectively. The analysis of the polysaccharide of WH revealed that it is a heteropolysaccharide, composed of D-xylose, L-galactose, and L-arabinose (Lalitha et al., 2012). The presence of glycosides was also detected in the distilled water and methanol extracts of WH (Kayathri et al., 2015). The presence of glycosides was also reported in the methanolic extract of WH (Baral et al., 2011).

6.3.1.7 STEROLS

Sterols can be found in various extracts of WH. The sterols reported from ethyl acetate and acetone extract include stigmasterol, 4-α-methyl-5-α-ergosta-8,14,24(28)-triene-3-β,4-β-diol,4-α-methyl-5-α-ergosta-8,24(28)-diene-3-β,4β-diol,β-sitosteroland4-α-methyl-5-α-ergosta-7,24(28)-diene-3-β,4-β-diol, campestral, and 6α-hydroxystigmata-4, 22-dien-3-one (Goswami et al., 1983; Della Greca et al., 1991; Hou Ming et al., 1991).

Plenty of organically-active compounds have been obtained from WH and *E. crassipes* are an interesting source of phytochemicals with potential therapeutic activities. The phytochemical composition of the extract of shoot and rhizome of *E. crassipes* showed the presence of cardiac glycosides, tannins, terpenoid, flavonoid, alkaloid phlobatannin, steroid, phenolics, quinone, and anthraquinone, (Joshi and Kaur, 2013). However, saponin was not detected in both the shoot and rhizome of the plant (Lata and Dubey, 2010). Phenolics, flavonoids, and tannin compounds were reported in the extract of WH (Rorong et al., 2012). Phenols, flavonoids, tannins, and alkaloids were detected in the leaf, petiole, and root of WH (Tulika et al., 2017).

Phytochemical analysis of WH detected coumarins, terpenoids, alkaloids, flavonoids, and, tannins whereas anthraquinones, cardiac glycosides, saponins, and, phlobatanins were absent in its extract (Khalid, 2020).

6.3.2 OTHER METABOLITES

Saponin, resins, and anthroquinone were detected in chloroform and methanol extracts of WH (Arayana et al., 1984; Aneke et al., 2007; Goud et al., 2009). The aqueous extract of shoots of WH showed the presence of quinone, anthraquinone, phlobatannin, and cardiac glycosides, while cardiac glycosides and phlobatanninare were absent in therhizome (Lata and Dubey, 2010). Anthocyanins and quinines have been found in the acid extract. Chloroform extract of WH leaves yielded Olefins (Do Amaral et al., 1990; Jayanthi et al., 2011).

6.4 PROBLEMS ASSOCIATED WITH WATER HYACINTH

WH is a foremost problem in waterways around the globe and thus is known as the nastiest aquatic plant in the world (Figure 6.2) (Mujere, 2017). It has not only occupied the water bodies but also leads to depletion of oxygen in the water, chokes the rivers, and provides a suitable environment for a breeding ground for mosquitoes, which has critically endangered the farmland irrigation, water transportation, drainage as well as human health (Su et al., 2018). It can also alter the transparency of the water, decrease the production of phytoplankton, and dissolve oxygen, phosphorous, nitrogen, heavy metals (Villamagna and Murphy, 2010). As it is a fast-growing plant, it builds up a large population and expands a dense mat over the water surface. This insidious weed possesses many risks ranging from ecological and economic to social. It leads to loss of biodiversity, causes eutrophication, clogs fresh waterways, affects agriculture and aquaculture, provides shelter for pests, and hampers shipping and recreational activities (Tulika et al., 2017). It also causes damage to vegetation, water pollution, causes irrigation problems, etc. (Yan et al., 2015; Das et al., 2015; Adanikin et al., 2017). Several problems caused by WH are as follows.

FIGURE 6.2 Problems due to water hyacinth.

6.4.1 BARRIER TO WATER TRANSPORT

Due to the widespread growth and population of WH, water transportation and shipping have been greatly affected. It is also getting dangerous to the transportation on lakes and rivers as they are covered with large floating islands of WH, whereas many domestic waterways of south-east Asia have been abandoned. Obstruction in navigation has also been faced, which is caused by mesh formed by the roots (Mujere, 2017). It can even get entangled with motorboat rotors (Bhattacharya and Kumar, 2010).

6.4.2 CHOKING OF WATER BODIES

In some places, WH can grow so thick that no one can even walk on it. Because of the overgrown population of WH, there is choking or blockage of various water bodies, thus also affecting the flow of water.

6.4.3 LOSS OF BIODIVERSITY

Due to the rapid growth of WH, the nutrients and oxygen of the water bodies get depleted, thus affecting other flora and fauna. It is a great challenge for any other species to survive in the surrounding water hyacinth, thus creating a threat to aquatic biodiversity (Guna et al., 2017). This can lead to an unevenness in the aquatic ecosystem and can also result in the extinction of a variety of fauna (Mujere, 2017). They also tend to take up nutrients rapidly thus making the ecosystems less fertile (Bhattacharya and Kumar, 2010).

6.4.4 PREVENTS PENETRATION OF SUNLIGHT

As a result of the high reproduction rate of WH species, they can proliferate until they cover the entire surface of the water in a short duration. The excessive growth of WH over the water surface restricts the dispersion of the sunlight and oxygen through the water surface and, as a result, the photosynthesis of submerged aquatic plants is retarded, thereby adversely impacting the oxygen generation required to support the aquatic life. Hence, it is very important to prevent the control of the growth of WH so that it doesn't cause any damage to the aquatic system (Ting et al., 2018; Priya and Selvan, 2014).

6.4.5 ADVERSE IMPACT ON FISHING INDUSTRY

Vegetative reproduction and vital growth rate of this plant impacts water flow, stops sunlight from reaching native aquatic plants, and diminishes dissolved oxygen of water, frequently slaughtering the fishes (Priya and Selvan, 2014), thus the fishermen get less or minimum chances of fishing which can also affect the economic status of their day to day life. It also affects fish reproduction, feeding, and productivity. When nets or lines get knotted in the roots of the plant, it can result in loss of fishing equipment which can lead to the subsequent loss of livelihood, and thus increasing poverty in the less developed places. This can even create serious socio-economic problems for the fishermen. The depletion of oxygen by the mats of hyacinth brings about lesser primary productivity and reduced biodiversity and greatly affects the whole food chain (Mujere, 2017).

6.4.6 BREEDING GROUND FOR MOSQUITOES AND PESTS

WH mat serves as a major platform for the breeding of mosquitoes (Priya and Selvan, 2014). They devastate the beauty of any place and also act as a reproduction ground for disease-causing pests and insects (Bhattacharya and Kumar, 2010). It also provides shelter to the poisonous snakes; reptiles, as well as crocodile attacks, have been increased due to this weed (Mujere, 2017).

6.4.7 EFFECTS ON HUMAN HEALTH

The excessive growth of WH leads to hygiene problems that affect human wellbeing (Santos et al., 2018). Skin rash, cough, malaria, cholera, gastro-intestinal disorders, bilharzia, encephalitis, and schistosomiasis are frequent diseases amongst the population close to water bodies covered by WH (Lalitha et al., 2012). The tangled WH weed serves as a habitat for the Snails, which hosts parasite *Schistosomiasis or bilharzia* (Borokini and Babalola, 2012). (Mujere, 2017).

6.4.8 CLOG IRRIGATION, WATER SUPPLY SYSTEMS, AND HYDROPOWER

Water hyacinth obstructs irrigation channels disturbing the current of water to fields, causing difficulty in fishing, and renders places inaccessible and unfit for human habitation. Several large hydropower systems are adversely affected by WH. It tends to obstruct hydroelectric turbines causing massive damage that impacts the environment and lead to major financial implication for the project (Bhattacharya and Kumar, 2010). It also clogs and obstructs the flow of water and damage several dams throughout the world (Mujere, 2017).

WH is placed on the quarantine list by many countries. WH is difficult to eliminate by biological, physical, and chemical methods, moreover, no long-term solutions were obtained using any of these techniques. Every year, a large sum is used for its management in the whole world. They compete with almost all other species growing in their locality thus decreasing the biodiversity. They also speed up the process of evaporation from water bodies (Bhattacharya and Kumar, 2010).

The effect of WH on zooplankton, phytoplankton, fish, birds, and macroinvertebrates is harmful. Since the preventive measures to control the development of WH are complex, the actions are taken to lower its growth, so that the economic and environmental effects can be reduced. In contrast, WH can be harvested and used in many beneficial ways for society (Lalitha et al., 2012).

6.5 BIOTECHNOLOGICAL INTERVENTIONS

Several biotechnological interventions for remediation of WH include production of biogas and ethanol, wastewater treatment (phytoremediation), sorbents for dyes and heavy metals, energy by direct combustion, reinforcement for composites, etc. (Guna et al., 2017). WH, as marine biomass has the benefit over global lignocellulosic biomass as it does not compete for land usage and food (Biswas et al., 2017).

6.5.1 USE OF WATER HYACINTH BIOMASS AS COMPOST

A controlled biological process where solid organic materials are converted into humic substances is known as composting. The decomposed form of WH help in the utilization of nutrients, and the organic compounds present in the compost can enhance the soil's organic carbon, phosphorus, nitrogen, and potassium, in addition to this, it also increases crop yield (Su et al., 2018). Its decomposed form can also be used as green manure. After the removal of hyacinth from water, it is left for a few days so that it gets dried before mixing with soil, ash, and several animal composts. Thus, decomposition by microbes breaks the lipids, fats, proteins, and sugar. These composts increase the fertility of soil and crop yield (Mujere, 2017). It is used as compost to get better soil structure and nutrients in waste biomass management and predominantly in emerging nations, where mostly artificial fertilizers are not affordable. It possesses a huge quantity of nutrients such as nitrogen, phosphorous, and potassium. Thus, WH based fertilizers are economical with high nutritive value and can be produced in less than 30 days) making it possible for farmers for improving their soil conditions (Newete and Byrne, 2016). The weed also contributes to protecting soil humidity and nutrient recovery (Rezania et al., 2017). For microbial decomposition, the best C/N ratio is between 20–40. Single WH composting is less efficient than the composting of leaves and manure and it may be because of the composition of leaves, i.e., cellulose and hemicelluloses, which can be easily degraded than lignin, the main component of WH (Polprasert et al., 1994).

6.5.2 TREATMENT OF WASTEWATER (PHYTOREMEDIATION) AND PURIFICATION

Contamination in water bodies can be reduced by growing aquatic plants in wastewater as a source of nutrients. Phytoremediation is an appropriate method for the economical treatment of groundwater, domestic wastewater, contaminated soils, and waste effluents of industries. It also has a greater capacity to remove nitrogen and phosphorus pollutants (Table 6.2). WH has a cyclable type of system and can be proved as a beneficial alternate for conventional treatment methods (Magar et al., 2018). It is capable of cleaning different polluted water bodies. The field and laboratory studies demonstrated that WH is capable of significantly reducing the contaminant load present in the swine wastewater (Rezania et al., 2017). Phosphorus contamination in the marine system comes mainly from three main sources: domestic sewage, agriculture, and industry (Rezania et al., 2015). In the early 1970s, WH was used to treat a variety of wastewater such as dairy waste, pulp and paper production, distillery, textile, electroplating, and heavy metals (Ting et al., 2018). The most important reason that it has been generally used in Southern France for the treatment of industrial wastewater is the high productivity rate of WH (Rezania et al., 2013). Phytoremediation is considered a green remediation technology for its eco-friendly method for the removal of pollutants from the water bodies. In a study, FTIR and ICP-Optical Emission Spectroscopy were used to check the ability of WH to remove heavy metals from the aqueous solution with more than 80% of the heavy metals efficiently removed by WH. Its roots were found to be more effective than the leaves and, was concluded that WH can be used as an agent for phytoremediation which can remove the dissolved pollutants from polluted water (Peng et al., 2020). WH was also found to be useful in the decolorization of harmful textile dye, Remazol Brilliant Red 3BS (RBR 3BS) discharged from industries (Bapat and Jaspal, 2020).

TABLE 6.2 Phytoremediation of Nutrition-Rich Wastewater by *E. crassipes*

Sources of Wastewater	Parameters	End result	Reference
Sewage treatment	Heavy metals, nitrates, phosphates, and BOD	Metal elimination (20–100%)	(Sinha and Sinha, 2000)
		Biological Oxygen Demand (97%)	
		Phosphates and nitrates (49%)	
Duck farmhouse	COD, TP, TN	Removal of COD 64.44 (%)	(Jianbo et al., 2008)
		Removal of TP 23.02 (%)	
		Removal of TN 21.78 (%)	
Swine wastewater	Phosphorus and Nitrogen	Adsorption efficiency up to 36% at saturation	(Chen et al., 2010)
Wastewater sample from Government College University, Lahore	Heavy metals, BOD, COD, Total Dissolved Solids (TDS), sulfates, and phosphates	BOD removal efficiency 66.66%	(Mahfooz et al., 2020)
		COD removal efficiency 66.67%	
		TDS reduction rate 911.5 g m^{-2} d^{-1}	

6.5.3 BIOETHANOL PRODUCTION

For bioethanol production, the utilization of WH seems to be an effective substitute for managing it and tackling the energy crisis and environmental pollution (Das et al., 2016; Sindhu et al., 2017). WH has to go through many pre-treatment steps like feedstock processing and saccharification for bioethanol production (Bote et al., 2020). The WH biomass can be useful in the production of biogas and its byproducts can be used for producing bioethanol and as organic manure by further decomposition of fermentable saccharides. The capacity of WH for bioethanol production is comparable to that obtained from agricultural waste; thus, is an important source for biofuel production (Bhattacharya and Kumar, 2010).

6.5.4 BIOFUEL OR BIOGAS PRODUCTION

The characteristics of an ideal biofuel crop are:

a) Naturally growing, and if the possible perennial plant.
b) High cellulose content with low lignin.
c) Easily degradable.
d) No competition with crop plants for light, space, and nutrients.
e) Pest, insect, and disease-resistant (Bhattacharya and Kumar, 2010).

WH is efficiently used for the production of biofuels because of its low content of lignin and fast growth rate (Bote et al., 2020). It has a high amount of cellulose (20%) and hemicellulose (33%) with low content of lignin (10%) thus, the cellulose and hemicelluloses can be easily converted to fermentable sugar for biofuel production (Bhattacharya and Kumar, 2010). The reason behind the low content of lignin is that it doesn't need land for its growth and grows well in water bodies on its own and thus, can be used as a source of effective biomass. Since it's very complex to remove the WH because the environment in which it grows can't be easily controlled; hence consumption of water hyacinth for biogas production can be beneficial for rural areas. As the water hyacinth is abundantly available, biogas is likely the best alternative to be used as biofuel (Bote et al., 2020). The thermal conversion of water hyacinth is at 400 °C, it contains 15% moisture and results in the production of biochar, biogas, and bio-oil. It has a high calorific value and is composed mainly of alkanes (26%), benzene derivatives (13%), nitro compounds (24%), phenol derivatives (27%), and other compounds (10%) (Santos et al., 2018). WH biomass when mixed with swine compost results in a better output of biogas compared with only using swine compost (Mujere, 2017). Using an anaerobic digestion process, biogas was also produced by a mixture of crushed WH and swine dung, and the digestate from the biogas fermenter was found to be an effective substitute for biofertilizer as it has high nutrient values and is eco-friendly (Unpaprom et al., 2020).

6.5.5 TEXTILE EFFLUENT TREATMENT

Harmful effluents discharge from different industries harmfully affects water resources, aquatic organisms, the richness of the soil, and the reliability of the ecosystem. Color dyes discharged from the textile industry are one of the major problems in the environment which pollutes the water bodies. The fabric wastewater is known as the most polluted amongst all the industrial sectors. Non-living biomass and, dried roots of WH were helpful in the removal of dyes, i.e., Methylene blue, and Victoria blue. Many parameters were considered, i.e., pH, contact time, sorbent dosage, and initial concentrations. Thus, the roots of WH were found as the cheaper source of biosorbent for basic dyes (Priya and Selvan, 2014). As WH is considered as a weed, so the roots of WH has been utilized as a substitute for dyes biosorption, BF-4B red reactive dye from an aqueous media was used for analysis where pseudo-second-order and Elovich models were used for the presentation of the experimental data and, the equilibrium state was achieved within 110 minutes, with 95% removal of dye (Rigueto et al., 2020).

6.5.6 ANTIMICROBIAL ACTIVITY

The antimicrobial activity of different extracts of WH has been evaluated. This activity was exhibited by the methanol extract and its fractions, as well as anti-algal activities (cyanobacteria and green microalgae) were also evaluated by the paper disc diffusion method. This analysis showed the presence of four phthalate derivatives and an alkaloid that has antimicrobial and anti-algal activities (Shanab et al., 2010).

The antibacterial activity against two bacterial strains, i.e., *Xanthomonas axonopodis* and *Bordetella pertussis,* and the antifungal activity against two fungi, i.e., *Penicillium italicum* and *Botrytis cinerea* were tested by disc diffusion method, and maximum potential was seen in leaf methanol extract against *B. pertussis* (17 mm) and, leaf chloroform extract against *B. cinerea* (29 mm) respectively (Khalid et al., 2020).

Against 7 pathogenic fungi and 11 human pathogenic bacteria, an antimicrobial assay was performed. The methanolic extract showed better antibacterial activity, whereas the aqueous portion attained by cold percolation showed comparatively better effects against different fungi (Baral et al., 2010).

The chloroform and ethanol extracts of WH were capable of preventing the growth of the test organisms, i.e., *Salmonella typhi*, *Acinetobacter* sp., *Schigella* sp., *Escherichia coli, Sclerotium rolfsii,* and *Fusarium proliferatum*. These results tend to provide validity to the conventional use of *E. crassipes* for microbial infection treatment (Baral et al., 2011)

6.5.7 ANTIOXIDANT ACTIVITY

WH exhibits effective antioxidant activity as demonstrated by DPPH scavenging method WH is rich in a wide variety of secondary metabolites. The antioxidant activity may be due to secondary metabolites containing unsaturated bonds and hydroxyl groups (Jayanthi et al., 2012; Khalid 2020). The DPPH free radical scavenging activity was reported for ethanol, methanol, and aqueous extract from the flowers of WH. The aqueous extract was found to be the most effective (Myint, 2018).

6.5.8 ANTITUMOR ACTIVITY

Methanol leaf extract (50%) of WH at different dosages, i.e., from 200 mg/kg body weight to 500 mg/kg body weight exhibited excellent reaction against B16F10 in-vivo melanoma tumor-bearing hybrid mice models (from Swiss albino female and C57BL male) (Ali et al., 2009).

6.5.9 WOUND HEALING ACTIVITY

The methanolic extract of WH leaves displayed significant wound healing activity in the rat wound experimental model system at 10% and 15% w/w concentration as a simple ointment. The treatments displayed better wound healing ability compared to that of the control (Ali et al., 2010).

6.6 VARIOUS OTHER APPLICATIONS OF WATER HYACINTH

WH is an abundantly available, inexpensive, sustainable, and renewable resource for various applications. It has a high water percentage, protein, higher energy, and fibrous tissues that can be used for various useful applications (Rezania et al., 2015). Several applications of WH include paper manufacturing, fiberboard production, basket, footwear and handicraft manufacturing, etc. (Figure 6.3).

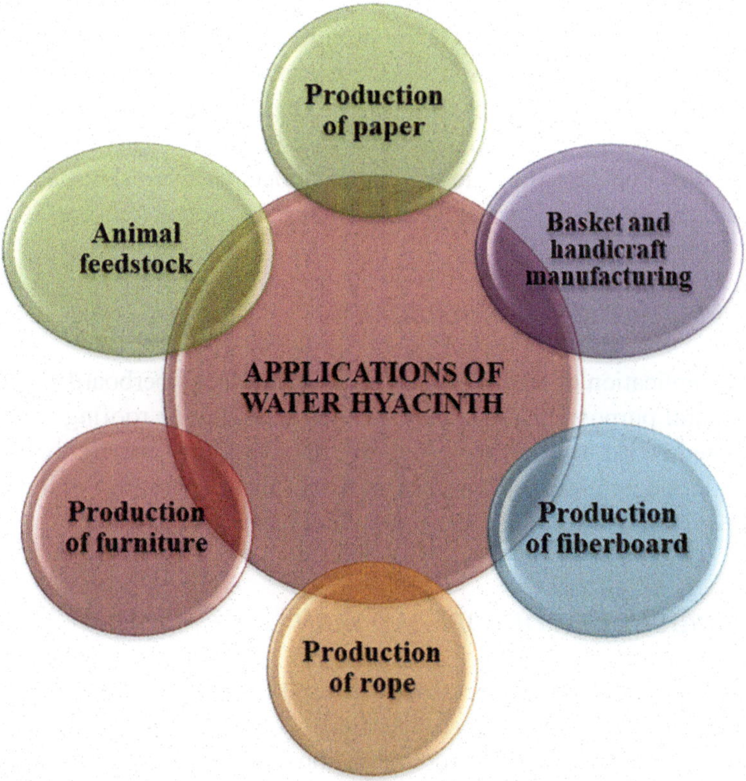

FIGURE 6.3 Various applications of water hyacinth.

6.6.1 PRODUCTION OF PAPER

WH is used for the production of paper but as its fiber length is very short, good quality paper can't be manufactured. So, to overcome this problem and improve the quality of the paper, the yarn is first combined with waste paper or jute to produce a good quality paper. The pulp is supplied with, calcium carbonate, bleaching powder, and sodium carbonate before it gets heated up (Mujere, 2017). In some countries like India, Indonesia, Bangladesh, and the Philippines, such small-scale cottage industry paper manufacturing projects are running successfully (Cho and Tifuh, 2012).

6.6.2 BASKET, FOOTWEAR, AND HANDICRAFT MANUFACTURING

The dried stalks and stems of WH are useful in manufacturing baskets, bags, footwear, and carpets for domestic use. For removal of moisture, the weed stalks are first desiccated which keeps them safe from getting rotten and ensure their long-term survival. The key to a good product is to make sure that the stalks are accurately dehydrated before use. Conventional weaving skills and basket manufacture are also done using the dried WH stalks heated (Mujere, 2017).

6.6.3 PRODUCTION OF FIBERBOARD

Another application of WH includes the production of fiberboards for using it for general purposes and can also be used as low-price roofing stuff, i.e., bituminized board. Both of these boards (fiber and bituminized boards) are made and used as a low-cost roofing material. These boards are useful in the making of ceilings and indoor partition walls (Mujere, 2017).

6.6.4 PRODUCTION OF ROPE AND YARN

The stem of WH is full of fibers and it is used for making ropes. The stalks are shredded along the length so that the fiber is exposed and then left for several so that it gets dried. The complete ropes are then treated with sodium metabisulphite to avoid it from getting rotten (Mujere, 2017).

6.6.5 PRODUCTION OF FURNITURE

The ropes produced by the stalks of WH are used for furniture manufacturing, where the ropes are winded around the bamboo frames to fabricate stylish, beautiful, and complete products (Borokini and Babalola, 2012).

6.6.6 *ANIMAL FEEDSTOCK*

WH is used as animal fodder, which solves some nutritional problems that persist in emerging countries. It can be fed to animals, which converts the nutrient into valuable foodstuffs for human utilization. WH feedstock is provided to animals with rice bran, vegetable waste, fish meal, and salt to make it an appropriate food for pond fish, geese, and pigs. Dried WH is very rich in vitamins, minerals, and proteins which serve as high-quality feedstock for fishes and animals in poultry (Mujere, 2017).

6.6.7 *FISH FEED*

Fishes like tilapia, silver carp, Chinese grass carp, sea cow, and the silver dollar fish are herbivores that consume floating or submerged flora and weeds. Such species of fish are used for the control of aquatic weeds and in addition to this; the weeds also serve as food for fish. A dehydrated form of WH has been included in the food of channel catfish fingerlings to enlarge their growth. It has been studied that the decaying matter of WH, following its chemical control, the nutrients are released that promote the growth of phytoplankton that also increases fish yield (Mujere, 2017).

6.7 CONCLUSION

WH is the most toxic floating marine plant on the earth, due to its capability to grow very fast over a vast area of any water body. It is known by many names around the whole world. It is native to South America located in the Amazon Basin. Water hyacinth is made up of cellulose, hemicelluloses, ash, lignin, and nitrogen. It has some negative as well as some positive effects on the environment. Its excessive growth can lead to many problems like choking of the water bodies, difficulty in navigation and transportation, loss of biodiversity, provides a favorable breeding ground for mosquitoes and other pests that causes many health-related problems and diseases, etc. It can even affect the livelihood of the local fishermen. Overall, these problems also have several applications like the production of paper, fiberboards, ropes, baskets, bags, footwear, furniture, etc. It also exhibits some therapeutic significance like anti-tumor activity, anti-oxidant activity, wound-healing property, etc. Many biotechnological interventions are also done for the remediation of this weed which includes utilization of WH biomass for manufacturing of compost, biofuel production, making of bioethanol, fertilizers, wastewater purification (phytoremediation), etc. Thus, water hyacinth can be both harmful as well as useful for human beings.

KEYWORDS

- **flavonoid**
- **phytochemicals**
- **phytoremediation**
- **water hyacinth**

REFERENCES

Adanikin, B. A., Ogunwande, G. A., & Adesanwo, O. O., (2017). Evaluation and kinetics of biogas yield from morning glory (Ipomoea aquatica) co-digested with water hyacinth (*Eichhornia crassipes*). *Ecological Engineering, 98*, 98–104.

Ali, H., Lata, N., Ahi, J., & Ganesh, N., (2010). Evaluation of wound-healing activity of *Eichornia crassipes:* A novel approach. *Drug Invention Today, 2*, 3.

Ali, H., Patel, M., Ganesh, N., & Ahi, J., (2009). The world's worst aquatic plant as a safe cancer medicine-antitumor activity on melanoma induced mouse by *Eichhornia crassipes*: *In vivo* studies. *J. Pharm. Res., 2*, 1365–1366.

Aneke, N. J., Oguzie, E. E., & Ukiwe, L. N., (2007). Physico-chemical properties of chloroform extract of water hyacinth (*Eichhoria crassipes*). *The African Journal of Plant Science and Biotechnology, 1*, 40–42.

Aravind, R. K., Rajan, D., Blesson, J., Chandran, S., Thampatty, A. R., & Veena, P. V., (2013). Detailed analysis on phytochemicals, antioxidants, antimicrobial activity of *Eichhornia crassipes*. *Inter. J. Sci. Res., 2*, 17–19.

Arayana, G. L., Rao, K. S., Pantulu, A. J., & Thyagarajan, G., (1984). Composition of lipids in roots, stalks, leaves and flowers of *Eichhornia crassipes* (Mart.) solms. *Aquatic Botany, 20*(3, 4), 219–227.

Bapat, S. A., & Jaspal, D. K., (2020). Surface-modified water hyacinth (*Eichhornia crassipes*) over activated carbon for wastewater treatment: A comparative account. *South African Journal of Chemistry, 73*, 70–80. https://doi.org/10.17159/0379-4350/2020/V73A11.

Baral, B., Vaidya, G. S., & Bhattarai, N., (2011). *Bioactivity and Biochemical Analysis of Water Hyacinth (Eichhornia crassipes)*, 33–39.

Baral, B., Vaidya, G. S., Bhattarai, N., & Baniya, R., (2010). Bioassay and screening of bio-active compounds of water hyacinth: An invasive aquatic weed of Fewa lake. In. *Proceedings of the International Conference on Biodiversity, Livelihood and Climate Change in the Himalayas*, 12–14.

Bayrakci, A. G., & Koçar, G., (2014). Second-generation bioethanol production from water hyacinth and duckweed in Izmir: A case study. *Renewable and Sustainable Energy Reviews, 30*, 306–316.

Bhattacharya, A., & Kumar, P., (2010). Water hyacinth as a potential biofuel crop. *Electron J. Environ. Agric. Food Chem., 9*(1), 112–122.

Biswas, B., Singh, R., Krishna, B. B., Kumar, J., & Bhaskar, T., (2017). Pyrolysis of azolla, *Sargassum tenerrimum* and water hyacinth for production of bio-oil. *Bioresource Technology, 242*, 139–145.

Borokini, T. I., & Babalola, F. D., (2012). Management of invasive plant species in Nigeria through economic exploitation: Lessons from other countries. *Management of Biological Invasions, 3*(1), 45.

Bote, M. A., Naik, V. R., & Jagadeeshgouda, K. B., (2020). Review on water hyacinth weed as a potential bio fuel crop to meet collective energy needs. *Materials Science for Energy Technologies, 3*, 397–406 https://doi.org/10.1016/j.mset.2020.02.003.

Breeding, T., (2015). In: Lenora, L. M., (ed.), *Anticancer Activity of Water Hyacinth [Eichhornia crassipes (mart) solms] on Human Cervical Cancer Cell Line, 3*(4), 327–331.

Chen, X., Chen, X., Wan, X., Weng, B., & Huang, Q., (2010). Water hyacinth (*Eichhornia crassipes*) waste as an adsorbent for phosphorus removal from swine wastewater. *Bioresource Technology, 101*(23), 9025–9030.

Cho, M. E., & Tifuh, J., (2012). Quantification of the impacts of water hyacinth on riparian communities in Cameroon and assessment of an appropriate method of control: the case of the Wouri River Basin. *World Maritime University Dissertations*. 29.

Das, S., Bhattacharya, A., Haldar, S., Ganguly, A., Gu, S., Ting, Y. P., & Chatterjee, P. K., (2015). Optimization of enzymatic saccharification of water hyacinth biomass for bio-ethanol: Comparison between artificial neural network and response surface methodology. *Sustainable Materials and Technologies, 3*, 17–28. https://doi.org/10.1016/j.susmat.2015.01.001.

Das, S., Gupta, A., Das, D., & Goyal, A., (2016). Enhanced bioethanol production from water hyacinth (*Eichhornia crassipes*) by statistical optimization of fermentation process parameters using Taguchi orthogonal array design. *International Biodeterioration and Biodegradation, 109*, 174–184. https://doi.org/10.1016/j.ibiod.2016.01.008.

Della, G. M., Monaco, P., & Previtera, L., (1991). New oxygenated sterols from the weed *Eichhornia crassipes* solms. *Tetrahedron, 47*(34), 7129–7134.

Dersseh, M. G., Melesse, A. M., Tilahun, S. A., Abate, M., & Dagnew, D. C., (2019). Water hyacinth: Review of its impacts on hydrology and ecosystem services—Lessons for management of Lake Tana. In: *Extreme Hydrology and Climate Variability* (pp. 237–251). Elsevier.

Do Amaral, M. D. C. E., Da Silva, A. J. R., & Salatino, A., (1990). Alkanes of surface waxes from eight species of aquatic angiosperms. *Aquatic Botany, 36*(3), 281–286.

Dölle, K., Tong, J., & Wang, Q., (2020). *Water Hyacinths – Biomass Production Potential at a Waste Water Treatment Plant., 23*(2), 1–10. https://doi.org/10.9734/JABB/2020/v23i230138.

Goswami, P. C., Nag, B., Sharma, A. K., Borthakur, A., Singh, H. D., & Baruah, J. N., (1983). Water hyacinth as a prospective source of stigmasterol. *Current Science, 52*(17), 806–808.

Goud, J. V., Suryam, A., & Charya, M. A. S., (2009). Biomolecular and phytochemical analyses of three aquatic angiosperms. *African Journal of Microbiology Research, 3*(8), 418–421.

Guna, V., Ilangovan, M., Prasad, M. G. A., & Reddy, N., (2017). *Water Hyacinth: A Unique Source for Sustainable Materials and Products, 5, 6*, 4478–4490. https://doi.org/10.1021/acssuschemeng.7b00051.

Hou, M. W. U., Hui, Q. Z., Sheng, Y. H. K., Xia, Fei, L. A. O., Shu, W. Y. U., Wen, H. S. U. N., & Zi, W. Y. U., (1991). A novel steroid from *Eichhornia crassipes*. *China Chemistry Letters, 2*(07), 509–512.

Jayanthi, P., Lalitha, P., & Aarthi, N., (2012). Larvicidal and pupicidal activity of extracts and fractionates of *Eichhornia crassipes* (Mart.) solms against the filarial vector *Culex quinquefasciatus* say. *Parasitology Research, 111*(5), 2129–2135.

Jayanthi, P., Lalitha, P., & Sripathi, S. K., (2011). *Phytochemical Investigation of the Extracts of Eichhornia Crassipes and its Solvent Fractionates, 4*(5), 1405–1406.

Jianbo, L. U., Zhihui, F. U., & Zhaozheng, Y. I. N., (2008). Performance of a water hyacinth (*Eichhornia crassipes*) system in the treatment of wastewater from a duck farm and the effects of using water hyacinth as duck feed. *Journal of Environmental Sciences, 20*(5), 513–519.

Joshi, M., & Kaur, S., (2013). *In vitro* evaluation of antimicrobial activity and phytochemical analysis of *Calotropis procera, Eichhornia crassipes* and *Datura innoxia* leaves. *Asian J. Pharm. Clin. Res., 6*(5), 25–28.

Kayathri, B., Kanimozhi, K., & Panneerselvam, A., (2015). Preliminary phytochemical analysis and *in vitro* investigation of antimicrobial activity of *Eichhornia crassipes* (mart.) solms. against poultry pathogens. *CIB Tech Journal of Microbiology, 4*(1), 19–27.

Khalid, S., Shaheen, S., Hussain, K., Shahid, M. N., & Sarwar, S., (2020). Pharmacological analysis of obnoxious water weed: *Eichhornia crassipes* (mart.) solms. *Journal of Animal and Plant Sciences, 30*(6), 1465–1475. https://doi.org/10.36899/JAPS.2020.6.0167.

Lalitha, P., Sripathi, S. K., & Jayanthi, P., (2012). Secondary metabolites of *Eichhornia crassipes* (waterhyacinth): A review (1949 to 2011). *Natural Product Communications, 7*(9), 1934578X1200700939.

Lata, N., & Dubey, V., (2010). Preliminary phytochemical screening of *Eichhornia crassipes*: The world's worst aquatic weed. *Journal of Pharmacy Research, 3*(6), 1240–1242.

Magar, R. B., Khan, A. N., & Honnutagi, A., (2018). Waste water treatment using water hyacinth waste water treatment using water hyacinth. *32nd Indian Engineering Congress*.

Mahfooz, Y., Yasar, A., Islam, Q. U., Rasheed, R., Naeem, U., & Mukhtar, S., (2020). Field testing phytoremediation of organic and inorganic pollutants of sewage drain by bacteria assisted water hyacinth. *International Journal of Phytoremediation*, 1–12. https://doi.org/10.1080/15226514.2020.1802574.

Mujere, N., (2017). *Water Hyacinth: Characteristics, Problems, Control Options, and Beneficial Uses*. https://doi.org/10.4018/978-1-4666-9559-7.ch015.

Muthunarayanan, V., Santhiya, M., Swabna, V., & Geetha, A., (2011). Phytodegradation of textile dyes by water hyacinth (*Eichhornia crassipes*) from aqueous dye solutions. *International Journal of Environmental Sciences, 1*(7), 1702–1717.

Myint, H. H., (2018). Antioxidant activity and detection of glutathione in flower extract of water hyacinth (*Eichhornia crassipes* (Mart.) solms. *3rd Myanmar Korea Conference Research Journal, 3*(1), 139–146.

Newete, S. W., & Byrene, M. J., (2016). The capacity of aquatic macrophytes for phytoremediation and their disposal with specific reference to water hyacinth. *Environmental Science and Pollution Research*. https://doi.org/10.1007/s11356-016-6329-6.

Nyananyo, B. L., Gijo, A. H., & Ogamba, E. N., (2007). The physico-chemistry and distribution of water hyacinth (*Eichhornia cressipes*) on the river nun in the Niger nelta. *Journal of Applied Sciences and Environmental Management, 11*(3).

Peng, H., Wang, Y., Tan, T. L., & Chen, Z., (2020). Exploring the phytoremediation potential of water hyacinth by FTIR Spectroscopy and ICP-OES for treatment of heavy metal contaminated water. *International Journal of Phytoremediation, 22*(9), 939–951. https://doi.org/10.1080/15226514.2020.1774499.

Polprasert, C., Kongsricharoern, N., & Kanjanaprapin, W., (1994). Production of feed and fertilizer from water hyacinth plants in the tropics. *Waste Management and Research, 12*(1), 3–11.

Priya, E. S., & Selvan, P. S., (2014). Water hyacinth (*Eichhornia crassipes*) – An efficient and economic adsorbent for textile effluent treatment – A review. *Arabian Journal of Chemistry*. https://doi.org/10.1016/j.arabjc.2014.03.002.

Rezania, S., Din, M. F., Mat, T. S., Ling, Y. E., Chellipan, S., & Kamyab, H., (2017). Beneficial Environmentally usage of water hyacinth: A mini review. *Recent Advances in Petrochemical Science, 1*(5).

Rezania, S., Din, M. F., Ponraj, M., & Sairan, F., (2013). Nutrient uptake and wastewater purification by water hyacinth and its effect on plant growth in a batch system nutrient uptake and wastewater purification with water hyacinth and its effect on plant growth in batch system. *Journal of Environmental Treatment Techniques, 1*(2), 81–85.

Rezania, S., Ponraj, M., & Fadhil, M., (2015). The diverse applications of water hyacinth with main focus on sustainable energy and production for new era: An overview. *Renewable and Sustainable Energy Reviews, 41*, 943–954. https://doi.org/10.1016/j.rser.2014.09.006.

Rigueto, C. V. T., Piccin, J. S., Dettmer, A., Rosseto, M., Dotto, G. L., De Oliveira, S. A. P., Perondi, D., et al., (2020). Water hyacinth (*Eichhornia crassipes*) roots, an amazon natural waste, as an alternative biosorbent to uptake a reactive textile dye from aqueous solutions. *Ecological Engineering, 150*(10), 105817. https://doi.org/10.1016/j.ecoleng.2020.105817.

Rorong, J. A., Prasetya, B., & Polii-mandang, J., (2012). Phytochemical analysis of water hyacinth (*Eichhornia crassipes*) of agricultural waste as biosensitizer for ferri photoreduction. *Agrivita, 34*(2), 152–160.

Santos, L. O., Silva, F. F., Santos, L. C., Carregosa, I. S. C., & Wisniewski, Jr. A., (2018). Potential bio-oil production from invasive aquatic plants by microscale pyrolysis studies. *Journal of the Brazilian Chemical Society, 29*(1), 151–158.

Shanab, S. M. M., Shalaby, E. A., Lightfoot, D. A., & El-Shemy, H. A., (2010). Allelopathic effects of water hyacinth [*Eichhornia crassipes*]. *PloS One, 5*(10), e13200.

Sindhu, R., Binod, P., Pandey, A., Madhavan, A., Alphonsa, J. A., Vivek, N., Gnansounou, E., Castro, E., & Faraco, V., (2017). Water hyacinth a potential source for value addition: An overview. *Bioresource Technology, 230*, 152–162.

Sinha, A. K., & Sinha, R. K., (2000). Sewage management by aquatic weeds (water hyacinth and duckweed): Economically viable and ecologically sustainable bio-mechanical technology. *Environmental Education and Information, 19*(3), 215–226.

Sircar, S. M., & Ray, A., (1961). Growth substances separated from the root of water hyacinth by paper chromatography. *Nature., 190*(4782), 1213–1214.

Solms, M., Lara-serrano, J. S., Rutiaga-quiñones, O. M., López-miranda, J., Fileto-pérez, H. A., Pedraza-bucio, F. E., & Rico-cerda, J. L., (2016). Physicochemical characterization of water hyacinth (*Eichhornia crassipes* (Mart.) solms). *Bioresources, 11*(3), 7214–7223.

Su, W., Sun, Q., Xia, M., Wen, Z., & Yao, Z., (2018). The resource utilization of water hyacinth (*Eichhornia crassipes* [Mart.] solms) and its challenges. *Resources, 7*(3), 46.

Thamaraiselvi, P., & Jayanthi, P., (2012). Preliminary studies on phytochemicals and antimicrobial activity of solvent extracts of *Eichhornia crassipes* (Mart.) solms. *Asian Journal of Plant Science and Research, 2*(2), 115–122.

Ting, W. H. T., Tan, I. A. W., Salleh, S. F., & Wahab, N. A., (2018). Journal of Water Process Engineering Application of water hyacinth (*Eichhornia crassipes*) for phytoremediation of ammoniacal nitrogen: A review. *Journal of Water Process Engineering, 22*(2), 239–249. https://doi.org/10.1016/j.jwpe.2018.02.011.

Tulika, T., Puneet, P., & Mala, A., (2017). Qualitative phytochemical analysis and antioxidant activity of methonolic extract of *Eichhornia crassipes* (Mart.) solms and *Pistia. International Journal of Pharmacognosy and Phytochemical Research, 9*(5), 632–636.

Tyagi, T., & Agarwal, M., (2017). Phytochemical screening and GC-MS analysis of bioactive constituents in the ethanolic extract of *Pistia stratiotes* L. and *Eichhornia crassipes* (Mart.) solms. *Journal of Pharmacognosy and Phytochemistry, 6*(1), 195–206.

Unpaprom, Y., Pimpimol, T., Whangchai, K., & Ramaraj, R., (2020). Sustainability assessment of water hyacinth with swine dung for biogas production, methane enhancement, and biofertilizer. *Biomass Conversion and Biorefinery, 2*. https://doi.org/10.1007/s13399-020-00850-7.

Villamagna, A. M., & Murphy, B. R., (2010). Ecological and socio-economic impacts of invasive water hyacinth (*Eichhornia crassipes*): A review. *Freshwater Biology, 55*(2), 282–298.

Yan, J., Wei, Z., Wang, Q., He, M., Li, S., & Irbis, C., (2015). Bioethanol production from sodium hydroxide/hydrogen peroxide-pretreated water hyacinth via simultaneous saccharification and fermentation with a newly isolated thermotolerant *Kluyveromyces marxianu* strain. *Bioresource Technology, 193*, 103–109.

CHAPTER 7

Advancements Towards Biomass Conversion for Sustainable Management of Solid Waste

AKANKSHA KULSHRESHTHA,[1] SOUMYA SASMAL,[1] MINAKSHI SAHU,[2] and O. N. TIWARI[2]

[1]*Department of Biological Sciences and Engineering, Netaji Subhas University of Technology Delhi–110078, India, E-mail: akankshak@nsut.ac.in*

[2]*Centre for Conservation and Utilization of Blue Green Algae, Division of Microbiology, ICAR- Indian Agricultural Research Institute, New Delhi–110012, India*

ABSTRACT

Solid waste management is a challenging process, especially in developing countries due to the continuous upsurge of population, urbanization, and lack of services and facilities, making solid waste management very difficult to handle. For our energy needs, we are dependent on fossil fuels like coal, petroleum, etc. Reserves for coal and petroleum are limited and subject to depletion depending on the usage. Various methods and techniques like pyrolysis, gasification, catalytic liquefaction, etc. are in use for solid waste conversion and energy/fuel generation. Furthermore, biochemical techniques like microbial fermentation and anaerobic digestion are of enormous use and have immense potential in the production of biogas and bioethanol/biobutanol, etc. Solid waste efficient treatment and conversion solves a two-way problem-one is waste disposal and another is energy generation. Here, we are discussing the major solid waste treatment process with an emphasis on biogas production by the anaerobic digestion method.

Biotechnology for Waste Biomass Utilization. Prakash Kumar Sarangi & Latika Bhatia (Eds.)

7.1 INTRODUCTION

India is the second-largest populated nation after China in the world with an approximate population of 138.20 crores. India occupies 2.4 % of the world's area but supports 17.7% of the world's population. Its urban population is continuously growing and facing a great problem between increasing urban population and available resources and services, there is a sharp decline in solid waste management, proper waste disposal system, and hygienic practices. Solid waste refers to unwanted, discarded useless waste arising from human and animal activities. Solid waste is generated from households, i.e., residential, commercial, and industrial setups. Solid waste management (SWM) is an important system dealing with the generation of waste, collection, transferring, and finally, disposal of waste in the best possible way after due consideration given to the environment, health, hygiene, conservation of resources, and aesthetics. Accordingly, SWM includes proper strategic planning, financial planning, engineering, and legal aspects. If SWM systems are not appropriate and waste is not handled properly, it can be a potential threat to the environment, it can lead to environmental pollution, climate change, adverse effect on natural resources, and human health. A solid waste dumping site is depicted in Figure 7.1.

FIGURE 7.1 Municipal corporation solid waste dumping site at Chennai, India.

7.1.1 THE CONCEPT OF SUSTAINABLE MANAGEMENT OF SOLID WASTE

"Transforming Our World" (United Nations 2015) is the report presented for the sustainable development goals during Rio+20 Conference, 20–22 June 2012, organized by the United Nations. Sustainable development goals are a collection of 17 global goals, to achieve a better, bright, and more sustainable future for all by 2030. Many sustainable goals are directly associated with solid waste accumulation, disposal, and recycling. Some of the associated goals are clean water and sanitation, affordable and clean energy, sustainable cities and communities, life on land, etc. To achieve these goals, we should have an integrated system of solid waste recycling and disposal in such a manner that solid waste impact can be reduced on the environment and life on the planet. Sustainable ways of solid waste management include the strategy, scientific principles, innovations, and advancements in handling solid waste in environment-friendly and sustainable ways.

7.1.2 SOLID WASTE CLASSIFICATION

Solid waste is unwanted, unusable, discarded waste from households, industries, hospitals, etc. Solid waste can be classified in various ways as follows:

A. *On the basis of their source of generation*
B. *On the basis of whether the waste is biodegradable/non-biodegradable*
C. *On the basis of hazardous/non-hazardous*

A. On the Basis of Source of Generation of Solid Waste

Based on their generative source, solid waste can be further classified into various types as shown in Figure 7.2.

1. *Household or domestic waste.* Solid waste is generated from households. The major part of the waste is organic waste, i.e., kitchen waste, plants, leaves, etc. Besides that, the e-waste, furniture waste, paper, glass, medicines, paint and varnishes, tin, aluminum, and other metal items, and majorly plastic items like bags, toys, toiletries, clothes, etc.
2. *Commercial waste.* Commercial waste is comprised of waste from markets, restaurants, hotels, sports, waste generated from institutions like school's research centers and non-profit organizations, waste from slaughterhouses, etc.

FIGURE 7.2 Classification of solid waste classification of solid waste.

3. *Agro and dairy waste*. Agro waste comprises mainly organic waste, i.e., crops, plants, etc. Apart from this, it also contains fertilizers and pesticides, dairy sludge, wastewater sludge, etc.
4. *Industrial waste.* Industrial waste consists of various types of waste depending on which industry the waste is being generated like power and chemical waste, tubes and tires, chemicals, dyes, paints, etc. Different types of batteries, equipment, food waste, etc.
5. *Municipal services.* This kind of service includes street cleaning, park sweeping, other recreational areas, water bodies, wastewater treatment plants, biogas treatment plants, etc.
6. *Construction and demolition.* Waste arises from new construction activities, renovation sites, construction or repair of roads, dams, etc., and demolition of buildings. The major type of waste produced from construction activities is steel, wood, concrete, dirt, and dust.
7. *Biomedical waste and hospital waste.* Waste generated from hospitals which include discarded medicines, dirty aprons, towels, bedsheets, pathological waste, etc. is known as biomedical waste products. Biomedical waste falls under the category of hazardous waste. Besides this, a lot of plastic waste, paper waste, and medical equipment are also generated in hospitals.

B. On the Basis of Biodegradable or Non-Biodegradable

On the basis of biodegradability, solid waste is segregated as biodegradable and non-biodegradable. Biodegradable is the term that refers to the substances that can be degraded. As per degradation time, the solid waste can be segregated and separated. Kitchen waste, paper waste, green waste, etc are examples. On the other hand, non-biodegradable substances are that which are not able to decompose in the natural environment like plastics, metals, paints, etc.

C. On the Basis of Hazardous/Non-Hazardous

Hazardous waste like biowaste /hospital waste, paints, flammable material are all toxic and fall under the category of hazardous waste. Non-hazardous waste is not toxic to humans, plants, and animals.

7.2 POTENTIAL OF A BIOMASS AS FUTURE ENERGY NEEDS

Biomass is the umbrella term that includes, plant, animal, algae, wood, food waste, or any organic matter that can be used for energy generation. Biomass is a renewable and potential source of future energy generation (Rajmohan and Varjani, 2019). Biomass generation is a kind of natural phenomenon. That's why it is a renewable source of energy generation. Modern sustainability lies in harnessing maximum energy and power generation from renewable sources of energy. The energy generation from biomass is a multistep process, it starts with the collection of municipal solid waste (MSW) from various sources of MSW generation like, households, institutions, treated biomedical waste, construction sites, etc. All the waste is collected at accumulation/dumping sites. After that, the waste is segregated on the basis of its physicochemical properties, further, the waste or biomass is pre-treated and later used for the generation of energy, either by the use of thermochemical conversion process or by the use of biochemical conversion methods (Chung, 2013) as depicted by the Figure 7.3 (flowchart).

7.2.1 BIOMASS PRE-TREATMENT METHODS

Solid organic waste products generated from our society can be treated as ample resources which can be converted to value-added different green chemicals such as biofuels. These solid organic wastes include food waste, animal manure, waste-activated sludge, yard waste, and agricultural waste products. Most of the organic waste materials or biomass materials are made of polymers of lignocellulose materials. Numerous solid organic waste products, particularly those with a relatively high lignocellulosic content pre-treatment are essential to disintegrate the polymers into their basic building block elements or comparatively less complex easily digestible polymers (Sasmal et al., 2012a).

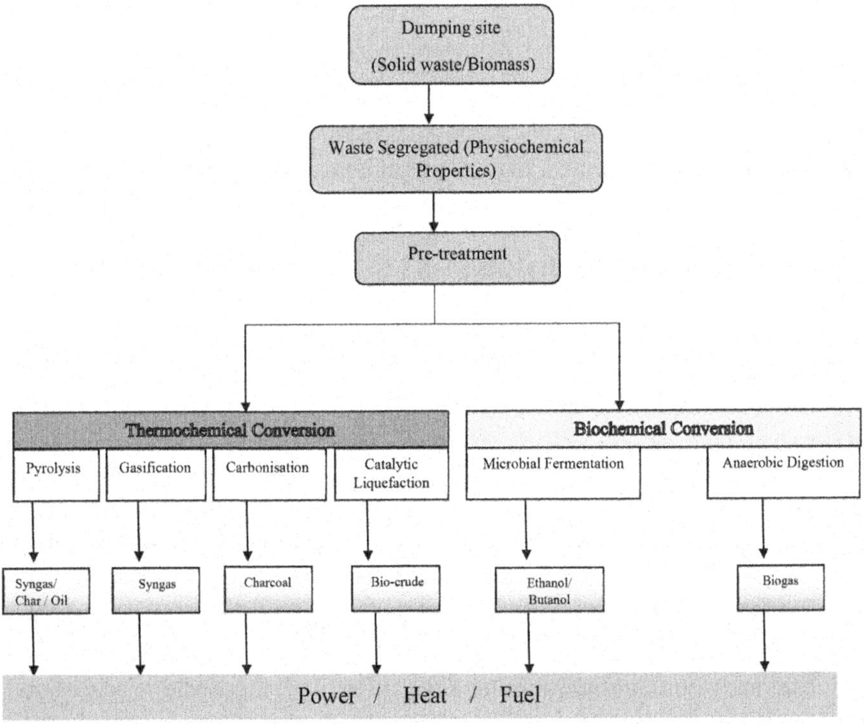

FIGURE 7.3 Flowchart representing the process of MSW treatment process.

There are various pre-treatment methods that exist as mentioned in Figure 7.4. Physical and chemical pre-treatment is the most common and rapid method but both the loss of materials and yield of hazardous chemicals as byproducts are more. Comparatively, the biological pre-treatment method is slow but it does not yield any hazardous product (Sasmal et al., 2018). The most commonly used chemicals in chemical pre-treatment methods are dilute sulfuric acid (Sasmal et al., 2013b) for acid pre-treatment, and sodium hydroxide, lime, or potassium hydroxides are used for the base pre-treatment method (Sasmal et al., 2013a). Apart from using acid or base, few researchers have proved the usage of solvents such as butanol, hexane, etc., which can also fulfill the purpose of the pre-treatment method. The objective of mechanical pre-treatment is to decrease the size of biomass materials to improve the surface area of the feedstock, as well as reduce the degree of polymerization and crystallinity of cellulose (Sasmal et al., 2012a). The crystallinity of lignocellulosic materials is also reduced through the mechanical pre-treatment process, which facilitates faster degradation of the basic sugar materials. Biological pre-treatment methods include enzymatic hydrolysis or the addition of suitable microorganisms into it for further digestion.

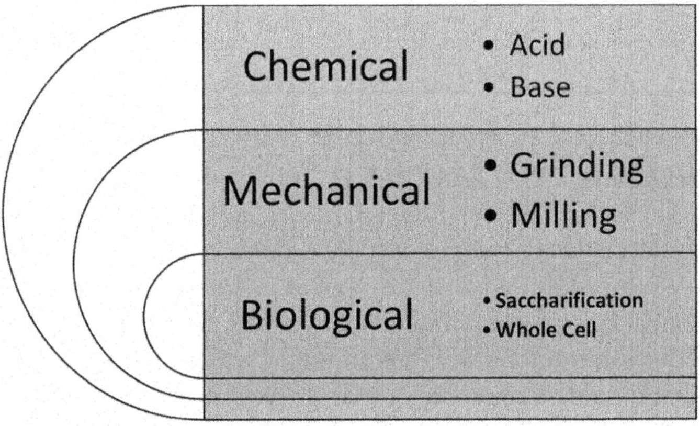

FIGURE 7.4 Major classification of pre-treatment methods.

The choice of pre-treatment to apply to biomass materials is not straightforward. Numerous factors including crystallinity and the degree of polymerization of the lignocellulosic biomass, obtainable surface area, and the number of acetyl groups present in lignin should be considered at the time of considering a particular pre-treatment process.

7.2.2 MECHANICAL PRE-TREATMENT

The mechanical pre-treatment process includes milling, steam explosion, high-pressure homogenization, etc. (Table 7.1). Mechanical pre-treatment methods are mostly preferred for both biogas and biofuel production plants for reduction of the size of biomass and reducing the degree of cellulose crystallinity.

TABLE 7.1 List of Various Mechanical Pre-treatment Methods

Type of Mechanical Pre-treatment	Pre-treatment Conditions	Feedstock	References
Milling	Particle lengths: 6–33 mm	Wheat straw	Dumas et al., (2015)
Thermal	Temperature: 70–121°C, Time: 0.5–15 h	Sewage sludge;	Ruffino et al., (2015)
Steam explosion	Temperature: 140°C, Time: 60 min	Wheat straw	Theuretzbacher et al., (2015)
High-pressure homogenization	Using Pressure 10 MPa	Yard Waste	Jin et al., (2015)
Hydrodynamic cavitation	Rotor speed: 2300–2700 rpm	Wheat straw	Patil et al., (2016)

7.2.3 CHEMICAL PRE-TREATMENT

The pre-treatment method in which different organic and inorganic chemicals are used collectively is known as the chemical pre-treatment method. Chemical pre-treatment methods are most robust among other established pre-treatment methods for solid waste disintegration or solid waste management. Various low-cost abundant chemicals are used for this purpose such as dilute acid, caustic soda, etc. Pre-treatment of areca nut husk (agricultural waste product generated after harvesting areca nut) was reported successfully using a combination of using lime and ultrasound (Sasmal et al., 2012b). Different chemical pre-treatment methods were used on different lignocellulosic biomass as tabulated in Table 7.2.

TABLE 7.2 List of Various Chemical Pre-Treatment Methods

Type of chemical pre-treatment	Pre-treatment conditions	Feedstock	References
Alkali (sodium hydroxide) hydrolysis	Base concertation 0.2–1% (w/w), Temperature: 50–121°C, Time 10–240 min	Sugarcane bagasse	Bolado-Rodríguez et al., (2016)
Acid hydrolysis (dilute sulfuric acid)	Temperature: 170°C, Acid concentration :1–5% (w/w)	Sunflower oil cakes	Monlau et al., (2013)
Solvent (Butanol)	Temperature: 121°C, Acid concentration: 0.015N–0.075N	Groundnut shell	Razi and Sasmal, (2020)

7.2.3 BIOLOGICAL TREATMENT

Processes for biological pre-treatment are moderate and environment-friendly. Biological treatment employs different micro-organisms, primarily bacteria, white and brown rot fungi, these micro-organisms produce lignin-degrading enzymes such as laccase and peroxidases to reduce or break down complex lignin into simplex components. An efficient biological pre-treatment requires no intervening mechanical size reduction, preserves the fractions of pentose (hemicellulose), prevents the development of degradation products that inhibit the growth of fermenting microorganisms, minimizes energy demands, and limits costs. Consequently, a biological pre-treatment is mainly aimed at breaking and eliminating lignin bonds and destroying the crystalline structure of cellulose to make it (more) vulnerable to enzymatic or microbial attacks, thus reducing the loss of carbohydrates for the anaerobic digestion (Howard et al., 2003).

An efficient biological pre-treatment process entails without intervening mechanical size reduction and conserves the fractions of pentose sugars (hemicellulose). It also prevents the formation of obnoxious products from biomass that inhibit the growth of fermenting microorganisms and minimizes energy demands (Zheng et al., 2014 and Sawatdeenarunat et al., 2017).

7.3 THERMOCHEMICAL CONVERSION METHODS

7.3.1 PYROLYSIS PROCESS

Pyrolysis is the process in which substrate (organic biomass) is heated at a temperature between 250–275°C. The exact temperature is dependent on the type of raw material used and reactor type in anaerobic conditions to break complex molecules into simpler forms (Sharma et al., 2014). The major product is bio-oil, syngas, and biochar, but primarily bio-oil production takes place (Pham et al., 2015).

Pyrolysis is a significant process, and by changing operating parameters, various combinations of solid, liquid, and gaseous products in different proportions can be generated (Czajczyńska et al., 2017). The transformation of low-energy molecules into high-energy biofuels takes place in pyrolysis (Chowdhury and Sarkar, 2012 and Biswal et.al., 2013). The major advantage of pyrolysis is that various types of waste can be used like kitchen waste, plastics, industrial waste, etc.

7.3.2 GASIFICATION

Gasification may be defined as a process of partial oxidation of biomass (Rajmohan et al., 2019). During gasification, syngas is synthesized primarily from biomass. Depending on the class of reactor, i.e., entrained bed, fixed bed, fluidized bed, moving grate furnace, vertical shaft, plasma reactor, and varying operating conditions, syngas for different compositions used for different applications can be produced (Arena, 2012; Young, 2010). Although any kind of organic waste can be used for gasification, the performance of the process for the generation of energy is significantly affected by the type of waste processed. Elemental composition, moisture content, ash content, bulk density, size, etc. are some of the important characteristics of the waste to be considered (Zevenhoven-Onderwater et al., 2001). The use of pre-treated waste in gasification is also advisable and important for the high performance of the reactor (Arena, 2012). Pyrolysis and gasification of biomass were evaluated by important parameters like the total yield of Syngas, hydrogen, and total energy. It was found that gasification is more suitable and better than pyrolysis in terms of the benchmarks chosen (Ahmed and Gupta, 2010).

7.3.3 CARBONIZATION

Carbonization is an established process used for the production of charcoal. In this process, the biomass materials are converted into charcoal-like carbonaceous material, it's a kind of slow pyrolysis process conducted in an oxygen-limiting or oxygen-free environment (Ronsse et al., 2015). Carbonization is dry wood distillation during which many volatile components are eliminated. Furthermore, hydrothermal carbonization (HTC) is getting attention nowadays, it's a wet process and converts food waste to energy-rich sources at low temperatures in between 180–350°C (Berge et al., 2011 and Liu et al., 2013). Some of the chemicals can be reused/recovered from HTC process water (Berge et al., 2011). Moreover, nutrients recovered from the liquids, mainly nitrogen species, can be used as a fertilizer (Li et al., 2013a). HTC technique is useful for the treatment of food waste as energy for drying is not required. During the digestion of food waste anaerobically, some of the CO_2 is released into the atmosphere. In the case of HTC, most of the carbon remains associated with the final char product. Catalytic HTC leads to the production of enhanced activated carbon (Watts et al., 2020).

7.3.4 CATALYTIC LIQUEFACTION

Catalytic liquefaction is the direct conversion of biomass into liquid or oily compounds in water. The solvent used in the process is most important to be considered as it can influence product yield, reaction operating conditions, and biocrude composition (Liu and Zhang, 2008; Zhang et al., 2010). Ethanol has several advantages, like cheap and easy availability. The critical pressure and temperature are lower than water. So, it has been used in the liquefaction process (Chumpoo and Prasassarakich, 2010). During the liquefaction process of biomass, ethanol-water, a dual solvent (co-solvent) is an efficient reaction medium (Chen et al., 2012; Yuan et al., 2007).

7.4 BIOCHEMICAL CONVERSION METHODS

The use of microorganisms for the production of ethanol is quite an old technique. It's significant that we have realized the true potential of microorganisms for the conversion of waste products into sources of energy. With the advent of recombinant DNA technologies and genetic engineering, there are a lot of opportunities and encouragement to use RDT for better prospects and efficient application of microorganisms in waste treatment.

7.4.1 MICROBIAL FERMENTATION

Owing to the rapid rise in oil consumption, energy demand is very difficult to be met by the number of current fossil fuels present. Continuous increase in usage of conventional diesel, depleting oil reserves, and levels of emissions of hazardous chemicals from burning of coal and petrol in vehicles and industries, compelled the research community to think and concentrate its attention towards development, design new and innovative methods and technologies for efficient production of biofuels and its judicious use in various sectors. Microbial fermentation is an old-school approach to convert organic materials into value-added chemicals using microorganisms as a catalyst. Production of different kinds of biofuels such as ethanol, butanol, from organic solid waste materials is the new normal in energy sectors. Overall, ethanol and butanol suggest a great promise as a petrol substitute.

The fermentation process is an important key phase in the production of bioethanol, where ethanol is produced directly from the fermentation agent's (yeast or bacteria) metabolic activity. Glucose-fermenting bacteria,

Zymomonas mobilis prefer to utilize glucose-rich substrates as their' preferred food' and produce ethanol. It is very difficult to expect that the pre-treated hydrolysate to be uniform or homogeneous in terms of the sugar monomer present. The hydrolysate always has different fractions of monomers and many other oligosaccharides which may act as inhibitors or indigestible substances for microbes. Theoretically, 0.51 kg/mol of ethanol along with 0.49 kg/mol of carbon dioxide is produced from one kg/mol of glucose or xylose.

Ethanol yield (fermented by *Saccharomyces cerevisiae)* from alkali pre-treated (12% NaOH) cotton linter and waste of blue jeans textiles resulted in around 0.48 g ethanol/g pretreated textiles used (Jeihanipour and Taherzadeh, 2009). Immobilization of Zymomonas *mobilis* on luffa (*Luffa cylindrica* L.) matrix used for ethanol production from mahula (*Madhuca latifolia* L.) flowers was investigated by Behera et al., 2011. The results showed a yield of ethanol around 247.9±0.08, g/kg flowers after 96 h fermentation. Ethanol production by the immobilized cells was 9.2% higher than that by free cells. The ethanol yield (*Yp/s*), and final sugar to ethanol conversion rate (%) obtained were 0.439±0.004 g/g, and 87.8%, respectively (Behera et al., 2011).

Apart from ethanol, another promising biofuel was bio-butanol which is four-carbon alcohol comprised of four isomeric structures, i.e., n-butanol (n-C_4H_9OH), iso-butanol (iso- C_4H_9OH), sec-butanol (sec- C_4H_9OH), and tert-butanol (tert- C_4H_9OH) (Grana et al., 2010).

Bio-butanol was traditionally produced through using process ABE (Acetone-Butanol- Ethanol), employing *Clostridium sp.* to ferment biomass sugars. The route is associated with a few drawbacks such as the low yield of butanol and the production of by-products (acetone and ethanol). However, butanol can be produced directly from ethanol by condensing aldol over metal oxides/ catalysts for hydroxyapatite. Various researchers across the globe have tried to improve the yield of butanol using a different kinds of lignocellulosic substrates as well as modifying the genes involved in butanol fermentation. The yield of butanol using genetically-modified *Clostridium* sp. may be high but the yield was not carried forward for a longer duration. The yield of butanol ranges from 6 gL^{-1} to 21 gL^{-1} (Table 7.3). Among various species of *Clostridium*, the yield of butanol by *C. beijerinckii*, was proved promising and high.

Although a good number of feedstocks and microorganisms are already successfully used to manufacture bio-butanol, there are multiple disadvantages associated with ABE fermentation that hinders it from competing with the petrochemical synthesis in economic terms.

TABLE 7.3 List of Biobutanol Production from Different *Clostridium* sp. and Their Yield

Microorganism	Feedstock	Yield	Reference
C. acetobutylicum (DSM 1731)	Oil palm sap	14.4 gL^{-1}	Komonkiat and Cheirsilp (2013)
C. beijerincki tyrobutyricum	Cassava starch	6.66 gL^{-1}	Li et al., (2013b)
C. beijerinckii, (BA101)	Maize meal	26 gL^{-1}	Ezeji et al., (2007)
C. acetobutylicum	Glycerol	13.57 gL^{-1}	Yadav et al., (2014)
C. acetobutylicum (ATCC 824)	Spoilage palm fruits	21.56 gL^{-1}	Abd-Alla et al., (2012)

7.4.2 ANAEROBIC DIGESTION

Anaerobic digestion is the process of converting organic waste into useful and usable products such as biogas and nutrient-enriched fertilizer under controlled anaerobic conditions (Chanakya et al., 2007; Guermoud et al., 2009). Anaerobic digestion is a potent method for the generation of renewable energy and nutrient recycling (Kosseva, 2011). The biogas produced can be further converted into compressed natural gas (CNG) and liquid petroleum gas (LPG) and these after further processing and purification can be used as transport fuel and household. The government of India has been emphasizing utilizing waste for the production of energy under the "waste-to-energy mission." It is having two advantages to generating energy and to solve the problem of waste disposal (Annual report, Ministry of Petroleum and natural gas, Govt. of India, 2020). Advancements in biogas technology have opened up new avenues for the utilization of different types of waste residues like crop waste and kitchen waste in an integrated manner (Bhatia et al., 2020). The advantage of anaerobic digestion is clean energy production, but the process is complex and tricky as different types of microorganisms are required for the process. For microbe's optimum temperature control and nutrient requirement, etc.to be considered for obtaining the maximum yield.

7.4.3 MICROBIAL COMMUNITIES IN ANAEROBIC DIGESTER

Overall process and reaction rate are dependent on the communities of microorganisms involved (Table 7.4). The dynamics between these microbes are complex and the quality and quantity of fuel produced are dependent on the interaction between these microorganisms (Bhatia et al., 2020). Total four different processes take place in biogas production. All these four processes; hydrolysis, acidogenesis, acetogenesis, and methanogenesis required different sets of microbes. Next-generation sequencing (NGS) and terminal restriction fragment length polymorphism (T-RFLP) are the techniques used for the detection of microorganisms in the anaerobic digestion process. Furthermore, metagenomics can also be used to find out microbial consortium in AD. 16S rRNA (Ribosomal RNA), denaturing gradient gel electrophoresis (DGGE) and polymerase chain reaction (PCR) are significant techniques to identify microbial communities.

TABLE 7.4 Important Species of Microbes Involved in Biogas Production (Bhatia et al., 2020)

Process	Microbial community involved
Hydrolysis	*Clostridium* sp., *Saccharomyces* sp., *Bidobacterium* sp., *Acetomyeces* sp. *Eubacterium* sp., *Corynebacterium* sp.
Acidogenesis	*Streptococcus* sp., *Staphylococcus* sp., *Pseudomonas* sp., *Fusobacterium* sp.
Acetogenesis	*Actinomyces* sp., *Succinomonas* sp., *Lachnospira* sp., *Propionibacterium* sp., *Anaerovibrio* sp.
Methanogenesis	*Methanobacterium* sp., *Methanogenium* sp., *Methanococcus* sp.

7.4.4 PROCESS IN ANAEROBIC DIGESTION (BIOGAS PRODUCTION)

There are four phases for the production of biogas from organic biomass materials (Voelklein et al., 2017). The very first step is hydrolysis in which the complex organic molecules like carbohydrates, fats, and proteins breaks down into simpler form and the main microbes involved in the process of hydrolysis are *Saccharomyces* sp., *Clostridium* sp., *Eubacterium* sp., etc. During the process of acidogenesis, the formation of acid, alcohol with some gases takes place by acetogenic bacteria, i.e., the third step. The organic acids and alcohols are transformed into acetic acid, carbon dioxide (CO_2), and hydrogen (H_2). After this, acetic acid, carbon dioxide, and hydrogen gas are converted into biogas. Flowchart (Figure 7.5) depicting the process of biogas production (Bhatia et al., 2018).

7.5 CASE STUDY ABOUT GOVARDHANNATHJI ENERGIES LLP BIOGAS PRODUCTION PLANT IN INDIA

For a better understanding of biogas production plants and their operational strategy, we are further discussing India's first biogas plant Govardhannathji

Energies LLP located at Kheda district, Gujarat with a capacity of 18,000 cubic meters per day. Central Financial Assistance (CFA) is being provided by the Ministry of New and Renewable Energy (MNRE) Government of India.

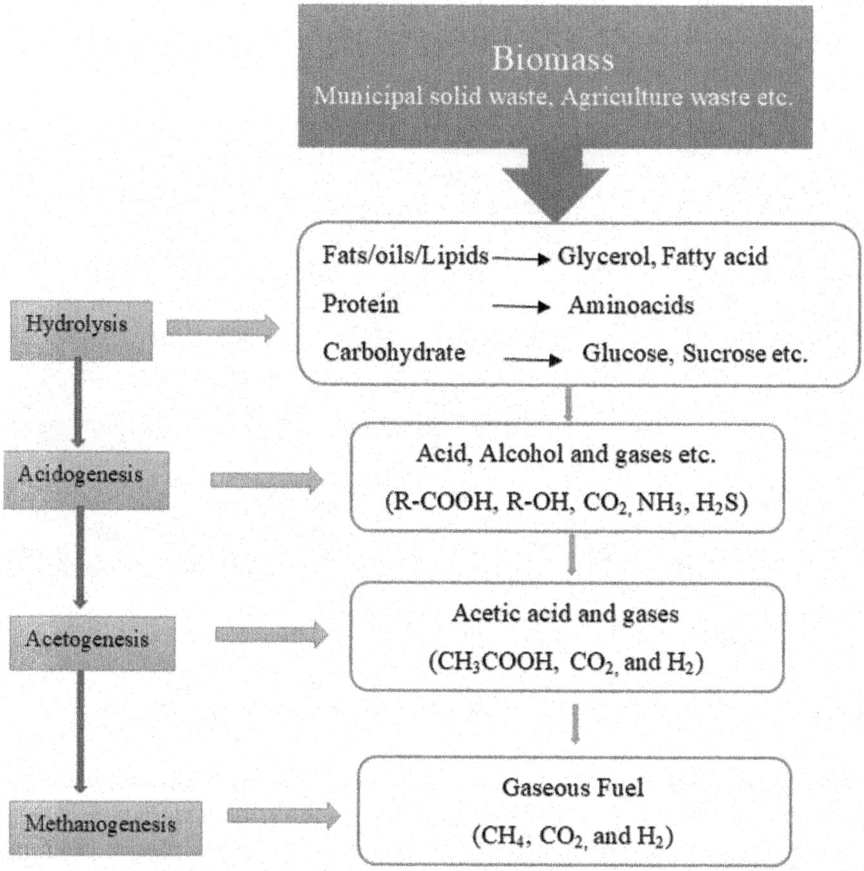

FIGURE 7.5 Flowchart representing the process of biogas generation.

The plant is designed with the following operating units:

1. Raw material storage
2. Raw material unloading and feed preparation
3. Biogas generation and storage
4. Biogas purification
5. Biogas compressing/bottling
6. Fertilizer production/packing

7.5.1 RAW MATERIAL STORAGE

MSW, food processing industry waste, cooked food waste, press mud, poultry waste, sugar mills, mandi waste, etc. are collected on a daily basis. After collection, these raw materials and various bio-degradable wastes are sorted and stored separately based on their various chemical and physical properties as depicted in Figure 7.6.

FIGURE 7.6 Waste segregation site for the process of biogas generation.

7.5.2 RAW MATERIAL UNLOADING AND FEED PREPARATION

After proper segregation and reaching the waste to the plant site, material unloading and feed processing take place. The mixture of biodegradable waste is fed into a feed preparation pit. The submersible agitator mixes different feedstock and brings it to a uniform, fluffy and homogenized form. From time to time it releases small quantities of feedstock into an open mouth pump. This open mouth pump joins an additional quantity of liquid with the biomass and pushes it forward. The liquid itself comes at the beginning of the daily preparation period from a fertilizer pit or sometimes directly from a digester. The mixture of feedstock and effluent is to be pumped into the digester.

The fertilizer pit and the feed preparation pit are fully equipped with high-quality submersible agitators. An additional chopper is placed in between the feed prep and the digester, for better refining, pumping, piping, and mixing of the biomass to attain better consistency and homogenization. The way of maintaining an acceptable fluid viscosity even of high dry matter containing mixtures reduces the demand for electrical self-consumption of the plant. Figure 7.7 depicting unloading of waste and the feeding process is depicted in Figure 7.8.

FIGURE 7.7 Waste unloading at the biogas generation plant site.

FIGURE 7.8 Waste raw material feeding tank.

7.5.3 BIOGAS GENERATION AND STORAGE

The substrate is used as feed for the anaerobic digestion (AD) process which takes place in the concrete digester as shown in Figure 7.9. The feeding of the AD works as a semi-automatic storage-flow process by which the biomass is guided into the digester. The digester is an advanced continuous stirred tank reactor (CSTR). All processes are fully automated and monitored by level switches, which are connected to the PLC-System, and control the pumping system. The digesters are equipped with a high-quality submersible agitator and operated at optimum (38°C ± 1°C) temperature range. This leads to a consistent process and economically-optimized demand for process heat.

FIGURE 7.9 Anaerobic digester.

The digester is comprised of a standing cylindrical tank of reinforced concrete with a volume of over 12,000 m^3 including a freeboard headspace for gas release. The digesters are covered with a double membrane gas roof for the storage of raw biogas. The solids are fed into the digestion system for decomposition or degradation of biomass or feed substrate. The degradation is done in one digester. The plant works on a storage-flow process. Biomass is fed into the single digester by the pumping system several times per day. Additionally, the re-circulated slurry is also pumped into the digester. The treated sludge is further pumped into the liquid fertilizer tank after due signs shown by the indicators present in the digester.

As described above (Figures 7.10 and 7.11, respectively), the digesters are equipped with high quality submerged agitator and operated in a mesospheric/moderate temperature range. These two parameters are significant and lead to an efficient and stable process of maximizing gas yield and minimizing greenhouse gases production.

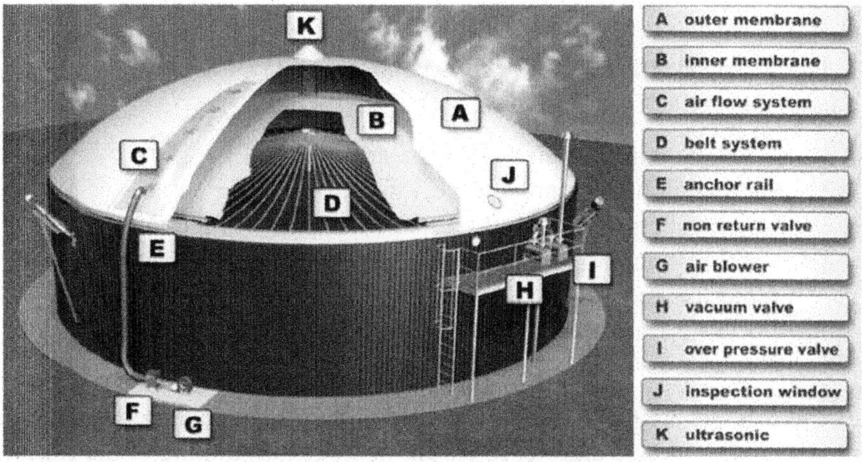

FIGURE 7.10 Anaerobic digester with double membrane biogas holder.

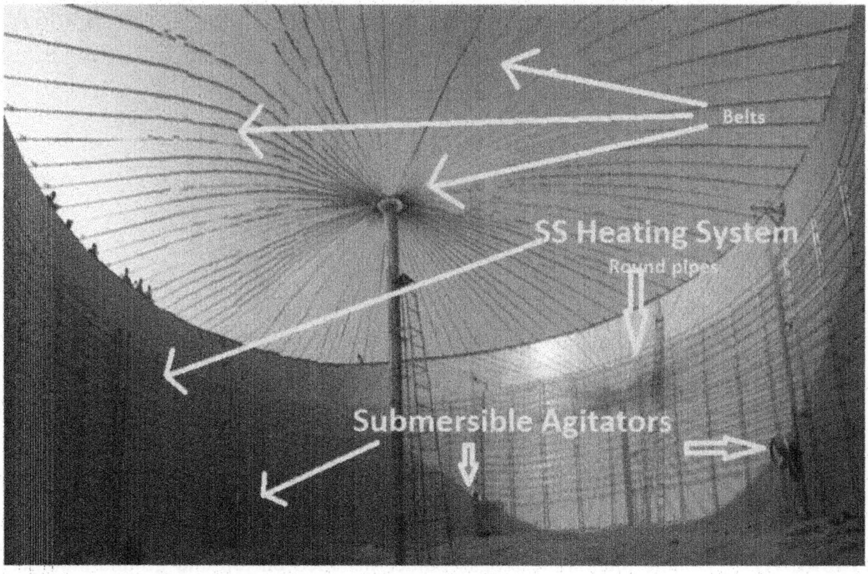

FIGURE 7.11 Anaerobic digester (inside picture).

7.5.4 BIOGAS STORAGE

The digester is covered with a flexible double membrane roof. Outer-membrane and inner membrane material: Euro-standard membrane, anti-UV, self-clean (PVDF coating), anti-corrosion, anti-aging, acid-proof, alkali proof, and high strength.

The double membrane roof is fully controlled by PLC/control panel with additional benefits"

1. Electric control cabinet display system.
2. Pressure/volume/control process display.
3. Function control system: LCC has more than 20 functions.
4. Sensor system: pressure/volume sensor made of stainless steel, explosion-proof.
5. Operation cabinet: Air boost blower, non-return valve Safety function, etc.

The reactors are CSTR and submerged agitators and high flow agitators are installed inside the digester to ensure homogenous mixing of the slurry. The digested organic matter produces a mixture of gases that comprises methane (CH_4) in high maximum concentration and the rest containing CO_2 and H_2S. This gas is called biogas. Biogas holders as shown in Figure 7.12. After digestion, the feed material is taken for further storage. This technology ensures maximum degradation of biodegradable feedstock and attaining high efficiency in biogas generation plants.

FIGURE 7.12 Double membrane biogas holder installed on top and ground-mounted holder.

7.5.5 BIOGAS UP-GRADATION/PURIFICATION/ENRICHMENT TECHNOLOGY

Pressure swing adsorption (PSA) technology for biogas purification has been used in this plant. Compressed biogas is separated from impurities like carbon dioxide with the help of PSA. At different pressure levels, carbon dioxide's adsorption/desorption on zeolites or activated carbon takes place. Hydrogen sulfide irreversibly gets adsorbed on adsorbing material and hence H_2S removing step is also an important part of the PSA process.

In this, the raw biogas passes through a gas washing process. The biogas is mixed with a special gas washing solution in an absorber unit. The carbon dioxide-loaded solution is then pumped into a stripping unit in which the solution unloads the carbon dioxide due to pressure and temperature changes. The solution is then recycled. In the next step, the biogas is pumped through a desulfurization unit to reduce the H_2S content to the limits of the gas compressor units. The desulfurization biogas is dried by cooling water and in a continuing step further dried by a chilling water system to minimize the risk of condensates, the biogas purification system as shown in Figures 7.13 and 7.14, respectively. The purified biofuel is bio-methane, CO_2, water vapor, and trace gases like H_2S must be removed before the further utilization of biogas due to the following reasons:

a) The corrosive nature of hydrogen sulfide gas.
b) Reduction of heating value as metal surfaces are sensitive to corrosion, due to the presence of water vapor and H_2S.

After purification of biogas, it goes to the filling unit via a high-pressure compressor.

7.5.6 BIOGAS COMPRESSING AND BOTTLING

The purified biogas to be compressed as depicted in Figures 7.15 and 7.16, respectively, bottled into CNG cascades/cylinder (metal and composite fiber cylinders) (Figures 7.17 and 7.18) at 200 bars for usage in the transportation sector with the help of a biogas compressor and gas filling manifold as shown in pictures below. The CBG (also known as BioCNG) is an excellent substitute for LPG/PNG in the commercial sector as a clean fuel for transportation, cooking, heating, and other related purposes.

FIGURE 7.13 Biogas purification system (dried by cooling water).

FIGURE 7.14 Biogas purification system (dried by chilling water).

FIGURE 7.15 Biogas compressor.

FIGURE 7.16 Gas filling manifold system.

FIGURE 7.17 Cylinder cascades during filling of CBG (metal fiber cylinders).

7.5.7 BIOFERTILIZER

The solids residuals from AD are separated from the liquid stream. The liquid phase after this separation unit is pre-treated and recycled. The separated solids are collected in the manure/fertilizer storage yard as shown in Figure

7.19 and sold as biofertilizers to the farmers/nurseries. Figures 7.20 and 7.21 is showing the fertilizer production and packaging zone, respectively.

FIGURE 7.18 Cylinder cascades during filling of CBG (composite fiber cylinders).

FIGURE 7.19 Solid-liquid separator for the production of biofertilizer.

FIGURE 7.20 Fertilizer production.

FIGURE 7.21 Fertilizer packaging zone.

7.6 CONCLUSION

Solid waste management can be portrayed as a challenge or problem in society. With the applications of scientific principles, organic waste or biomass can be converted into a valuable energy resource. Anaerobic digestion is a biological approach for the generation of biogas and remnants can be used as biofertilizers. Anaerobic digestion is a four-step process in which various microorganisms are playing a significantly important role in biomass to biogas conversion. It is one of the most preferred, efficient, economical, clean, and green methods for biogas generation or solid waste treatment. A lot of opportunities and scope still lie for the development of active and more useful microbial consortia, pre-treatment methods, digester design, etc. To conclude, the implication of strict government policies for solid waste management, disposal, and generation of energy by organic waste will be a step forward towards the sustainable generation of fuel and energy and solves many future problems in society.

ACKNOWLEDGMENTS

The authors are very much thankful to the authority of NSUT, New Delhi and ICAR-IARI, New Delhi for moral support and encouragement. We are also grateful to Govardhannathji Energies LLP, a biogas production plant for providing information and required figures and pictures regarding biogas production in India

KEYWORDS

- **anaerobic digestion**
- **bioethanol**
- **biofuel**
- **biogas**
- **solid waste**

REFERENCES

Abd-Alla, M. H., & El-Enany, A. W., (2012). Production of acetone-butanol-ethanol from spoilage date palm (*Phoenix dactylifera* L.) fruits by mixed culture of *Clostridium acetobutylicum* and *Bacillus subtilis*. *Biomass and Bioenergy*, 172–178.

Ahmed, I. I., & Gupta, A. K., (2010). Pyrolysis and gasification of food waste: Syngas characteristics and char gasification kinetics. *Applied Energy*, 101–108.

Arena, U., (2012). Process and technological aspects of municipal solid waste gasification. A review. *Waste Management*, 625–639.

Behera, S., Mohanty, R. C., & Ray, R. C., (2011). Ethanol production from mahula (*Madhuca latifolia* L.) flowers using free and immobilized (in *Luffa cylindrical* L. sponge discs) cells of *Zymomonas mobilis* MTCC 92. *Annals of Microbiology*, 469–474.

Berge, N. D., Ro, K. S., Mao, J., Flora, J. R., Chappell, M. A., & Bae, S., (2011). Hydrothermal carbonization of municipal waste streams. *Environmental Science & Technology*, 5696–5703.

Bhatia, R. K., & Ramadoss, G., (2020). Jain, A. K., Dhiman, R. K., Bhatia, S. K., Bhatt, A. K., Conversion of waste biomass into gaseous fuel: Present status and challenges in India. *BioEnergy Research*.

Bhatia, S. K., Joo, H. S., & Yang, Y. H., (2018). Biowaste-to-bioenergy using biological methods–a mini-review. *Energy Conversion and Management*, 640–660.

Biswal, B., Kumar, S., & Singh, R. K., (2013). Production of hydrocarbon liquid by thermal pyrolysis of paper cup waste. *Journal of Waste Management*.

Bolado-Rodríguez, S., Toquero, C., Martín-Juárez, J., Travaini, R., & García-Encina, P. A., (2016). Effect of thermal, acid, alkaline, and alkaline-peroxide pretreatments on the biochemical methane potential and kinetics of the anaerobic digestion of wheat straw and sugarcane bagasse. *Bioresource Technology*, 182–190.

Chanakya, H. N., Ramachandra, T. V., Vijayachamundeeswari, M., (2007). Resource recovery potential from secondary components of segregated municipal solid wastes. *Environmental Monitoring and Assessment*, 119–127.

Chen, Y., Wu, Y., Zhang, P., Hua, D., Yang, M., Li, C., Chen, Z., & Liu, J., (2012). Direct liquefaction of *Dunaliella tertiolecta* for bio-oil in sub/supercritical ethanol–water. *Bioresource Technology*, 190–198.

Chowdhury, R., & Sarkar, A., (2012). Reaction kinetics and product distribution of slow pyrolysis of Indian textile wastes. *International Journal of Chemical Reactor Engineering*.

Chumpoo, J., & Prasassarakich, P., (2010). Bio-oil from hydro-liquefaction of bagasse in supercritical ethanol. *Energy & Fuels*, 2071–2077.

Chung, J. N., (2013). Grand challenges in bioenergy and biofuel research: Engineering and technology development, environmental impact, and sustainability. *Frontiers in Energy Research, 1*, 4.

Czajczyńska, D., Anguilano, L., Ghazal, H., Krzyżyńska, R., Reynolds, A. J., Spencer, N., & Jouhara, H., (2017). Potential of pyrolysis processes in the waste management sector. *Thermal Science and Engineering Progress*, 171–197.

Dumas, C., Damasceno, G. S., Barakat, A., Carrère, H., Steyer, J. P., & Rouau, X., (2015). Effects of grinding processes on anaerobic digestion of wheat straw. *Industrial Crops and Products*, 450–456.

Ezeji, T., Qureshi, N., & Blaschek, H. P., (2007). Production of acetone–butanol–ethanol (ABE) in a continuous flow bioreactor using degermed corn and *Clostridium beijerinckii*. *Process Biochemistry*, 34–39.

Grana, R., Frassoldati, A., Faravelli, T., Niemann, U., Ranzi, E., Seiser, R., Cattolica, R., & Seshadri, K., (2010). An experimental and kinetic modeling study of combustion of isomers of butanol. *Combustion and Flame*, 2137–2154.

Guermoud, N., Ouadjnia, F., Abdelmalek, F., & Taleb, F., (2009). Municipal solid waste in Mostaganem city (Western Algeria). *Waste Management*, 896–902.

Howard, R. L., Abotsi, E. L., Van, R. E. J., & Howard, S., (2003). Lignocellulose biotechnology: Issues of bioconversion and enzyme production. *African Journal of Biotechnology*, 602–619.

Jeihanipour, A., & Taherzadeh, M. J., (2009). Ethanol production from cotton-based waste textiles. *Bioresource Technology*, 1007–1010.

Jin, S., Zhang, G., Zhang, P., Jin, L., Fan, S., & Li, F., (2015). Comparative study of high-pressure homogenization and alkaline-heat pretreatments for enhancing enzymatic hydrolysis and biogas production of grass clipping. *International Biodeterioration & Biodegradation,* 477–481.

Komonkiat, I., & Cheirsilp, B., (2013). Felled oil palm trunk as a renewable source for biobutanol production by *Clostridium* spp. *Bioresource Technology,* 200–207.

Kosseva, M. R., (2011). Management and processing of food wastes. In: *Reference Module in Earth Systems and Environmental Sciences Comprehensive Biotechnology* (2nd edn., pp. 557–593).

Li, L., Ai, H., Zhang, S., Li, S., Liang, Z., Wu, Z. Q., Yang, S. T., & Wang, J. F., (2013a). Enhanced butanol production by coculture of *Clostridium beijerinckii* and *Clostridium tyrobutyricum*. *Bioresource Technology*, 397–404.

Li, L., Diederick, R., Flora, J. R., & Berge, N. D., (2013b). Hydrothermal carbonization of food waste and associated packaging materials for energy source generation. *Waste Management*, 2478–2492.

Liu, Z., & Zhang, F. S., (2008). Effects of various solvents on the liquefaction of biomass to produce fuels and chemical feedstocks. *Energy Conversion and Management*, 3498–3504.

Liu, Z., Quek, A., Hoekman, S. K., & Balasubramanian, R., (2013). Production of solid biochar fuel from waste biomass by hydrothermal carbonization. *Fuel*, 943–949.

Monlau, F., Latrille, E., Da Costa, A. C., Steyer, J. P., & Carrère, H., (2013). Enhancement of methane production from sunflower oil cakes by dilute acid pretreatment. *Applied Energy*, 1105–1113.

Patil, P. N., Gogate, P. R., Csoka, L., Dregelyi-Kiss, A., & Horvath, M., (2016). Intensification of biogas production using pretreatment based on hydrodynamic cavitation. *Ultrasonics Sonochemistry,* 79–86.

Pham, T. P., Kaushik, R., Parshetti, G. K., Mahmood, R., & Balasubramanian, R., (2015). Food waste-to-energy conversion technologies: Current status and future directions. *Waste Management*, 399–408.

Rajmohan, K. S., & Varjani, S., (2019). Trends and advances in bioenergy production and sustainable solid waste management. *Energy & Environment*.

Razi, M. O., & Sasmal, S., (2020). Organosolv pre-treatment of groundnut (*Arachis hypogaea*) shell and its upshot. *Biomass Conversion and Biorefinery*, 1–8.

Ronsse, F., Nachenius, R. W., & Prins, W., (2015). Carbonization of biomass. In: *Recent Advances in Thermo-chemical Conversion of Biomass* (pp. 293–324). Elsevier.

Ruffino, B., Campo, G., Genon, G., Lorenzi, E., Novarino, D., Scibilia, G., & Zanetti, M., (2015). Improvement of anaerobic digestion of sewage sludge in a wastewater treatment plant by means of mechanical and thermal pre-treatments: Performance, energy and economical assessment. *Bioresource Technology*, 298–308.

Sasmal, S., & Mohanty, K., (2018). Pretreatment of lignocellulosic biomass toward biofuel production. In: *Biorefining of Biomass to Biofuels* (pp. 203–221). Springer, Cham.

Sasmal, S., Goud, V. V., & Mohanty, K., (2012a). Characterization of biomasses available in the region of North-East India for production of biofuels. *Biomass and Bioenergy,* 212–220.

Sasmal, S., Goud, V. V., & Mohanty, K., (2012b). Ultrasound assisted lime pretreatment of lignocellulosic biomass toward bioethanol production. *Energy & Fuels*, 3777–3784.

Sasmal, S., Goud, V. V., & Mohanty, K., (2013a). Delignification kinetics of lime pretreatment—An ineluctable tread for augmenting saccharification. *Journal of Biobased Materials and Bioenergy*, 660–664.

Sasmal, S., Goud, V. V., & Mohanty, K., (2013b). Dilute Sulfuric Acid Pretreatment of Bon Bogori (Ziziphus rugosus): Proselyte to Amorphous Biomass for Biofuel Production. *Journal of Bioprocess Engineering and Biorefinery*, 225–229.

Sawatdeenarunat, C., Sung, S., & Khanal, S. K., (2017). Enhanced volatile fatty acids production during anaerobic digestion of lignocellulosic biomass via micro-oxygenation. *Bioresource Technology*, 139–1345.

Sharma, S., Meena, R., Sharma, A., & Goyal, P., (2014). Biomass conversion technologies for renewable energy and fuels: A review note. *IOSR J. Mech. Civ. Eng.*, 28–35.

Theuretzbacher, F., Blomqvist, J., Lizasoain, J., Klietz, L., Potthast, A., Horn, S. J., Nilsen, P. J., et al., (2015). The effect of a combined biological and thermo-mechanical pretreatment of wheat straw on energy yields in coupled ethanol and methane generation. *Bioresource Technology*, 7–13.

Voelklein, M. A., O'Shea, R., Jacob, A., & Murphy, J. D., (2017). Role of trace elements in single and two-stage digestion of food waste at high organic loading rates. *Energy*, 185–1892.

Watts, M. K., Pradhan, R., & Dutta, A., (2020). Catalytic hydrothermal carbonization treatment of biomass for enhanced activated carbon: A review. *Waste and Biomass Valorization*, 1–6.

Yadav, S., Rawat, G., Tripathi, P., & Saxena, R. K., (2014). A novel approach for biobutanol production by *Clostridium acetobutylicum* using glycerol: A low cost substrate. *Renewable Energy*, 37–42.

Young, G. C., (2010). *Municipal Solid Waste to Energy Conversion Processes: Economic, Technical, and Renewable Comparisons*. John Wiley & Sons.

Yuan, X. Z., Li, H., Zeng, G. M., Tong, J. Y., & Xie, W., (2007). Sub-and supercritical liquefaction of rice straw in the presence of ethanol–water and 2-propanol–water mixture. *Energy*, 2081–2088.

Zevenhoven-Onderwater, M., Backman, R., Skrifvars, B. J., & Hupa, M., (2001). The ash chemistry in fluidized bed gasification of biomass fuels. Part I: Predicting the chemistry of melting ashes and ash–bed material interaction. *Fuel*, 1489–1502.

Zhang, L., Xu, C. C., & Champagne, P., (2010). Overview of recent advances in thermo-chemical conversion of biomass. *Energy Conversion and Management*, 969–982.

Zheng, Y., Zhao, J., Xu, F., & Li, Y., (2014). Pretreatment of lignocellulosic biomass for enhanced biogas production. *Progress in Energy and Combustion Science*, 35–53.

CHAPTER 8

Downstream Processing of Waste Biomass: From Biophysical Aspects of Biomass Yield to Engineered Microbial Cells for Better Harvesting

SURAJIT DEBNATH

Department of Medical Laboratory Technology, Women's Polytechnic, Tripura, India, E-mail: Surajitdebnath9@gmail.com

ABSTRACT

Industrial manufacturing of high-value products through bio-processes and minimizing waste through optimized downstream processing is imperative for sustainable biotechnology. The high-value biomass can have multiple origins, either a peptide therapeutics expressed from cell culture or a bio-product converted by microbial cell mass. Finally, the product of interest should be isolated in a highly purified and appropriately concentrated form so that the product could be ready for final formulation and packaging. The process for biomass production for therapeutics, biofuels, industrial chemicals such as alcohols, etc., is well developed. However, for a better industrial aspect, the processes do require rigorous standardization. Thus, for cost-effective biomass production, the phase of downstream processing of bio-manufacturing typically starts with the harvesting of bioreactor biomass. This chapter focuses on a molecular biological intervention that can be attempted to effectively harvest biomass by engineering the phenotype of microbial cells. Moreover, the concepts involved in downstream processing steps are explained. Finally, the application of bioinformatics and computational biology to minimize costs for molecular biology and genetic engineering workflow is discussed. This chapter for the first time proposes

Biotechnology for Waste Biomass Utilization. Prakash Kumar Sarangi & Latika Bhatia (Eds.)

the concept of "*Industrial Yield and Environmental Sustainability Protocol through Effective Downstream Processing.*"

8.1 INTRODUCTION

Concerns on availability and production of resource like fuel, food, and consumables to meet the demand of the constantly increasing global population require substantial innovation in industrial techniques. To ensure this, an inter-disciplinary scientific endeavor is the need of the hour. Improved downstream processing, implementation of the stringent protocol of recycling and reuse of resources, use of alternative building blocks as raw materials or carbon sources, strategies for waste minimization, optimum use of microbial cell factories are some of the vital steps that can be standardized to ensure value-added compounds. These steps can not only ensure industrial yield enhancement but also environmental sustainability through aspects like rapid wastewater treatment and so on. This chapter for the first time proposes the *concept "Industrial Yield & Environmental Sustainability Protocol through Effective Downstream Processing*"(Figure 8.1).

Reduction of the operational costs of bioprocesses is a major anticipation to bring down the costs of value-added bioproducts. To do so, using less refined substrates or low-value raw materials for microbial bioprocesses is opted. This would also result in potentially less dependence on fossil fuels for the conventional production of industrial chemicals, reagents, and bioproducts. The other advantage of enhanced downstream processing is also substantial reduction of pre-treatment (Baruah et al., 2018).

Various factors like effective removal of biomass to ensure continuous production of the target product, strain selection of organisms and culture conditions, abiotic factors, etc. have an effect on the plasticity of product yields. Although there are various protocols that are being studied experimentally throughout the world, a fraction of them has been scaled up to a pilot stage. The technical bottlenecks in this regard are the inefficient removal of biomass or the overall downstream processing. In this chapter, a combinatorial approach for genetic manipulation of a target organism for enhancing efficient downstream processing for biomass conversion is described. For sustainability and availability of environmental resources in the future, the waste products from various industries must be regarded as nutrient sources for another bioprocess through using microbial cell factories. To execute this in a sustainable way paradigm shift in the downstream processing technology, harvesting of biomass, and ensuring the optimal culturing ability of strains should be ensured.

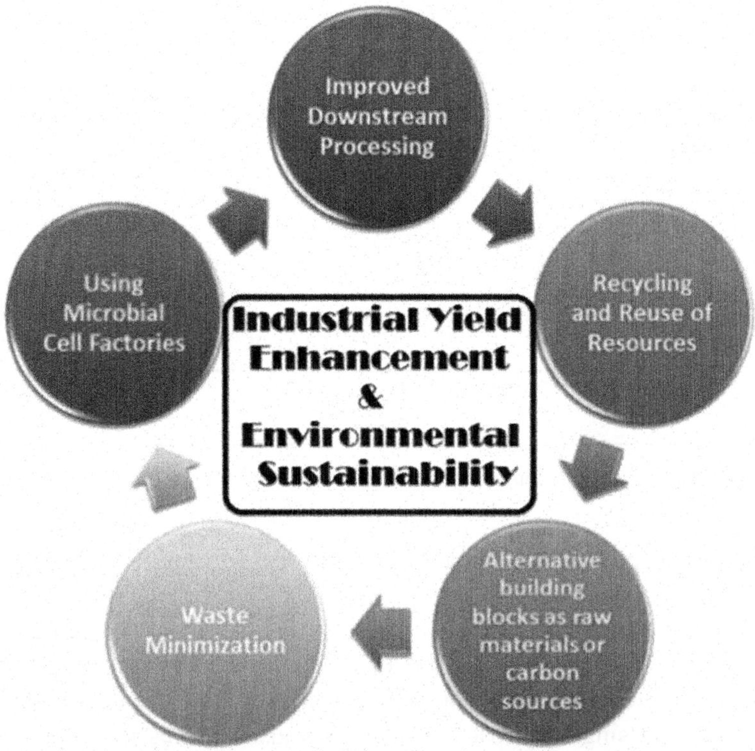

FIGURE 8.1 Industrial yield and environmental sustainability protocol through effective downstream processing.

8.2 BIOPROCESS TECHNOLOGY FOR SUSTAINABLE DOWNSTREAM PROCESSING OF BIOMASS

Initially, for this research, a publicly available dynamic bibliographic database was analyzed with specific MeSH terms to understand the recent trend and development in bioprocess biotechnology. Interestingly, *"Bioprocess Technology and Biomass"* research was initiated as early as 1976. In the timeline of *"Bioprocess Technology and Biomass"* research till 2020, three different phases can be seen. After 2009, there has been very rapid growth in the interest in *"Bioprocess Technology and Biomass"* research. In the year 2019 alone, more than 50 research papers were published in top peer-reviewed journals according to the database (Figure 8.2).

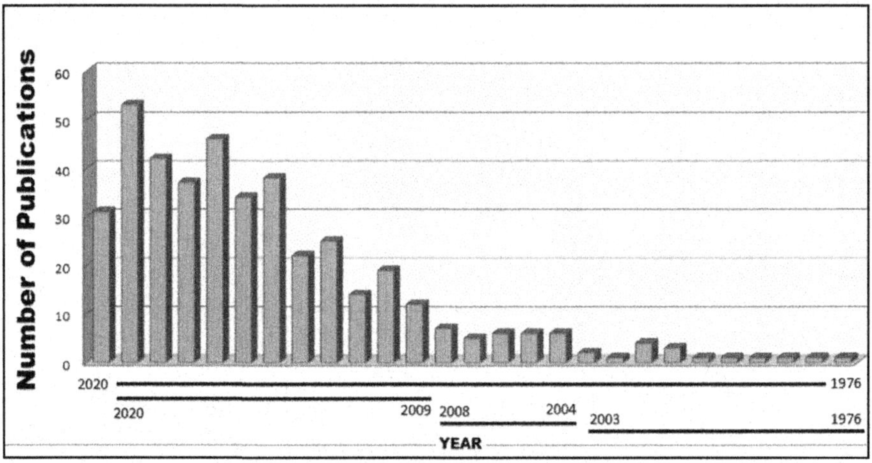

FIGURE 8.2 Research trend in "bioprocess technology and biomass."

The philosophy of the research in bioprocess technology has been in mainly three aspects which can be classified as innovations in (1) biofuels, (2) fine chemicals and drugs, and finally (3) secondary environmental innovations. However, the use of bioprocess technology and downstream processing has been multifactorial. Value addition of the end product is already there. Using low-cost raw materials also ensures the economy of these processes (Figure 8.3).

In the above schematic representation, biofuels essentially refer to bio-diesel and the bio-gas concepts. Biogas production, a highly sustainable bioprocess harnessing microbe cells has been a popular topic of interest for researchers (Lora Grando et al., 2017). In this bioprocess, microbe-based anaerobic digestion is harnessed in reactors for the production of methane from bio-waste. This is an excellent example of bio-conversion of low-value raw materials into a high-value product in the renewable energy landscape. According to studies (Hanemaaijer et al., 2015), community systems biology and quantitative modeling may aid to enhance our comprehension of the complex interplay of microbial communities in biogas production. Therefore, by using molecular biology and genetic engineering there is huge scope to augment the sustainability of this bio-process.

Similarly, the production of fine chemicals and drugs as well as the secondary environmental applications such as wastewater treatment also harness substantial contribution from the interplay of microbial communities and these aspects will be discussed in later sections. To understand the dynamic behavior of the microbial communities in the bio-process, the following concepts of the next section are imperative.

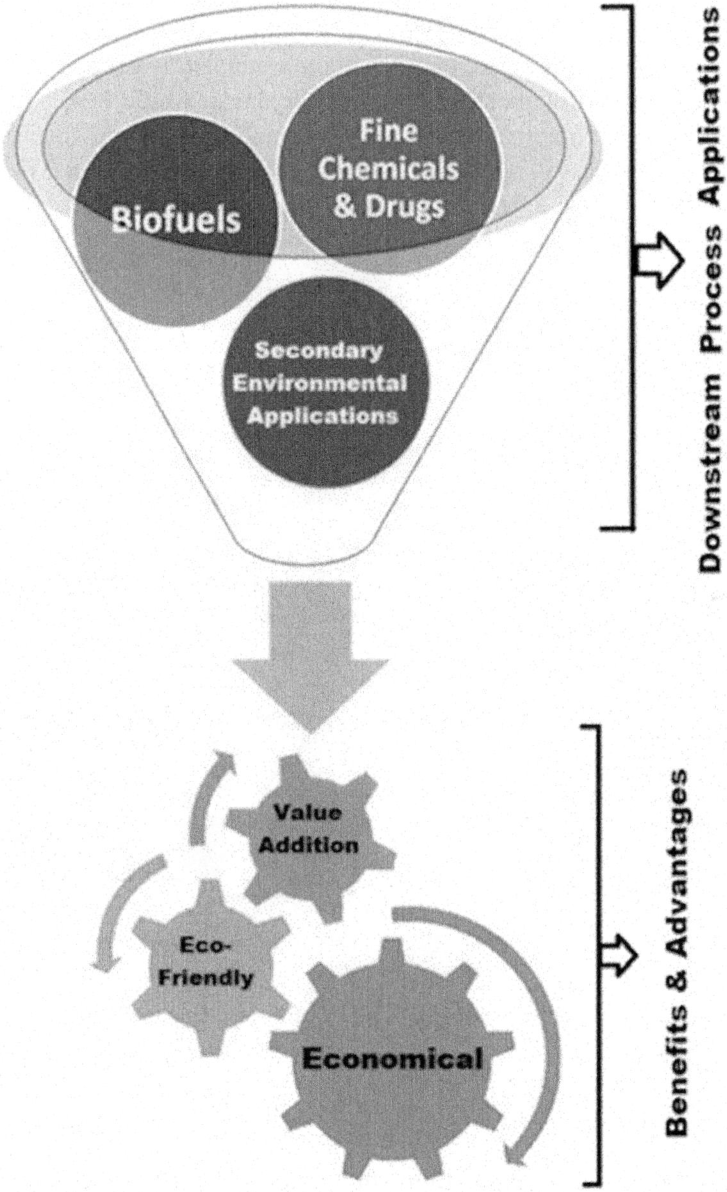

FIGURE 8.3 Downstream processing applications and advantages.

8.2.1 CONCEPTS AND CONSIDERATIONS OF BIOPROCESS TECHNOLOGY

Exploration of novel property of a natural biological resource is the first step for developing a bioprocess strategy. For microbe-based bio-conversion (Figure 8.4) and bio-process: (1) the source organism is generally isolated from its habitat. Subsequently, the source organism would be (2) cultured in a controlled environment in the lab. The analysis and assessment of the proposed bio-process using the isolated organism would be under (3) scrutiny through lab-scale R&D. Finally, upon successful assessment (4) scaling up of the process for industrial-scale would be done.

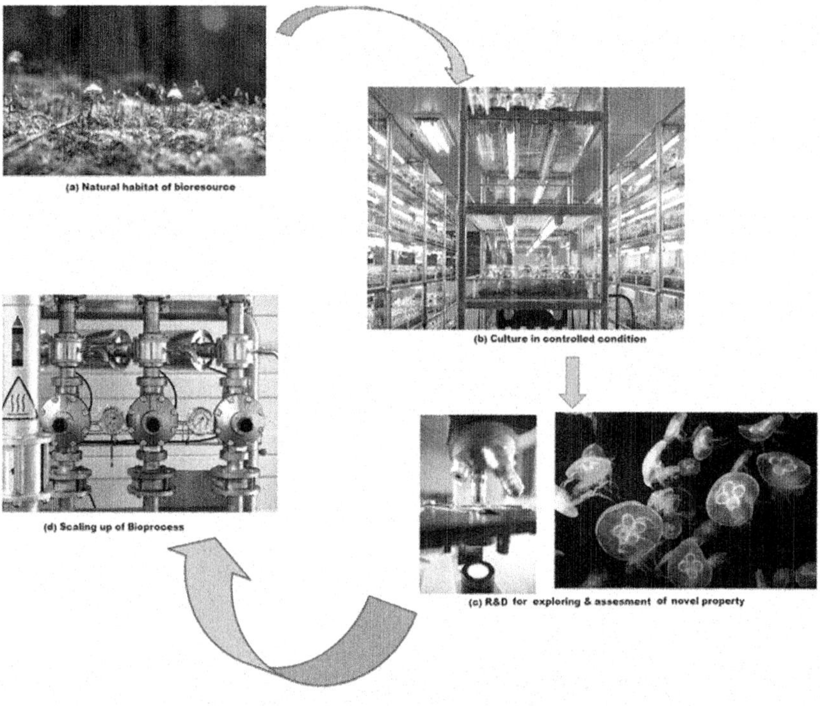

FIGURE 8.4 Steps in establishing novel bioprocess for value-added end products.

8.2.1.1 DYNAMICS OF CELLULAR BIOMASS YIELD

Lab-scale analysis of microbial growth aids in the scale-up process of bio-conversion or biomass production. Therefore, for microbe-based bioprocesses, it is imperative to comprehend the dynamics of biomass yield.

During microbial cell growth in a medium, a general trend shown above (Figure 8.5) is generally found. Interestingly, this trend is found in-vitro as well as a large-scale industrial setup as well (Figure 8.5).

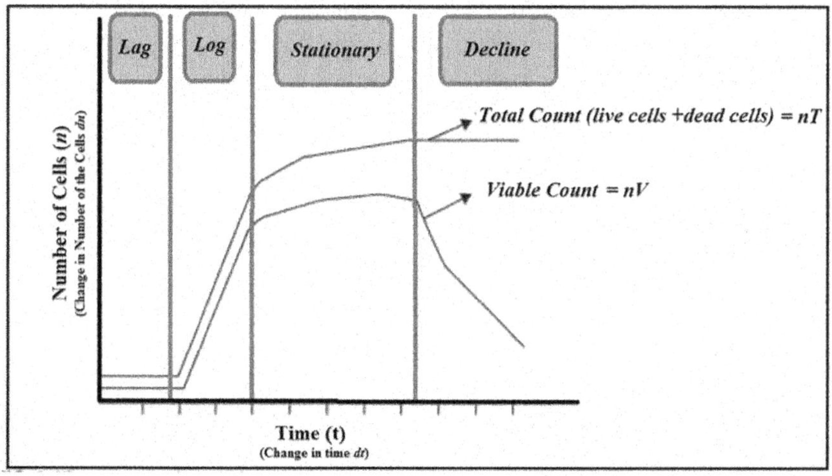

FIGURE 8.5 Steps in establishing novel bioprocess for value-added end products.

8.2.1.1.1 Initial Lag

The initial lag phase is where the specific growth rate is not at maximum level and this lag of growth is mainly due to the spores as well as the dead cells present. The inoculum takes time to acclimatize to the environment through adjustment in physiological conditions. Culture conditions and changes in nutrients are responsible for the lag. An example of this adjustment in physiology or growth lag is seen in shade-adapted cells that are exposed to light. The lag phase is minimized once at a later exponential growth phase, the cells themselves become the inoculum.

8.2.1.1.2 Log

Once the cells have adjusted at the late lag phase, they begin to grow in the new environment and multiply. Eventually, the cells enter the logarithmic growth phase or the exponential phase. As resources are available in optimum concentration at the latter phase the cells grow as an exponential function of time and divide.

8.2.1.1.3 Linear Growth

When everything is ok in the medium growth goes on, but a steady state is observed where there is no net growth of biomass. This is because the growing cells do not outnumber the cells that are being inhibited due to the limitation of resources. The toxic metabolic waste buildup is another reason other than the resource limitation that leads to linear growth and subsequent decline in numbers of viable cells.

8.2.1.2 KINETICS OF LARGE-SCALE BIOMASS YIELD IN BIOGAS PLANTS

Microbial community-driven process is harnessed in biogas plants. It is for converting organic waste material to methane and CO_2, i.e., biogas production, the naturally occurring anaerobic digestion. According to various research (Sören et al., 2019), Anaerobic Digestion Model No. 1 (ADM1) is the standard kinetic model of anaerobic digestion. It couples to Flux-Balance Analysis (FBA) of methanogenic species. This will be discussed in detail in a later section.

8.2.1.3 KINETICS OF SMALL-SCALE BIOMASS YIELD: LAB SCALE BIOMASS YIELD

With respect to lab-scale biomass growth kinetics, Solid-State Fermentations are regularly used for different biomass production in the lab scale. In this regard, lab-scale Mycelium Growth Rate is used by several researchers (Dantigny et al., 2004; Baranyi et al., 1993). To examine biomass growth of fungal biomass even in the petri plate *in-vitro* relationship of radius growth of mycelium with time can be an effective estimate of biomass production on the lab scale. This equation can represent biomass yield in a laboratory petri dish, where at every 12 or 24 hrs the measurements of colony diameter in two perpendicular directions were taken on the surface of the solid-state fermentations (SSF) petri dishes.

8.2.1.4 INDUSTRIAL ASPECTS OF WASTE BIOMASS RE-UTILIZATION

Microbe-based systems are an efficient converting strategy for low-value to high-value conversions. For example, in land-based integrated systems, the organisms such as microalgae together with bacteria aids in clearing aquaculture wastes. The resultant biomass of the microalgae may be utilized as feed for herbivores and filter-feeders. The microbe cells grow under stress and also lead to bio-remediation or detoxification of wastes.

By clear understanding of nutrient recycling processes waste biomass re-utilization becomes highly beneficial. For example, recirculating water from mariculture ponds (Neori and Shpigel, 2002) through the suitably designed phytoplankton or seaweed containing algal bio-filter. Excess CO_2 and dissolved nutrients can efficiently be extracted by this which reduces BOD. This, in turn, recharges the dissolved oxygen and subsequently makes the treated water suitable for reuse for a clean discharge as well as for mariculture. In a similar context, it is seen that deep wastewater reservoirs are becoming popular with the concept that agricultural irrigation should be met with treated wastewater that can be stored and reused. Once the biology of an organism is well understood, it can be harnessed for an associated benefit. Another example is the re-use of metals that are sorbed by microbial systems (Kotrba and Ruml, 2000; Klimmek et al., 2001). Microalgal biomass has been successfully used as a sorbing material. The sorbing material can be restored for reuse while these metals can be recovered from cell wall surfaces of algal biomass.

Another dimension of biomass reuse is sustainability. Such as the rapidly increasing global population and associated concerns over food availability and production it seems imperative to implement recycling of resources towards minimization of waste. The nutrient-rich brewery wastewater is recycled to enhance biomass growth that may be used for some other purpose (Simate et al., 2011). In research it was also seen that re-use even of culture media can be executed for biomass growth, which again can be a sustainable scenario. The growth of *Scenedesmus* sp was seen to be well maintained even with metallic salt (Kim et al., 2011) and bio-flocculant in a reused medium. Similarly, in biofuel synthesis using enzymatic bioprocess reuse of enzymes makes this a sustainable option (Jegannathan et al., 2008). Lipase enzyme reuse makes the process a sustainable option for the production of bio-diesel. Similarly, re-use of the medium was also reported in the harvesting of microalgae (Wu et al., 2012).

However, the issue of biomass recovery is a hindrance in many large-scale bioprocesses. The issues of biomass recovery will be dealt with in the next sections. Among the organic wastes that can be efficiently re-used for generating valuable bio-resources are (i) hydrophobic substrates, (ii) crude glycerol, and (iii) alternative saccharides. However, these components are further sub-categorized, such as in hydrophobic substrates, there are (a) pure oil, (b) used oil and industrial fats, and (c) oily wastewater. The crude glycerol is a by-product from various industrial processes such as the saponification process in oleochemical industries and biodiesel and bioethanol production. Among the alternative saccharides, the major components are (a) monosaccharides such as xylose and galactose, (b) several disaccharides and polysaccharides such as inulin and molasses containing sucrose. However, recalcitrant plant components, as well as food wastes, are also some of the reusable carbon sources that can be converted (Figure 8.6).

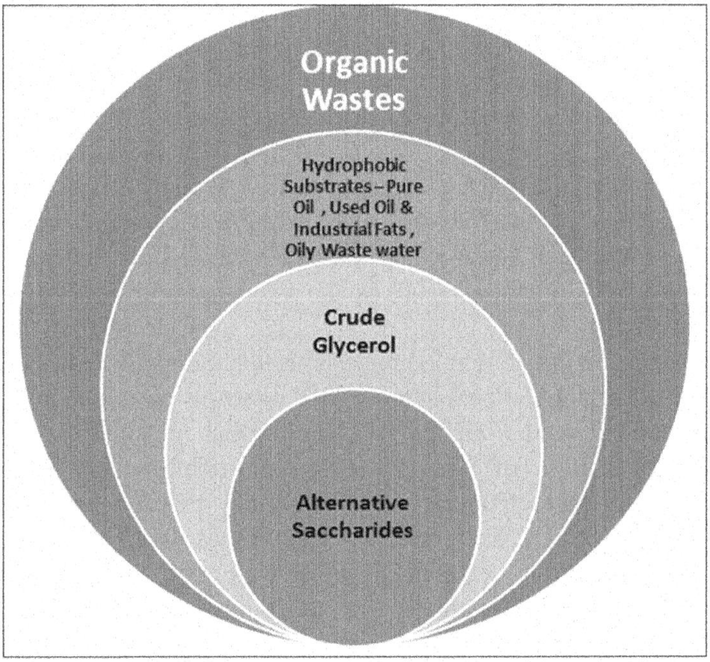

FIGURE 8.6 Components of re-useable carbon sources for bio-conversion.

8.2.2 BIOPROCESS TECHNOLOGY FOR DOWNSTREAM PROCESSING OF BIOMASS.

During an industrial bioprocess among several steps, the harvesting step is crucial. Efficiency and economics both can be improved by proper harvesting methods in microbe-based bio-process. According to the microbe cell biology & desired products the harvesting process can be standardized.

Among the harvesting techniques employed for biomass recovery, some are technology-intensive and some are easy to deploy. (a) Flocculation, (b) flotation, and (c) filtration are some of the methods that are less demanding in terms of technology use. On the other hand, several other technology-intensive methods of harvesting are: (i) centrifugation, (ii) magnetic separation, (iii) electrolysis, (iv) ultrasound, (v) immobilization, and so on.

Now, if the bio-product of interest is an intracellular entity, the complete isolation of the product of interest would involve more than aggregation of the microbe cell. In those cases, the cells should be disintegrated so that the product of interest is released. The product harvesting in these cases is mainly of two types, physical method & Chemical method (Tables 8.1 and 8.2).

TABLE 8.1 Post-Harvest Product Recovery for Intra Cellular Bio-Product: Chemical Means

Method of Product Recovery – Intra-cellular bio-product	
S. No	**Chemical and Biochemical Means**
1	Enzymatic methods
2	Chemical treatment.
3	Osmotic shock

TABLE 8.2 Post-Harvest Product Recovery for Intra Cellular Bio-Product: Physical Means

Method of Product Recovery – Intra-cellular bio-Product	
S. No.	**Physical Means**
1	Ultrasonication by an Ultrasonic homogenizer
2	Bead-mill
3	High-pressure homogenization (HPH) by a French press
4	Electroporation

Extraction and release of target biomass from bulk are done by a variety of means (Table 8.1). The research found that the combinatorial approach could enhance recovery to a great extent. For example, the enzymatic method and the microwave approach shown to enhance the lipid extraction from *Rhodosporidium toruloides* Y4 (Jin et al., 2012). In some cases for cell disruption to release cellular components, enzymatic methods showed good results (Zheng et al., 2011) in comparison to mechanical methods. There are studies that utilize enzymes for releasing cellular components, for example, various enzymes such as cellulase, pectinase, protein hydrolyzing papain, snailase, etc., have been used for arachidonic acid extraction (Young et al., 2011). The alcalase and neutrase have also been used for the degradation of the cell wall of *Mortierella alpine*. However, as biological methods phages or autolysis (Geciova et al., 2002) are some other biological methods that are in use. One of the disadvantages of the enzyme system may be its repeated re-use. A study showed that a serious problem could be the reduced efficacy of recycled immobilized enzymes (Zhang et al., 2010). However, it is also seen that even in tough cells with rigid coats such as chlorella, a mixture of enzymes can even destabilize it (Braun and Aach, 1975).

The method of ultrasonication (González-Fernández et al., 2012) has been used in a variety of bioprocesses including for *Scenedesmus sp.* increasing the biogas fermentation yield up to 4 g/L. According to research ultrasonication generates heat, releases free radicals, and generates shock waves that in cell envelopes of microorganisms can cause damage the leading to the release of intracellular products (Miller et al., 2002). The high-intensity ultrasound aids to form microbubbles. These microbubbles undergo compression and implodes. Fermentation yield is significantly increased by ultrasound up to a frequency of 40 kHz. It was seen in the case of biomass production from *S. obliquus*YSW15 (Yoo et al., 2012). Ultrasonic induced cavitation is also affected by the temperature of the liquid media as well as the viscosity. The low temperature of the system is also favorable for sonolysis in an effective way for micro-organisms (Jiang et al., 2006). Standardization of the ultrasonic cavitation process showed that at the low frequency, it is significantly more intense than that of a high frequency (Cravotto et al., 2008). Interestingly, utilizing simulations of microbubble sizes and microturbulence velocity of ultrasonication it was found that ultrasonication itself increases the extraction yield in comparison to the Soxhlet methods (Ranjan et al., 2010). Ultrasonication in a "solvent-free" extraction system was also successful in separating lipids in the absence of a solvent (Adam et al., 2012). Compared with other methods, when it comes to cell disruption methods ultrasonication can be the best (Prabakaran and Ravindran, 2011) in some algal species. For biodiesel-based biofuel research, a separation method has been standardized by Marrone et al., (2011) which is for the extraction using ultrasounds that include cell residues and lipids.

8.2.2.1 BIOPHYSICAL ASPECTS OF BIOMASS YIELD

During a microbe-based bio-factory or reactor, several biophysical principles govern the yield. These concepts are imperative for the sustainable and optimum output of any bioprocess.

According to growth kinetics in microbe-based bioprocess for biomass production, the "doubling time *(t_d)*" or generation time is an important concept that is explained as the time that is required for doubling the number of cells. Now in a microbial culture where the cells are exponentially growing, the mathematical expression for the number of cells is represented as

$$2^0N_0 \rightarrow 2^1N_0 \rightarrow 2^2N_0 \rightarrow 2^3N_0 \rightarrow 2^nN_0$$

where, "N_0" – initial number of cells; "*n*" – number of generations.

Now, in the standard kinetic model of anaerobic digestion that is coupled with flux-balance analysis (FBA) models of methanogenic species (Anaerobic Digestion Model No. 1 – ADM1) the assessment of microbial growth and decay is described by a Monod-type equation as follows (Sören et al., 2019):

$$\frac{dX}{dt} = \mu(S).X\ .I - kd\ .X, \tag{1}$$

In Eq. (1), X indicates microbial biomass concentration (in gm/Lt); whereas, μ(S) – specific growth rate (substrate concentration-dependent). It is further derived as follows:

$$\mu = \mu_{max} \cdot \frac{\mathrm{S}}{\mathrm{Ks} + \mathrm{S}}$$

Here the terms are: "μ_{max} *(1/d)*" – maximal specific growth rate.

Substrate concentration – "*S*" *(gm/Lt)*. Another term "*Ks* (gm/Lt)" represents a half-saturation constant. In this equation, the product of inhibitory terms is "*I*" have values ranging from 0 (complete inhibition) to 1 (no inhibition). The specific biomass decay rate is "k."

With respect to a lab-scale scenario biomass growth can be assessed by the radius growth rate. The term (*Kr*), expressed in mm/day) is the radius growth rate. This can be an effective estimate of biomass production on the lab scale. It can be expressed as follows:

$$r = Kr.t + r_0, \tag{2}$$

where "*r*" represents colony radius at a time "*t*"; and "r_0" represents colony radius at a time "t_0." The constant growth rate is "*Kr.*"

Once the growth of the biomass reaches a specific threshold, the next step is product recovery. Now if the product of interest is trapped inside the cell then it should be liberated. It has been seen that among several methods to procure the product from biomass, ultrasonication is a good option. The energetics of ultra-sonication again is an important aspect in this context.

The quantitative analysis on the energy supplied to the biomass that needs to be treated is estimated (González-Fernández et al., 2012). Ultrasonic energy that is consumed in this analysis has been estimated as follows:

$$E_S = \frac{P X t}{V X TS_0}$$

where, E_s – the amount of ultrasonic energy consumed; P – ultrasonic power; t – treatment time; V – volume; and TS_0 – concentration of biomass.

Another tool to liberate intracellular contents from the bulk of the biomass can be the use of the electromagnetic field (EF) (Sheng et al., 2011). The authors used the application of pulsed electric field (EF) to release the intracellular contents which can be calculated by the following equation:

$$TI = K\frac{V^2 Df\sigma HRT}{L^2}$$

In this equation, the terms are as follows: TI – the intensity of the PEF, V – the applied voltage, D – the pulse width, f – the pulse frequency, σ – the sample conductivity, L – the distance between the electrodes, HRT – the residence time of the liquid media, and K – a constant for unit conversion. Subsequent purification strategies that have been standardized for biomass capture and to make it into a finished product are relevant to this discussion in the chapter. However, due to the limited scope of the work, "*Purification Principle, Methodologies & Recent Development in Bioprocess Technology*" I will be discussed in future work.

8.2.3 INTERVENTION FOR DESIRED TRAIT TO ENHANCE DOWNSTREAM PROCESSING

The desired trait in a microbe for enhanced bioprocess is the most anticipated goal in industrial biotechnology. If we look at the history of genetic engineering and recombinant DNA technology several notable breakthroughs are noteworthy. For example, Zaslavskaia et al., (2001) in diatom *P. tricornutum* the trophic conversion and transformation of the alga with a single gene that encodes a glucose transporter was achieved. In this context, it can be noted that the cultivation of heterotrophic microalgae in fermenters is of limited application. It is because the microalgae are by and large obligate photoautotrophs. One cannot grow them in a fermenter. But once you can change its trait it can be grown in a fermenter as well. By changing the nature of the photoautotrophs and by enhancing the capacity to use organic molecules as energy and carbon sources the trait of the autotrophs can be changed. By doing so the microalgae may now be used for biosynthesis of many high-value products even by heterotrophic cultivation of these transgenic microorganisms in fermenters. In recent times (2010–2020), there has been rapid development in genetic engineering and recombinant DNA technology.

8.2.3.1 PROTEIN ENGINEERING AND DESIGNER PHENOTYPE FOR MICROBE-BASED BIOMASS PROCESSING

Designer phenotypes in bioprocess technology have huge potential for various applications. From waste biomass, value-added end products are a highly anticipated outcome. An important industrial product is polyunsaturated fatty acids (PUFAs) which is a commercially important nutraceutical with high demand. According to recent research, a novel anaerobic polyketide synthase-like pathway has been discovered in marine organisms (Napier, 2002) that aids in the biosynthesis of polyunsaturated fatty acids (PUFAs). Now the genetic machinery of the polyketide synthase-like pathway can be engineered into an organism for the direct production of PUFA. Generally, for the production of PUFA more than five desaturase and elongase genes are required for the typical aerobic pathway. However, these novel genes of the polyketide synthase-like pathway can be used to direct the synthesis of PUFAs in heterologous hosts. Introducing and regulating the gene cluster of polyketide synthase-like pathway is relatively simple compared to the more than five desaturase and elongase genes required for the typical aerobic pathway. This designer trait in a microorganism makes the synthesis of PUFAs via transgenic organisms practically possible.

However, if we look at it, the phenotypic trait modulation, initially it was in agriculture and biomedical applications. For example (Peters et al., 1981), the genetic engineering of symbiotic N2-fixation and conservation of fixed nitrogen. Several examples for poultry and veterinary feed-related application are also available. Recombinant DNA technologies now allow for the economic production of a variety of feed supplements as well such as microbial phytases. Similarly, applications in other industries also can be mentioned. Increased production of zeaxanthin and other pigments (Lagarde et al., 2000) by application of genetic engineering techniques in *Synechocystis* sp. is also there. As early as 1997, the genetic engineering of resistant traits to bleaching herbicides affecting phytoene desaturase and lycopene cyclase in cyanobacterial carotenogenesis was reported (Windhoevel et al., 1997).

In bioprocess technology and waste recycling, fungi-based methods are very popular. However, removing the biomass from the broth is generally done by energy-intensive methods. The methods are gravity sedimentation, filtration, and flotation as well as centrifugation, and finally flocculation. For large-scale harvesting, it is still a demanding process when it comes to the separation of biomass. Post-recovery extraction of biomass too is a costly affair. Several options have been studied with much detail to reduce the cost of removing the accumulated biomass from the treated water where biofloculation seems to be the most promising option. Here the accumulated biomass forms a 3d structure or flocs. This clump-like entity can be segregated easily. According to reports (Govender et al., 2008) flock formation is inhibited by the negatively charged microalgal surfaces under normal growth conditions that prevent their self-flocculation. To address this by altering the surface charges various approaches are harnessed (Gultom, 2013). Among fungi-based co-culture of microalgae, yeast is also an alternative since the physiology, biochemistry and molecular biology of yeast flocculation is well understood. It has a track record of efficient flocculation and there is a scope that molecular biology and genetic engineering are harnessed in this regard. Already cell flocculation has been widely employed as a simple and cheap method to harvest yeast biomass from the fermentation broth (Li et al., 2015) and *S. cerevisiae* has also been used in environmental applications like industrial wastewater treatment (Machado et al., 2008). *Saccharomyces cerevisiae* yeast cells constitute a good alternative for wastewater treatment, mainly because of its availability (can be obtained in a high amount at low cost as a by-product of fermentation industries) and capacity to accumulate a broad variety of heavy metals under a wide range of external conditions (Manheim & Nelson, 2013).

8.2.3.2 MOLECULAR BIOLOGY AND GENETIC ENGINEERING FOR MICROBE-BASED BIOMASS PROCESSING

In microbe-based biomass processing, molecular biology and genetic engineering interventions can enhance biomass separation. In this regard fungi and algae-based genetic engineering can open a new vista for enhanced downstream processing.

In a pilot study in chloroplast engineering, it was reported that(Mayfield et al., (2003) attempts of protein expression based on *Chlamydomonas reinhardtii*, the unicellular green alga. In this work, the used chloroplast-targeted transgenes to express an antibody. The replacement of the chloroplast *psbA* gene with the chimeric *psbA-transgene Chimera was also reported (*Manuell et al., 2007*). This intervention* results in the accumulation of recombinant protein more than that of the total protein up to above 5% which approaches that of other robust expression systems such as bacteria. Here, replacement of the chloroplast *psbA* gene with the chimeric *psbA-transgene Chimera* results in the accumulation of recombinant protein above 5% of total protein. This level of recombinant protein accumulation approaches that of other robust expression systems, including bacteria.

According to research studies high copy number of the trans-gene in a typical photosynthetic cell can be expressed in operons (Fischer et al., 2004). As there is little or no post-translational modification that may alter a protein's functional aspect chloroplast-expressed protein, contains the correct amino-terminal sequence (Manuell et al., 2007).

The SDN Lab of Department of Medical Laboratory Technology, Women's Polytechnic, Govt. of Tripura has proposed a strategy to enhance the flocculation phenotype by a transgene integration through homologous recombination of a FLO Cassette (an operon of Flocculation genes from *Saccharomyces* to *Chlamydomonas reinhardtii.* In this work (Debnath, 2019), a detailed strategy is discussed for this genetic engineering intervention for enhanced flocculation phenotype. The specific steps that should be emphasized may be noted as below (Figure 8.7):

8.2.4 COMPUTATIONAL BIOLOGY FOR COST-EFFECTIVE STANDARDIZATION OF WET LAB PROTOCOLS ON ENGINEERING OF DESIRED TRAIT TO ENHANCE DOWNSTREAM PROCESSING

Due to the near precise prediction capabilities of a bio-molecular phenomenon and its economic operational costs, computational simulations in the microprocessor environment become a method of choice (Debnath and Addya, 2015). Computational simulations are easy to perform. Also, it requires minimal investments in terms of infrastructure. However, they are sensitive enough to differentiate the intricacies among protein candidates that may aid to prioritize the suitable proteins for molecular biology-genetic engineering experiments. Therefore, this process helps to minimize the cost of the overall trial and error expenditure. Figure 8.8a depicts the major steps that are required for a general pipeline to use computational simulations for cost-cutting of genetic engineering and molecular biology experimental trial and errors.

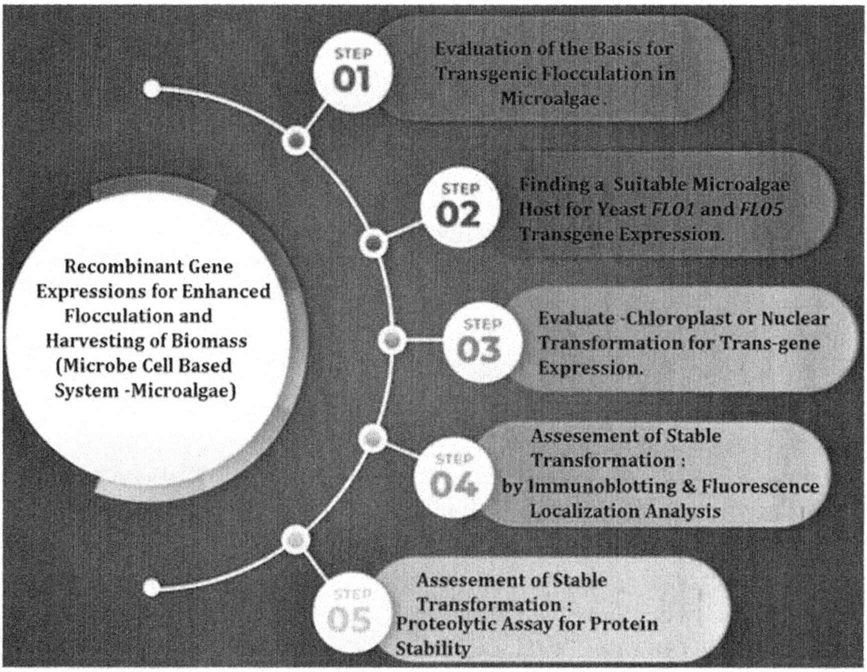

Figure 8.7. Strategy for engineering enhanced flocculation in microbe-based bioprocess (Debnath, 2019).

Among computational simulation, structural analysis of proteins is a vital technique. Structure analysis of proteins is done by X-ray crystallography or NMR spectroscopy, both of which are methods that require substantial skill, infrastructure as well as a financial investment. In 1958, the first crystal structure of a protein was reported. The structure for myoglobin in 3D was resolved at 6 A° (23). Macromolecular crystallography has come a long way in the half-century and advancement in instrumentation as well as computational processing capacity of a large amount of data was the key factor for this. Protein Data Bank (PDB) global repository for crystal structures (and later structural models obtained by NMR spectroscopy, fiber diffraction, electron microscopy, and some other techniques) provided a unique resource for the scientific community.

Due to the advantage in terms of structure resolution, X-Ray crystallography is the preferred method for deducing protein structure. To make out the operational complexity in structural studies by X-ray crystallography, we shall briefly see the principle involved in X-Ray crystallography.

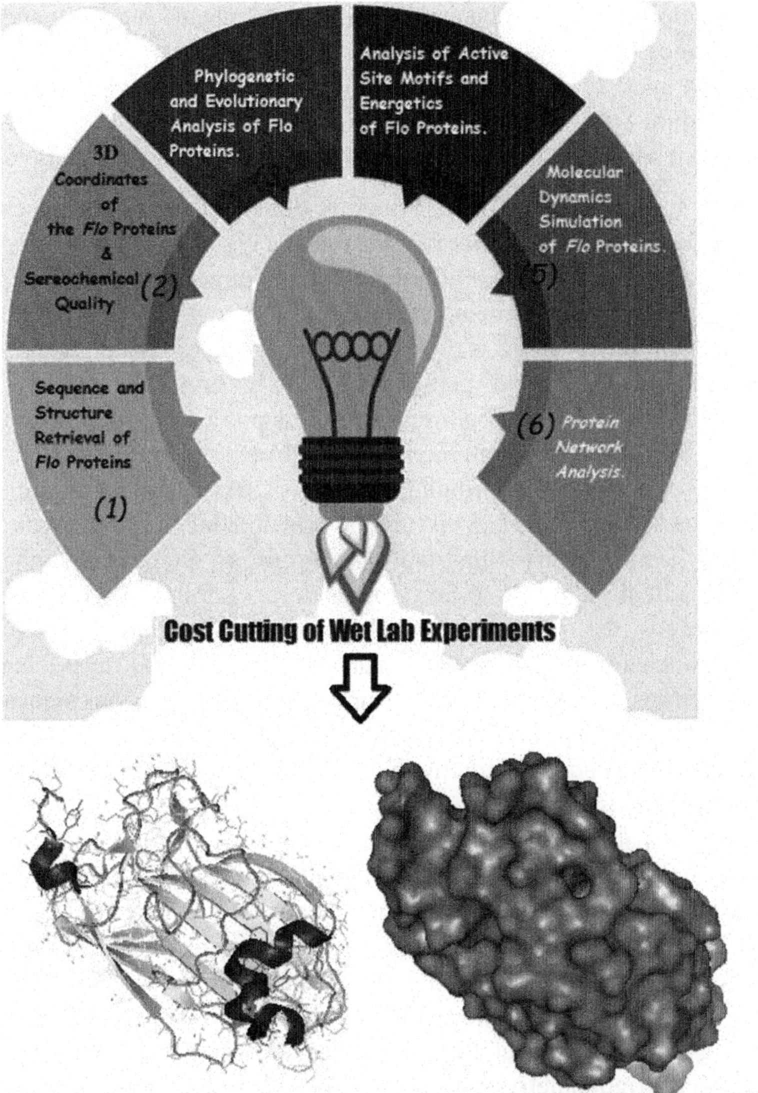

(2D Structure of *FLO1* in Helix-Sheet) (*FLO1* 3D Surface Structure)

FIGURE 8.8a Cost-cutting through computational simulation for enhanced flocculation in microbe-based bioprocess.

X-Ray crystallography is performed in crystals (Figure 8.8b, A) and the criteria for a good crystal is:

i) If it is organized in an ordered fashion in a crystal lattice. Generally, crystals are regular arrays of atoms (Figure 8.8b, B).

ii) A good crystal will be arranged in crystal planes and the planes are formed by an arrangement of atoms in the crystal molecule.
iii) In a good crystal the crystal planes are separated by a distance "d" (Figure 8.8b, B). The crystal planes are also called the scatterers as these are the planes from which the incident X-ray gets scattered.

Atoms scatter X-ray waves, primarily through the atoms' electrons. Therefore in a good and symmetrical and ordered crystal, the scattering of the X-ray will also be symmetrical or ordered. This symmetry or the order can be easily analyzed. For the protein, crystallography X-rays are chosen because the wavelength (λ) of X-rays is typically the same order of magnitude (1–100 angstroms) as the spacing "d" between planes in the crystal. In principle, any wave impinging on a regular array of scatterers produces diffraction. But to produce significant diffraction, the spacing between the scatterers (i.e., "d") and the wavelength (λ) of the impinging wave should be similar in size.

In this background, we can discuss the principle behind X-ray scattering and its application in crystallography. According to Bragg's law when an X-ray is incident on a regular array of scatterers (atoms on the plane that scatter X-ray waves, primarily through the atoms' electrons), the X-ray is scattered much like reflection. The scattered (or reflected) rays generated by the electrons of the atoms present on the symmetric crystal planes will cancel one another out, in most directions through destructive interference. However, they can add constructively in a few specific directions if specific conditions are satisfied.

When the spacing between two successive planes of atoms (or electrons) is "d" (Figure 8b, B) and the X-Ray is incident (Black ray, Figure 8b, C) on the plane of atoms is "θ" the X-Ray is reflected at an angle "θ." (Red ray, Figure 8.8b, C). The hypothetical extended straight line (Black broken lines in Figure 8.8b, C) of the incident ray makes an angle "2θ" with the reflected ray. The extended straight line (Red broken line) makes an angle "θ" with that of the spacing "d" (Blue Straight Line, Figure 8.8b, C). From the "Sin" convention of trigonometry, we find "$dSin\theta$" (green, Figure 8b, C).

Now, interference of the scattered rays will take place (Figure 8.8b, D). This interference is possible if the path-length difference of the reflected or scattered rays is " 2dsin θ."

Moreover, constructive interference is only possible if the " $2dsin\ \theta$" is equal to an integral number of wavelengths of the incident ray. Therefore according to Bragg's law, constructive and analyzable interference of the reflected or scattered X-Rays can be obtained if "$n\lambda = 2\ d\ sin\theta$," where "$n$" is an integral number of wavelengths "λ."

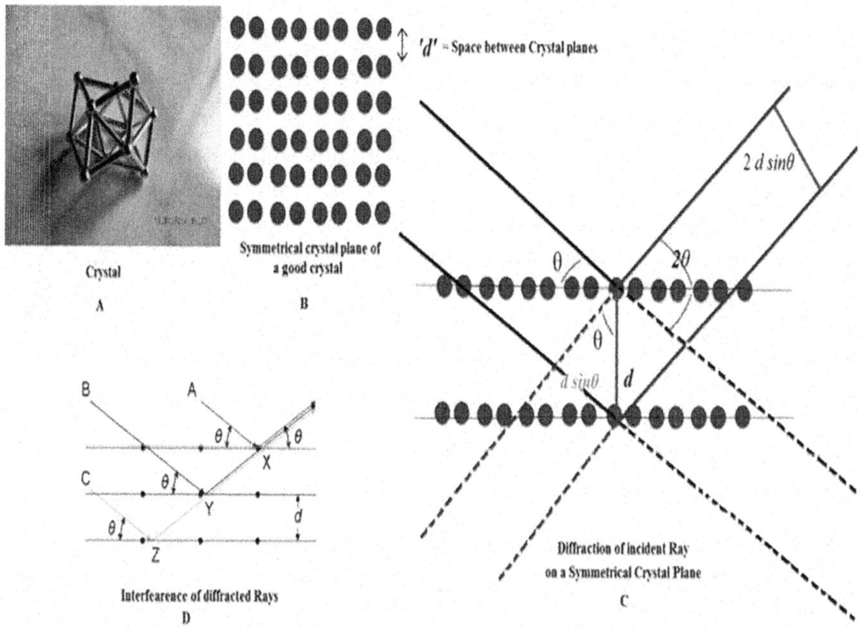

FIGURE 8.8b Cost-cutting through computational simulation–principle of X-Ray crystallography.

8.3 RECENT DEVELOPMENT IN THE APPLICATIONS OF BIOTECHNOLOGY FOR VALUE-ADDED BIOMASS YIELD

Biotechnology for value-added biomass yield has several aspects; for example, microbial cell factories. Biochemical production of chemicals and value-added compounds by microbial cell factories requires carbon sources. Some non-conventional microbes such as *Yarrowia lipolytica,* are able to catabolize non-conventional carbon sources such as alkanes by alkane monooxygenases (Takai et al., 2012). The yeast *Yarrowia lipolytica* can even use hydrophobic substrates (Fickers et al., 2005). For a variety of consumer products, bioprocess engineering of microalgae is in use (Harun et al., 2010). In biotechnological processes, yeast flocculation is also important among microbe-based technologies (Domingues et al., 2000). In earlier studies of microbe-based systems, continuous ethanol production by flocculating strains was a significant bioprocess (Teixeira et al., 1990). Alcoholic fermentation from cellulose biomass was a breakthrough (Okonkwo et al., 2016). Along with cellulose, uncommon carbohydrates such as Inulin were also attempted for bioconversion (Chi et al., 2011). The use of recombinant DNA technology for microbe-based bioprocess opened a whole new dimension (Darvishi et al., 2017). Fungi-based treatment of various spent mediums and recycling is one of the applications of bioprocess technology (Hultberg et al., 2017). Recycling of wastes from industry has been a primary motive of this technology (Kerby, 2017). Environmental application of bioprocess technology also involves remediation process and bio-leaching (Zhang and Elser, 2017; Munir et al., 2017). Another aspect of bioprocess technology and downstream processing for industrial application has been linked with nutrition and public health (Ween et al., 2017). Therefore, it can be opined that biotechnology for biomass recovery and downstream processing for efficient product recovery may impact mankind in several ways, and it's up to us to harness this technology for the sustainable and sound development of the whole biosphere.

8.4 CONCLUSION

Innovation in biotechnology today has become more imperative than ever in human history. When the first fermentation was done we recognized its importance only for food and beverages, but today biotechnology became necessary even for surviving in the ever-changing planet. In this chapter, my focus has been an overall discussion about bioprocess technology and some necessary mention of important concepts associated with it. It is also worth mentioning that today the academic fraternity is constantly being challenged to bring about impactful innovation even without having proper resources. Therefore, a constant remolding of one's capabilities and coping with challenges in a resource-limited setup has become mandatory. In this chapter, some of the concepts and considerations of bio-process technology has been discussed. The dynamics of cellular biomass yield with some light on the basic understanding of kinetics of large-scale and small-scale biomass yield has been attempted. Aspects of industrial waste biomass re-utilization was also discussed. A significant chunk of the discussion is on biophysical aspects of biomass yield. Later on, the text describes the Interventions for desired trait to enhance downstream processing. The topic of protein engineering and designer phenotype for microbe-based biomass processing has also been touched. Discussion was also made on the molecular biology and genetic engineering for microbe based biomass processing. For cost effective experimentation application of Computational Biology has been advocated for standardization of wet lab protocols for enhance downstream processing. Computational biology fundamentally simulates real-life bio-molecular phenomena of diverse kinds (Debnath and Addya, 2015). In this context, the principle of X-ray crystallography a method that is used for the analysis of enzymes and proteins in terms of its 3D structure was discussed as well. If the chapter helps the learners to appreciate the overall view of bioprocess technology in simple terms the endeavor will be fruitful.

ACKNOWLEDGMENT

The associates, interns, and students who are trained in my lab for their courses, dissertation as well as internships are acknowledged. The researchers who have cited my works and whom I have cited for my literature were major motivation for me to continue my research in a resource limited set-up.

KEYWORDS

- **bio-processes**
- **bioinformatics**
- **biotechnology**
- **desired phenotype**
- **downstream processing**
- **harvesting**
- **high-value biomass**
- **molecular biology**

REFERENCES

Adam, F., Abert-Vian, M., Peltier, G., & Chemat, F., (2012). "Solvent-free" ultrasound-assisted extraction of lipids from fresh microalgae cells: A green, clean and scalable process. *Bioresource Technology, 114*, 457–465.

Baranyi, J., Roberts, T. A., & McClure, P., (1993). A non-autonomous differential equation to model bacterial growth. *Food Microbiology, 10*(1), 43–59.

Baruah, J., Nath, B. K., Sharma, R., Kumar, S., Deka, R. C., Baruah, D. C., & Kalita, E. (2018). Recent trends in the pretreatment of lignocellulosic biomass for value-added products. *Front Energy Res., 6*, 1–19.

Braun, E., & Aach, H. G., (1975). Enzymatic degradation of the cell wall of chlorella. *Planta, 126*(2), 181–185.

Chi, Z. M., Zhang, T., Cao, T. S., Liu, X. Y., Cui, W., & Zhao, C. H., (2011). Biotechnological potential of inulin for bioprocesses. *Bioresource Technology, 102*(6), 4295–4303.

Cravotto, G., Boffa, L., Mantegna, S., Perego, P., Avogadro, M., & Cintas, P., (2008). Improved extraction of vegetable oils under high-intensity ultrasound and/or microwaves. *Ultrasonics Sonochemistry, 15*(5), 898–902.

Dantigny, P., Guilmart, A., & Bensoussan, M., (2005). Basis of predictive mycology. *International Journal of Food Microbiology, 100*(1–3), 187–196.

Darvishi, F., Moradi, M., Madzak, C., & Jolivalt, C., (2017). Production of laccase by recombinant *Yarrowia lipolytica* from molasses: Bioprocess development using statistical modeling and increase productivity in shake-flask and bioreactor cultures. *Appl. Biochem. Biotechnol., 181*, 1228–1239.

Debnath, S., & Addya, S., (2015). In-silico modelling of SERAC1: Protein involved in a developmental neural disorder MEGDEL syndrome characterized by 3-methyl glutaconic aciduria type IV with sensory-neural deafness, encephalopathy and Leigh-like syndrome. *International Journal of Developmental Neuroscience, 47*, 1–2.

Debnath, S., & Addya, S., (2015). Mis-sesnse mutations in tafazzin (TAZ) that escort to mild clinical symptoms of Barth syndrome is owed to the minimal inhibitory effect of the mutations on the enzyme function: In-silico evidence. Interdisciplinary Sciences. *Computational Life Sciences, 7*(1), 21–35.

Debnath, S., (2019). Characterization of extracellular proteins to explore their role in bio-flocculation for harvesting algal biomass for wastewater treatment. *The Role of Microalgae in Wastewater Treatment*, 229–266. Springer, Singapore.

Domingues, L., Vicente, A. A., Lima, N., & Teixeira, J. A., (2000). Applications of yeast flocculation in biotechnological processes. *Biotechnology and Bioprocess Engineering, 5*(4), 288–305.

Fickers, P., Benetti, P. H., Waché, Y., Marty, A., Mauersberger, S., Smit, M. S., & Nicaud, J. M., (2005). Hydrophobic substrate utilization by the yeast *Yarrowia lipolytica*, and its potential applications. *FEMS Yeast Res., 5*, 527–543.

Fischer, R., Stoger, E., Schillberg, S., Christou, P., & Twyman, R. M., (2004). Plant-based production of biopharmaceuticals. *Current Opinion in Plant Biology, 7*(2), 152–158.

Geciova, J., Bury, D., & Jelen, P., (2002). Methods for disruption of microbial cells for potential use in the dairy industry—A review. *Int. Dairy J., 12*, 541–553.

González-Fernández, C., Sialve, B., Bernet, N., & Steyer, J. P., (2012). Comparison of ultrasound and thermal pretreatment of Scenedesmus biomass on methane production. *Bioresource Technol., 110*, 610–616.

Govender, P., Domingo, J. L., Bester, M. C., Pretorius, I. S., & Bauer, F. F., (2008). Controlled expression of the dominant flocculation genes FLO1, FLO5, and FLO11 in *Saccharomyces cerevisiae. Applied and Environmental Microbiology, 74*(19), 6041–6052.

Grando, R. L., De Souza, A. A. M., Da Fonseca, F. V., Sánche, A., & Barrena, R., (2017). Font X. Technology overview of biogas production in anaerobic digestion plants: A European evaluation of research and development. *Renewable and Sustainable Energy Reviews., 80*, 44–53.

Gultom, S. O., & Hu, B., (2013). Review of microalgae harvesting via co-pelletization with filamentous fungus. *Energies, 6*(11), 5921–5939.

Hanemaaijer, M., Röling, W. F., Olivier, B. G., Khandelwal, R. A., Teusink, B., & Bruggeman, F. J., (2015). Systems modeling approaches for microbial community studies: From metagenomics to inference of the community structure. *Frontiers in Microbiology, 6*, 213.

Harun, R., Singh, M., Forde, G. M., & Danquah, M. K., (2010). Bioprocess engineering of microalgae to produce a variety of consumer products. *Renewable and Sustainable Energy Reviews, 14*(3), 1037–1047.

Hultberg, M., & Bodin, H., (2017). Fungi-based treatment of brewery wastewater—biomass production and nutrient reduction. *Appl. Microbiol. Biotechnol., 101*, 4791–4798.

Jegannathan, K. R., Abang, S., Poncelet, D., Chan, E. S., & Ravindra, P., (2008). Production of biodiesel using immobilized lipase: A critical review. *Crit. Rev. Biotechnol., 28*, 253–264.

Jiang, Y., Petrier, C., & Waite, T. D., (2006). Sonolysis of 4-chlorophenol in aqueous solution: Effects of substrate concentration, aqueous temperature and ultrasonic frequency. *Ultrason. Sonochem., 13*, 415–422.

Jin, G., Yang, F., Hu, C., Shen, H., & Zhao, Z. K., (2012). Enzyme-assisted extraction of lipids directly from the culture of the oleaginous yeast *Rhodosporidium toruloides*. *Bioresource Technol., 111*, 378–382.

Kerby, C., & Vriesekoop, F., (2017). An overview of the utilization of brewery by-products as generated by British craft breweries. *Beverages, 3*, 24.

Kim, D. G., La, H. J., Ahn, C. Y., Park, Y. H., & Oh, H. M., (2011). Harvest of Scenedesmus sp. with bioflocculant and reuse of culture medium for subsequent high-density cultures. *Bioresour. Technol., 102*, 3163–3168.

Klimmek, S., Stan, H. J., Wilke, A., Bunke, G., & Buchholz, R., (2001). Comparative analysis of the biosorption of cadmium, lead, nickel, and zinc by algae. *Environmental Science & Technology, 35*(21), 4283–4288.

Kotrba, P., & Ruml, T., (2000). Bioremediation of heavy metal pollution exploiting constituents, metabolites and metabolic pathways of livings. A review. *Collection of Czechoslovak Chemical Communications, 65*(8), 1205–1247.

Lagarde, D., Beuf, L., & Vermaas, W., (2000). Increased production of zeaxanthin and other pigments by application of genetic engineering techniques to *Synechocystis* sp. strain PCC 6803. *Applied and Environmental Microbiology, 66*(1), 64–72.

Li, T., Lin, G., Podola, B., & Melkonian, M., (2015). Continuous removal of zinc from wastewater and mine dump leachate by a microalgal biofilm PSBR. *Journal of Hazardous Materials, 297*, 112–118.

Machado, M. D., Santos, M. S., Gouveia, C., Soares, H. M., & Soares, E. V., (2008). Removal of heavy metals using a brewer's yeast strain of *Saccharomyces cerevisiae*: The flocculation as a separation process. *Bioresource Technology, 99*(7), 2107–2115.

Manheim, D., & Nelson, Y., (2013). Settling and bioflocculation of two species of algae used in wastewater treatment and algae biomass production. *Environmental Progress & Sustainable Energy, 32*(4), 946–954.

Manuell, A. L., Beligni, M. V., Elder, J. H., Siefker, D. T., Tran, M., Weber, A., McDonald, T. L., & Mayfield, S. P., (2007). Robust expression of a bioactive mammalian protein in *Chlamydomonas* chloroplast. *Plant Biotechnology Journal, 5*(3), 402–412.

Marrone, B. L., Coons, J., Dale, T., & Kalb, D., (2011). Extraction of lipids from algae using ultrasonic fields. *Algae Biomass Summit*. Minneapolis, Minnesota, United States.

Mayfield, S. P., Franklin, S. E., & Lerner, R. A., (2003). Expression and assembly of a fully active antibody in algae. *Proc. Natl. Acad Sci. USA, 100*, 438–442.

Miller, D., Pislaru, S., & Greenleaf, J., (2002). Sonoporation: Mechanical DNA delivery by ultrasonic cavitation. *Somat. Cell Molec. Gen., 27*, 115–134.

Miller, M. W., Miller, D. L., & Brayman, A. A., (1996). A review of *in vitro* bioeffects of inertial ultrasonic cavitation from a mechanistic perspective. *Ultrasound Med. Biol., 22*, 1131–1154.

Munir, M. T., Li, B., Boiarkina, I., Baroutian, S., Yu, W., & Young, B. R., (2017). Phosphate recovery from hydrothermally treated sewage sludge using struvite precipitation. *Biores. Technol., 239*, 171–179.

Napier, J. A., (2002). Plumbing the depths of PUFA biosynthesis: A novel polyketide synthase-like pathway from marine organisms. *Trends in Plant Science, 7*(2), 51–54.

Neori, A., & Shpigel, M., (2002). Mariculture sustainability by integration of algae/algivores to fish/shrimp ponds. In: *The 1st Congress of the International Society for Applied Phycology* (Vol. 26, p. 30). Aquadulce, Roquetas de Mar, Almeria (Spain). Universidad de Almeria Servicio de Publicaciones.

Okonkwo, C. C., Azam, M. M., Ezeji, T. C., & Qureshi, N., (2016). Enhancing ethanol production from cellulosic sugars using scheffersomyces (*Pichia*) stipitis. *Bioprocess. Biosyst. Eng., 39*, 1023–1032.

Peters, G. A., Ito, O., Tyagi, V. V., & Kaplan, D., (1981). Physiological studies on N 2-fixing azolla. In: *Genetic Engineering of Symbiotic Nitrogen Fixation and Conservation of Fixed Nitrogen* (pp. 343–362). Springer, Boston, MA.

Prabakaran, P., & Ravindran, A. D., (2011). A comparative study on effective cell disruption methods for lipid extraction from microalgae. *Lett Appl. Microbiol., 53*, 150–154.

Ranjan, A., Patil, C., & Moholkar, V. S., (2010). Mechanistic assessment of microalgal lipid extraction. *Ind. Eng. Chem. Res., 49*, 2979–2985.

Simate, G. S., Cluett, J., Iyuke, S. E., Musapatika, E. T., Ndlovu, S., Walubita, L. F., & Alvarez, A. E., (2011). The treatment of brewery wastewater for reuse: State of the art. *Desalination., 273*, 235–247.

Sören, W., Sabine, K., Fabian, B., Denny, P., Dirk, B., Steffen, K., & Florian, C., (2019). Augmenting biogas process modeling by resolving intracellular metabolic activity. *Front Microbiol., 10*, 1095. doi: 10.3389/fmicb.2019.01095.

Takai, H., Iwama, R., Kobayashi, S., Horiuchi, H., Fukuda, R., & Ohta, A., (2012). Construction and characterization of a *Yarrowia lipolytica* mutant lacking genes encoding cytochromes P450 subfamily 52. *Fungal Genet. Biol., 49*, 58–64.

Teixeira, J. A., Mota, M., & Goma, G., (1990). Continuous ethanol production by a flocculating strain of *Kluyveromyces marxianus*: Bioreactor performance. *Bioprocess Engineering, 5*(3), 123–127.

Ween, O., Stangeland, J. K., Fylling, T. S., & Aas, G. H., (2017). Nutritional and functional properties of fishmeal produced from fresh byproducts of cod (*Gadus morhua L.*) and saithe (*Pollachius vir ens*). *Heliyon, 3*, e00343.

Windhövel, U., Sandmann, G., & Böger, P., (1997). Genetic engineering of resistance to bleaching herbicides affecting phytoene desaturase and lycopene cyclase in cyanobacterial carotenogenesis. *Pesticide Biochemistry and Physiology, 57*(1), 68–78.

Wu, Z., Zhu, Y., Huang, W., Zhang, C., Li, T., & Zhang, Y., (2012). Evaluation of flocculation induced by pH increase for harvesting microalgae and reuse of flocculated medium. *Bioresour. Technol., 110*, 496–502.

Yoo, G., Park, W. K., Kim, C. W., Choi, Y. E., & Yang, J. W., (2012). Direct lipid extraction from wet *Chlamydomonas reinhardtii* biomass using osmotic shock. *Bioresource Technol., 123*, 717–722.

Young, G., Nippen, F., Titterbrandt, S., & Cooney, M. J., (2011). Direct transesterification of biomass using an ionic liquid co-solvent system. *Biofuels., 2*, 261–266.

Zaslavskaia, L. A., Lippmeier, J. C., Shih, C., Ehrhardt, D., Grossman, A. R., & Apt, K. E., (2001). Trophic conversion of an obligate photoautotrophic organism through metabolic engineering. *Science, 292*(5524), 2073–2075.

Zhang, J., & Elser, J. J., (2017). Carbon:Nitrogen:Phosphorus stoichiometry in fungi: A meta-analysis. *Front Microbiol., 8*, 1281.

Zhang, X., Hua, Q., Sommerfeld, M., Puruhito, E., & Chen, Y., (2010). Harvesting algal biomass for biofuels using ultrafiltration membranes. *Bioresour. Technol., 101*, 5297–5304.

Zheng, H., Yin, J., Gao, Z., Huang, H., Ji, X., & Dou, C., (2011). Disruption of *Chlorella vulgaris* cells for the release of biodiesel-producing lipids: A comparison of grinding, ultrasonication, bead milling, enzymatic lysis, and microwaves. *Appl. Biochem. Biotech., 164*, 1215–1224.

CHAPTER 9

Production of Commercial Products by Vermiculture and Vermicomposting

DOWLURU S. V. G. K. KALADHAR[1] and TANTRAVAHI SRINIVASAN[2]

[1]*Department of Microbiology and Bioinformatics, Atal Bihari Vajpayee University, Bilaspur, Chhattisgarh, India, E-mail: dkaladhar@gmail.com*

[2]*Department of Botany, Indira Gandhi National Tribal University, Amarkantak, Madhya Pradesh, India*

ABSTRACT

Vermiculture is the cultivation of annelid worms like earthworms or bloodworms that are used for the purpose of bait or in composting. Vermicomposting is the process of production of vermicompost that contains several water-soluble nutrients, excellent natural nutrient-rich organic fertilizer, and a soil conditioner. Production systems like batch reactors (containers on legs or the ground), stacked bins or containers, windrow systems on the concrete or the ground, and continuous flow reactors are generally designed for the bulk production of vermiculture and vermicompost. Some of the commercial products obtained in vermiculture and vermicomposting are the production of vermicompost and its products, plastic degradation, bioturbation, worms for bait or feed, and medicines.

9.1 INTRODUCTION

Vermicompost (or vermicompost/casting) is the product formed due to the process of decomposition of vegetable or food waste, vermicast, and bedding materials, by several types of species of earthworms like white worms, red

Biotechnology for Waste Biomass Utilization. Prakash Kumar Sarangi & Latika Bhatia (Eds.)

wigglers, etc. (Kalam and Ahmad, 2017; Yuvaraj and Mahendran, 2020; Yadav et al., 2014). The process of production of vermicompost is called vermicomposting. Vermicomposting is a biotechnological mesophilic process that improves the conversion of the waste material to produce better and useful end products. The process of rearing worms for vermicomposting is called vermiculture (Lim et al., 2015). During these processes, several commercial products will be formed that is very much useful for the daily living process for humans. Worms need minerals, vitamins, protein, and carbohydrates and most of the feeds should contain cellulose and carbohydrates than the protein (Ng et al., 2001; Musyoka et al., 2019; Calvert, 1979). A carbon-to-nitrogen ratio should be 30:1 parts for the good growth of worms in the production system. Earthworms play an essential role in chemical, biological, and physical processes in the maintenance of ecological parameters.

Castings are the pure worm waste produced from earthworms (Hidalgo and Harkess, 2002). Vermicompost includes the castings, bedding, and organic material decomposition, along with the living process of worms, and other worm bin creatures. The species used in vermicomposting are *Eisenia fetida* (tiger worm or red wiggler), *Eisenia hortensis* (European nightcrawlers), *Lumbricus rubellus*, *Eudrilus eugeniae* (African Nightcrawlers), and *Perionyx excavatus* (Blue worms) (Figure 9.1). The most common worms that are used in composting systems, red worms (*Eisenia andrei, Lumbricus rubellus*, and *Eisenia fetida*) feed most rapidly at temperatures between 15–25°C (about 59–77°F). The worms can survive from 10°C (or 50°F) to 30°C (or 86°F). Worms are less susceptible to diseases and very sensitive to environmental conditions. The worms like *Perionyx excavates, Lumbricus rubellus*, etc., are suitable for the warmer climates. A worm bin should be kept in a protected area without direct sunlight when kept outside and should be shielded against frostiness in the winter. The ecosystem for the growth of worms in vermiculture and vermicomposting require digesting organic wastes like eggshells, fruits, fruit peels, vegetables, cracker, cereal, tea bags, coffee grounds, coffee filters, grains like bread, leaves, grass clippings, newspapers, and/or paper toweling (Singh, 2020; Saba et al., 2019; Kaladhar, 2018; Yuvaraj and Mahendran, 2020).

The agricultural fields treated with the fungicides like thifluzamide and excess pesticides will cause changes in the soil ecosystem and damage fertility due to the killing of surface living earthworms (Kumar et al., 2020; Yao et al., 2020) (Figure 9.2). Hence, vermiculture and vermicomposting have gained importance to dominate agricultural toxins and increase fertility for the development of the productivity of agriculture. If the worms die due to fungicides, pesticides, or herbicides, the bed becomes acidic and the gases like methane, NO_2, etc., are released into the bedding.

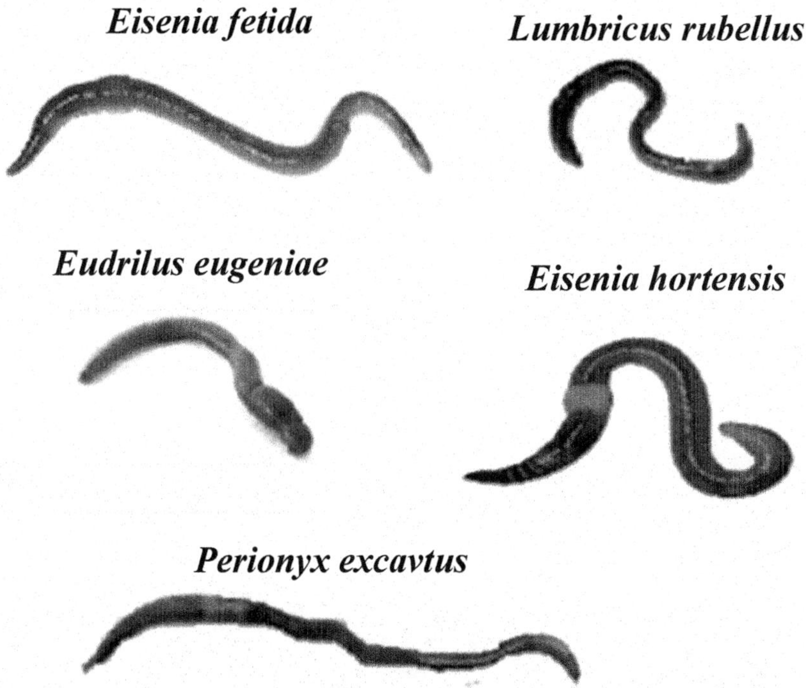

FIGURE 9.1 Some species of worms used in vermiculture and vermicomposting.

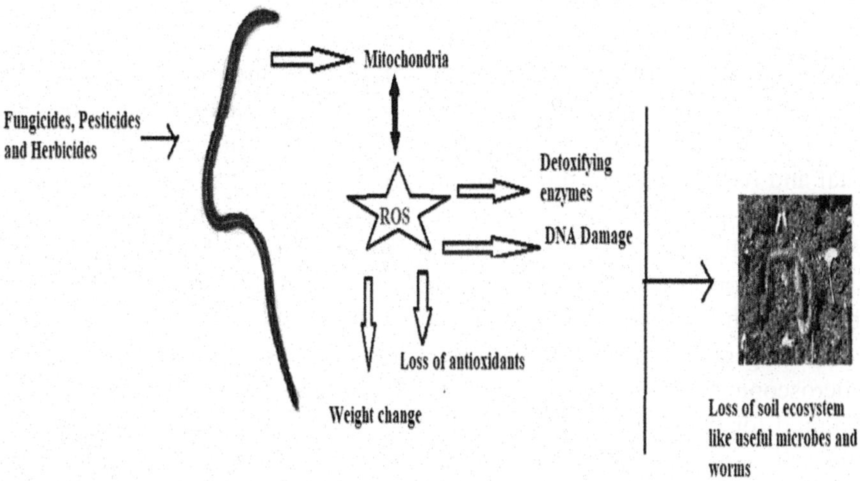

FIGURE 9.2 Process of losing soil ecosystem.

9.2 LIFE CYCLE OF WORMS

Earthworms show direct development and do not undergo metamorphosis (Adrianov and Maiorova, 2010; Venter and Reinecke, 1988). There are a few stages in the life cycle of an earthworm. The fertilized eggs are present in a protective form called a cocoon. A fertilized cocoon may hold around one to about 20 juvenile earthworms, which varies according to the type of species and external conditions. The hatchlings (or baby worms) emerge from the cocoon and burrow into the soil, where baby worms grow into the juvenile and then convert to mature worms about 10 weeks (Figure 9.3).

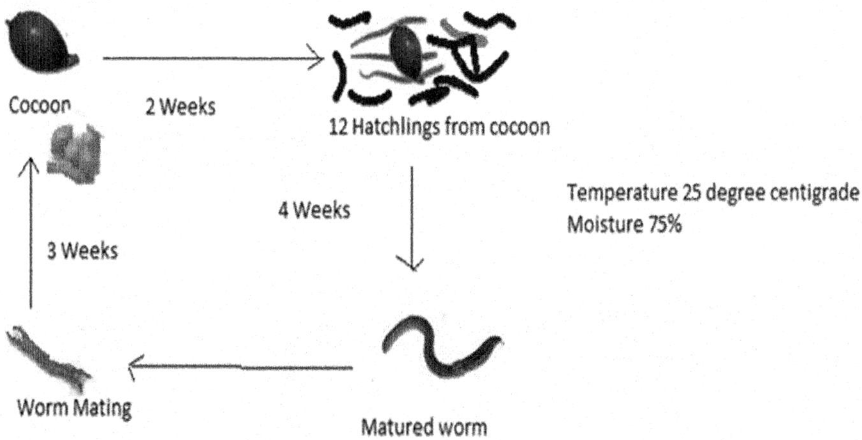

FIGURE 9.3 Lifecycle of earthworms.

Worms can survive a wide range of temperatures between 55- and 77-degrees Fahrenheit (or 13 to 25°C). The worms need organic substrate or moist bedding and convert it into castings. Moisture and oxygen are vital and require good bedding for the growth of worms. The worm's skin is generally photosensitive, and hence, worms require a dark environment. Worms in the containers should include higher nutrient breakdown, better air circulation, less soil compaction, and maintenance of slightly acidic soil that is beneficial for most plants like herbs, shrubs, and trees. Worms do not live in separation but are mutualism or interspecies with microorganisms. Microscopic organisms like bacteria and a few fungi are compulsorily required for the living conditions of earthworms. The most common pests present in worm bins are ants, fruit flies, snails, spiders, pill bugs, soldier fly larvae, centipedes, salamanders, frogs, and rodents.

9.3 PRODUCTION PROCESS FOR VERMICULTURE AND VERMICOMPOST

A large wooden box or pit of 30 cm deep has to be kept or made in a shady area (Gajalakshmi and Abbasi, 2004). A net or a mesh has to be made for culturing worms. Wastes like fruit peels, dried animal dung, vegetable or fruit wastes, husk, coffee powder, tea powder, weeds, old newspapers have to be kept and should spread a layer of soil on the above wastes. Water should be sprinkled to make a wet base and should avoid excess water. Some redworms should be placed in the box or a pit containing wastes and soil. After 30 days, mix and move the top layer and the pit should be covered with gunny bags or grass. The compost will be finally ready after 4 weeks (Figure 9.4). The food wastes should be kept in corners so that all the worms move towards corners and compost in the middle. The vermicompost should be removed and made dry in the sun. The worms can be reused for vermiculture or can be used as bait or feed and medicines.

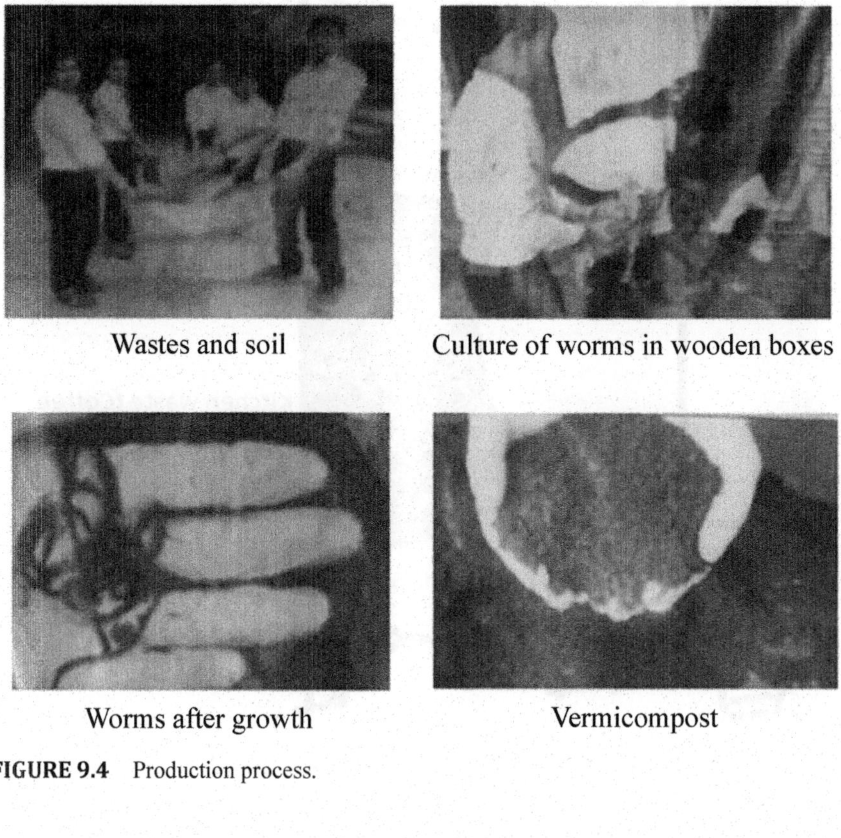

Wastes and soil — Culture of worms in wooden boxes

Worms after growth — Vermicompost

FIGURE 9.4 Production process.

In-vessel continuous flow closed systems have been designed for the production of a bulk product of vermicompost through vermiculture (Dominguez and Edwards, 2010). Food and additional bedding like newspaper, coffee grounds, peelings, leaves, fruit and vegetable cores, tea, bread, spoiled rice, etc. from kitchen wastes can be added from the above to encourage the worms to continue the feeding from the upwards layer in the vessel. The casting harvest and drain mechanism are removed from the bottom of the system or vessel. About one pound of earthworms that contain approximately 1,000 adults can multiply fast up to 2,000–3,000 in favorable conditions like space, bedding material, moisture, temperature, pH, and amount of food available. Production systems are grouped as batch reactors (containers on legs or the ground), stacked bins or containers, windrow systems on the concrete or the ground, and continuous flow reactors (Figure 9.5).

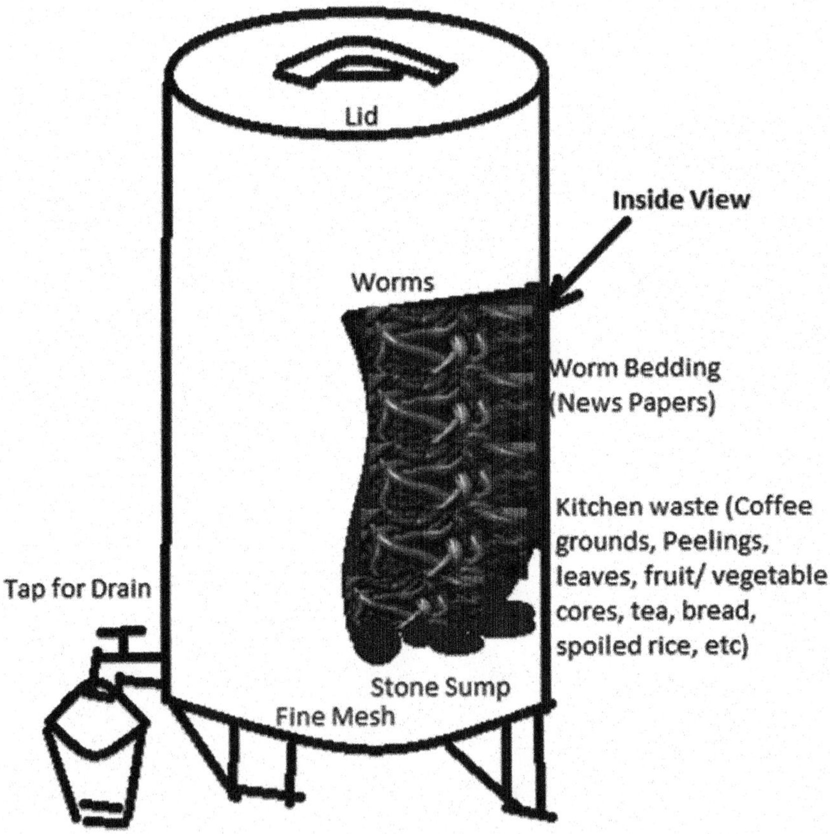

FIGURE 9.5 In-vessel continuous flow closed system for vermicompost.

9.4 COMMERCIAL PRODUCTS PRODUCE IN VERMICULTURE AND VERMICOMPOST

Vermiculture is a new emerging industry that provides a variety of commercial products from environmental protection (Simsek-Ersahin, 2011), disease management to industrial products like medicines, food, etc. Worms have been added as important sources as fishing bait, an enzyme and protein source for several products like animal food, a medical source, and biodegradable cleansers. Some of the commercial products obtained in vermiculture and vermicomposting are as follows:

a. vermicompost and its products;
b. plastic degradation;
c. bioturbation;
d. worms for bait or feed; and
e. medicines.

a. Vermicompost and its products

Vermicomposting is a cold composting process as heat will not be generated by the bacteria during the decomposition of organic materials into nutrient-rich compost (Sánchez et al., 2017; Rajiv et al., 2010; Insam and De Bertoldi, 2007). Vermicompost provides many beneficial microorganisms like bacteria, fungi, and protozoa, and nutrients like calcium, nitrogen, potassium, phosphorus, and magnesium to the soil and make the land fertile for the growth of soil biota. Livestock manure mixed with organic materials is good for a worm business as the management cost will be less and also can be used in other fields like feeds, bioremediation, etc.

Vermicompost improves the soil aeration or bioturbation and water holding capacity and enriches the soil with nutrients like NKP [i.e., Nitrogen (N), Potassium (K) and Phosphorus (P)], organic matter, plant growth hormones, enzymes, and micro-organisms (Sinha et al., 2009). The regular composting procedures are sufficient to decompose organic matter into water and carbon dioxide but they could partially degrade plastic polymers and release residual forms of the polymers which are still toxic to the soil environment. The earthworms can perform more effective decomposition of these plastics in two ways: (1) gut-associated process; and (2) cast-associated process. Complex symbiotic interactions between the earthworms, microorganisms, organic matter, and soil invertebrates provide benefits like bio-oxidation, fragmentation, and stabilization of organic matter.

The ingested organic matter along with the plastics are physically degraded by grinding in the gizzard and later the biochemical transformation is initiated by the enzymes released by the gut and its associated microbes (Pathma and Sakthivel, 2012). The nutrients thus formed are absorbed whereas the degraded plastic along with the reminiscences of organic matter, mucus, enzymes, etc. are egested as casts. The casts that are released attract microorganisms and soil fauna that further decompose the plastic. As this process continues for a prolonged period, the plastic would be completely degraded in due course of time. The rate of degradation is directly dependent on the richness and amount of the organisms present.

The study of microbiology and chemistry of vermicompost extract are complex. The vermicompost (humified material) is used in treating wastewaters, improving agricultural productivity, remediating polluted soils, and protecting crop production. Humic substances like fulvic and humic acids are helpful in the development of the natural environment (Yang et al., 2019; Peña-Méndez et al., 2005; Benedetti et al., 1996). The molecular weights of humic substances for fulvic acids range from 500 to 5,000 Daltons and for humic acids range from 3,000 to 1,000,000 Daltons (Figure 9.6). These substances provide benefits in the conversion of nitrogen to stable organic forms, destroying harmful pathogens, decreasing the volume of wastes in soil, and improving the physical and chemical characteristics of the soil residues. Vermicomposts with organic amendments release weak acids like citric acid, lactic acid, maleic acid, oxalic acid, propanoic acid, butyric acids, etc., and minerals like nitrogen, potassium, magnesium, calcium, and phosphorus into soils. Vermicomposts are important for stimulating the production of antifungal substances in the soil by organisms like earthworms.

Vermicompost producing and augmented with enriching minerals like iron, manganese, and phosphorus promotes the development of microbes like *Pseudomonas syringae, Pseudomonas mosselii, Fusarium oxysporum* f. sp. *Ciceris,* etc., act as ecofriendly bio-fungicides (Sarma et al., 2010). Vermicompost extracts can be applied in foods, pharmaceuticals, and fuels. Applications of vermicompost have been used in agricultural practices, growth of phytopathogenic microorganisms, antimicrobial compounds, digested foods, pharmaceuticals, and biological atmosphere. Vermitechnology is an important method for improving the nutrient status of fly ash and reducing the toxic heavy metal content.

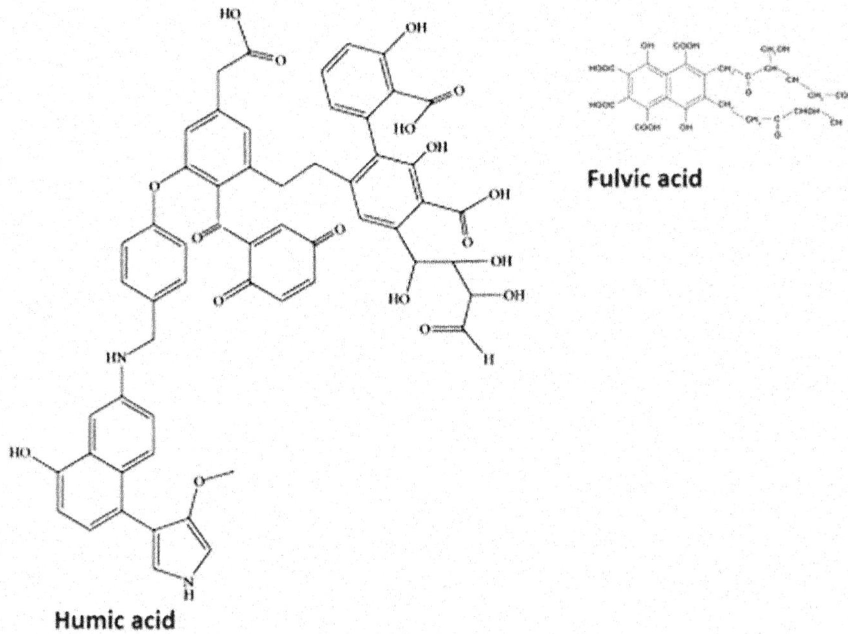

FIGURE 9.6 Humic and fulvic acids.

b. Plastic degradation

The anthropocene period is proposed to be the plastic age due to the role of the usage of plastic in every human activity (Halden, 2010). The improper disposal of plastic has contaminated both soil and water to the level of affecting and altering the ecosystems. It is estimated that annually the land is dumped up to 23 times with more plastic than the oceans and polyethylene-based mulching is one of the major contributors. Plastic waste is ubiquitous in the world is affecting the environment and health of every organism. To alleviate this condition the biodegradable plastic is replacing non-biodegradable plastic but the rate of biodegradation of any plastic is lower than the estimates of laboratory studies. The standardized physicochemical conditions of optimal biodegradation are not available in the environment and hence, the routine biodegradation rate predicted by the microorganisms is very poor. In this regard, the earthworms with better survival and biodegradation capabilities can also create optimal habitats for microbial proliferation (Masciandaro et al., 2013). The earthworms primarily modify the biological, physical, and chemical properties of the soil through bioturbation. The earthworms will also decompose solid wastes like plastic (Table 9.1).

TABLE 9.1 Worms as Sources in Plastic Degradation

Worm	Plastic and materials	Importance	Reference
Mealworms (*Tenebrio molitor*)	Polystyrene	Support plant growth	Koh et al., (2020)
Superworms (*Zophobasmorio*)	Polystyrene	Support plant growth	Koh et al., (2020)
Achroiagrisella (lesser waxworm)	Polyethylene	Unceasing environmental pollution	Kundungal et al., (2019)
Lumbricus terrestris	Polyethylene films, biodegradable plastics (polylactic acid and polyhydroxy alkanoate), biodegradable paper, and litter	Agriculture and in control of plastic pollution	Zhang et al., (2018)
Elsiniafetida, Eudriluseuginae, and *Perionyx excavatus*	Plastic, hard fruit, and eggshells and bones	Waste management, soil detoxification, and regeneration, and sustainable agriculture	Sinha et al., (2002)
Eisenia andrei	Solid paper-pulp mill sludges and primary sewage sludge	C/N ratio and the degree of extractability of heavy metals were lower; increased the humification rate	Elvira et al., (1996)
Eisenia fetida	Low-density polyethylene microplastics	Control of plastic pollution	Rodríguez-Seijo et al., (2018)

c. Bioturbation

The mobility of the earthworms through the soil and its burrowing activity changes the pattern of soil component distribution and this phenomenon is known as bioturbation (Lavelle, 1988; Binet et al., 2006; Chan, 2001). The process enhances the aeration and water percolation of the soil (Figure 9.7). Therefore, the loosening of the soil and its components will increase the oxidoreduction of the components, and also the imbibition of water leads to further decomposition. Thus, bioturbation permits quicker degradation. Moreover, earthworms cause multiple indirect chemical and biological effects. The biostructures like middens, casts, etc. are rich in microbial and fauna diversity and are also sites of extracellular enzymes. The middens being nutrient-rich are abundant in mesofauna diversity and microorganisms. Casts and other drilosphere components also host diverse fauna and microorganisms especially arthropods, bacteria that also contribute to the further degradation of all components including plastic.

The mobility of the worms through the plastic waste facilitates the binding of meso and micro-sized plastic to its mucous skin and facilitates their transportation to lower layers of soil leading to enhanced accessibility of the plastic to the fauna and microbial communities (Slizovskiy et al., 2011). The bioturbation also creates microhabitats for plastic degradation. The middens are the accumulation of litter, casts, and plastics, together and casts contain the partially degraded ingested plastic along with the left out the organic matter. The furrows accommodate plastic in addition to organic composites of mucus, cast, and litter. These nutrient-rich zones attract diverse organisms and due to their mechanical and biological activity, the plastic gets degraded.

d. Worms for bait or feed

The worms are mostly used as food for livestock and poultry farms, fish, rabbitries, horse stables, etc. (Sherman, 2018) (Figure 9.8). The nightcrawler (or *Lumbricus terrestrius*) is commonly harvested and will be marketed as fishing bait. *Eisenia fetida* and various species of earthworms can break down wastes like cattle waste, poultry waste, and pig waste. These wastes rapidly under controlled conditions are used in the production of commercial products like vermicomposts, high-grade proteins, animal feed, and medicines. Separation of earthworms from humus can be done by manual, behavioral, mechanical, chemical, and electric methods. Due to high nutritional value, fish meal, increased digestibility, and deliciousness, worms are been widely used as an animal protein source especially in the aquaculture industry

(Jabir et al., 2012). From ancient times, the worms like nightcrawlers and marshmallows are mostly used in feeds in trout fishing. *Eudrilus eugeniœ* is highly demanded as fishing bait that is adapted to tropical climates possess high economic value (Figure 9.8).

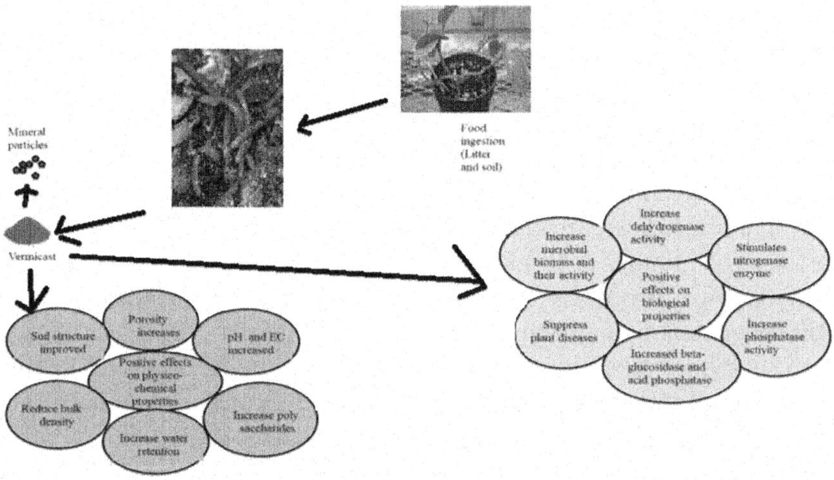

FIGURE 9.7 Bioturbation and importance.

FIGURE 9.8 Worms as bait and feed.

Night Crawlers, *super worms*, red worms, or red wiggler mix are commonly used as bait in fishing (Appelhof and Olszewski, 2017) with below mixing products like cornmeal, wheat bran, powdered milk, rice bran, corn flour, purina worm chow, rabbit manure or chicken mash. Night crawlers and red wigglers are the live bait for small crappies and sunfish, catfish, walleye, bass, perch, bluegill, etc. Some processed food of worms for feeding poultry, fish, rabbitries, horses, birds, and tortoise also contains preservatives (Table 9.2).

TABLE 9.2 Worms as Bait and Feed

Worm	Place	Fish type	Reference
West African nightcrawler (*Eudrilus eugeniae*)	North America	Ornamental fish like *Mexico sturgeon* (*Acipenser oxyrhynchus desotoi*)	Mason et al., (1992)
brandling worm (*Eisenia fetida*)	North America	Ornamental fish like *Mexico sturgeon* (*Acipenser oxyrhynchus desotoi*)	Mason et al., (1992)
Marine worm *Perinereisbrevicirrus*	Shizuoka	Red sea bream *Chrysophrys major*	Shinya et al., (1981)
Red earthworm (*Eisenia fetida*)	India	Aquafeeds like cat fish, eel, Catla, etc.	Tacon, (1995)
Meal worm (*Tenebrio molitor*)	Africa	African catfish (*Clarias gariepinus*)	Ng, (2001)

e. Medicines

The earthworm's used in medicine from ancient days for various remedies in the treatment of diseases and nutrition (Cooper et al., 2012a). Folk medicine from Burma and India used to eat earthworms in the treatment of diseases like pyorrhea, hepatoprotective, antioxidant, anti-inflammatory, antipyretic, etc. The Divine Farmer's Materia Medica Classic was the first documented book from china of around 200 B.C–200 A.D was documented with 365 drugs (Shen, 2010). Traditional Chinese medicine (TCM), Ayurveda, practices in Vietnam, Japan, and Korea have focused on earthworms as sources of human food due to their non-toxic nature (Table 9.3).

9.5 MERITS AND DEMERITS OF VERMICULTURE AND VERMICOMPOSTING

9.5.1 *MERITS OF VERMICULTURE AND VERMICOMPOSTING*

Earthworms are commercially used for human benefit as they are not grown as pests, vectors of disease, or parasites, not expensive and time-consuming as related to ethical parameters (Finke, 2002). Vermiculture and vermicomposting have a lot of advantages like providing nutrients to the soil and improving soil structure, improving the ability to hold nutrients, improving aeration and water holding ability, and providing beneficial microorganisms for the growth of plants.

TABLE 9.3 Medicinal Properties of Earthworms

Medicinal property	References
Hepatoprotective and antioxidant properties	Balamurugan et al., (2008); Prakash et al., (2008); Jamshidzadeh et al., (2018)
Anti-inflammatory and antipyretic property	Omar et al., (2012); Mathur et al., (2011)
Stop spasms, reduce heat toxins, treat a high fever with convulsions, swollen and painful joints, settle wheezing, promote urination, long-term cough, and high blood pressure.	Shen, (2010)
Promote dilong neuron regeneration.	Cooper et al., (2012a); Cooper et al., (2012b)

Production of earthworms is having much demand in the organization of solid wastes, production of vermicompost, and production of animal feeds, commercialization of bait, improving soil fertility, bioturbation, and soil detoxification. Vermicompost may be also used in the preparation of paper, cardboard, toys, several side materials in household items, etc. (Gupta and Garg, 2009). The vermicompost may also hold heat more compare to normal soil and can use as warmers.

9.5.2 DEMERITS OF VERMICULTURE AND VERMICOMPOSTING

The demerits of vermiculture and vermicomposting are the system is complex, depends on the choice of species, cost-intensive, deposition of heavy metals in soil, bad odor problems, greenhouse gas (GHG) Emissions like methane, NO_2, and CO_2 (Lim et al., 2016; Nigussie et al., 2016, 2017). The disadvantages of vermicompost are less nitrogen loss, complex management process, do not kill weed seeds, and may leachate for spreading plant pathogens (Figure 9.9).

9.6 CONCLUSION

Vermicomposting is an efficient natural technique in converting milk processing industry sludge into nutrient-rich biomanure. Vermiculture operations are very important to *produce* live poultry feed, *bait worms, or medicines.*

Vermiculture and vermicompost production is very much beneficial in vermicompost production, plastic degradation, bioturbation, worms for bait or feed, and medicines. Vermicomposting plays an important role in diminishing public health problems due to bioaccumulation of heavy metals and reduction of environmental impacts. The commercial production of vermiculture and vermicompost may provide good benefits for future entrepreneurships.

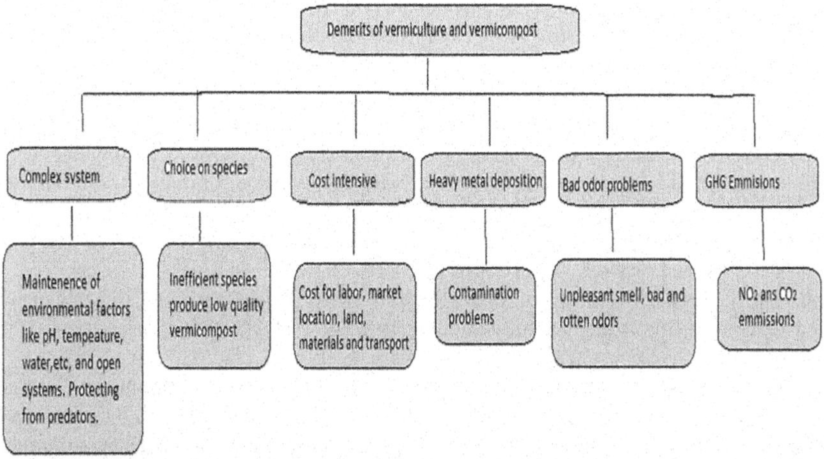

FIGURE 9.9 Demerits of vermiculture and vermicomposting.

ACKNOWLEDGMENTS

The authors would like to thank the administration of Atal Bihari Vajpayee University, Bilaspur (CG) and IGNTU, Amarkantak (MP) for providing assistance, encouragement, and kind support.

KEYWORDS

- **vermiculture**
- **vermicompost**
- **worms**
- **production**
- **commercial products**

REFERENCES

Adrianov, A. V., & Maiorova, A. S., (2010). Reproduction and development of common species of peanut worms (Sipuncula) from the sea of Japan. *Russian Journal of Marine Biology, 36*(1), 1–5.

Appelhof, M., & Olszewski, J., (2017). *Worms Eat My Garbage: How to Set up and Maintain a Worm Composting System: Compost Food Waste, Produce Fertilizer for Houseplants and Garden, and Educate Your Kids and Family*. Storey Publishing.

Balamurugan, M., Parthasarathi, K., Ranganathan, L. S., & Cooper, E. L., (2008). Hypothetical mode of action of earthworm extract with hepatoprotective and antioxidant properties. *Journal of Zhejiang University Science B., 9*(2), 141.

Benedetti, M. F., Van, R. W. H., & Koopal, L. K., (1996). Humic substances considered as a heterogeneous donnan gel phase. *Environmental Science & Technology, 30*(6), 1805–1813.

Binet, F., Kersanté, A., Munier-Lamy, C., Le Bayon, R. C., Belgy, M. J., & Shipitalo, M. J., (2006). Lumbricid macrofauna alter atrazine mineralization and sorption in a silt loam soil. *Soil Biology and Biochemistry, 38*(6), 1255–1263.

Calvert, C. C., (1979). Use of animal excreta for microbial and insect protein synthesis. *Journal of Animal Science, 48*(1), 178–192.

Chan, K. Y., (2001). An overview of some tillage impacts on earthworm population abundance and diversity—Implications for functioning in soils. *Soil and Tillage Research, 57*(4), 179–191.

Cooper, E. L., Balamurugan, M., Huang, C. Y., Tsao, C. R., Heredia, J., Tommaseo-Ponzetta, M., & Paoletti, M. G., (2012a). Earthworms dilong: Ancient, inexpensive, noncontroversial models may help clarify approaches to integrated medicine emphasizing neuroimmune systems. *Evidence-Based Complementary and Alternative Medicine, 2012*. Article ID 164152.

Cooper, E. L., Hirabayashi, K., & Balamurugan, M., (2012b). Dilong: Food for thought and medicine. *Journal of Traditional and Complementary Medicine, 2*(4), 242–248.

Dominguez, J., & Edwards, C. A., (2010). Relationships between composting and vermicomposting. *Vermiculture Technology Earthworms, Organic Wastes, and Environmental Management, 1*, 11–26.

Elvira, C., Goicoechea, M., Sampedro, L., Mato, S., & Nogales, R. J., (1996). Bioconversion of solid paper-pulp mill sludge by earthworms. *Bioresource Technology, 57*(2), 173–177.

Finke, M. D., (2002). Complete nutrient composition of commercially raised invertebrates used as food for insectivores. *Zoo Biology, 21*(3), 269–285.

Gajalakshmi, S., & Abbasi, S. A., (2004). Earthworms and vermicomposting. *Indian Journal of Biotechnology, 3*, 486–494.

Gupta, R., & Garg, V. K., (2009). Vermiremediation and nutrient recovery of non-recyclable paper waste employing *Eisenia fetida*. *Journal of Hazardous Materials, 162*(1), 430–439.

Halden, R. U., (2010). Plastics and health risks. *Annual Review of Public Health, 31*, 179–194.

Hidalgo, P. R., & Harkess, R. L., (2002). Earthworm castings as a substrate amendment for chrysanthemum production. *Hortscience, 37*(7), 1035–1039.

Insam, H., & De Bertoldi, M., (2007). Microbiology of the composting process. *Compost Science and Technology, 8*(1), 25–48.

Jabir, M. D., Razak, S. A., & Vikineswary, S., (2012). Chemical composition and nutrient digestibility of super worm meal in red tilapia juvenile. *Pakistan Veterinary Journal, 32*(4), 489–493.

Jamshidzadeh, A., Dabagh, F., Farshad, O., Ommat, M. M., Mahdavinia, A., Azarpira, N., Shahbazi, M., et al., (2018). Hepatoprotective properties of the glycolipoprotein extract from Eisenia foetida. *Trends in Pharmaceutical Sciences, 4*(3), 149–160.

Kaladhar, D. S. V. G. K., (2018). *Vermiculture and Organic Farming*. Paradise Press.

Kalam, A., & Ahmad, S. R., (2017). An outlook of vermicomposting and its Scope in future. *J. Chem. Biol. Phys. Sci., 7*, 49–65.

Koh, D. W., Ang, B. Y., Yeo, J. Y., Xing, Z., & Gan, S. K., (2020). *Plastic Agriculture Using Worms: Augmenting Polystyrene Consumption and Using Frass for Plant Growth Towards a Zero-Waste Circular Economy*. bioRxiv.

Kumar, R., Sharma, P., Gupta, R. K., Kumar, S., Sharma, M. M., Singh, S., & Pradhan, G., (2020). Earthworms for eco-friendly resource efficient agriculture. In: *Resources Use Efficiency in Agriculture* (pp. 47–84). Springer, Singapore.

Kundungal, H., Gangarapu, M., Sarangapani, S., Patchaiyappan, A., & Devipriya, S. P., (2019). Efficient biodegradation of polyethylene (HDPE) waste by the plastic-eating lesser waxworm (*Achroiagrisella*). *Environmental Science and Pollution Research, 26*(18), 18509–18519.

Lavelle, P., (1988). Earthworm activities and the soil system. *Biology and Fertility of Soils, 6*(3), 237–251.

Lim, S. L., Lee, L. H., & Wu, T. Y., (2016). Sustainability of using composting and vermicomposting technologies for organic solid waste biotransformation: Recent overview, greenhouse gases emissions and economic analysis. *Journal of Cleaner Production, 111*, 262–278.

Lim, S. L., Wu, T. Y., Lim, P. N., & Shak, K. P., (2015). The use of vermicompost in organic farming: Overview, effects on soil and economics. *Journal of the Science of Food and Agriculture, 95*(6), 1143–1156.

Masciandaro, G., Macci, C., Peruzzi, E., Ceccanti, B., & Doni, S., (2013). Organic matter–microorganism–plant in soil bioremediation: A synergic approach. *Reviews in Environmental Science and Bio/Technology, 12*(4), 399–419.

Mason, W. T., Rottmann, R. W., & Dequine, J. F., (1992). *Culture of Earthworms for Bait or Fish Food*. Florida Cooperative Extension Service, Institute of Food and Agricultural Sciences, University of Florida.

Mathur, A., Verma, S. K., Singh, S. K., Prakash, A., Prasad, G. B., & Dua, V. K., (2011). Anti-inflammatory activity of earthworm extracts. *International Journal of Pharmaceutical Sciences and Research, 2*(2), 278.

Musyoka, S. N., Liti, D. M., Ogello, E., & Waidbacher, H., (2019). Utilization of the earthworm, *Eisenia fetida* (Savigny, 1826) as an alternative protein source in fish feeds processing: A review. *Aquaculture Research, 50*(9), 2301–2315.

Ng, W. K., Liew, F. L., Ang, L. P., & Wong, K. W., (2001). Potential of mealworm (*Tenebrio molitor*) as an alternative protein source in practical diets for African catfish, *Clarias gariepinus*. *Aquaculture Research, 32*, 273–280.

Nigussie, A., Bruun, S., De Neergaard, A., & Kuyper, T. W., (2017). Earthworms change the quantity and composition of dissolved organic carbon and reduce greenhouse gas emissions during composting. *Waste Management, 62*, 43–51.

Nigussie, A., Kuyper, T. W., Bruun, S., & De Neergaard, A., (2016). Vermicomposting as a technology for reducing nitrogen losses and greenhouse gas emissions from small-scale composting. *Journal of Cleaner Production, 139*, 429–439.

Omar, H. M., Ibraheim, Z. Z., El-Shimy, N. A., & Ali, R. S., (2012). Anti-inflammatory, antipyretic and antioxidant activities of the earthworms extract. *Journal of Biology and Earth Sciences, 2*(1), 10–17.

Pathma, J., & Sakthivel, N., (2012). Microbial diversity of vermicompost bacteria that exhibit useful agricultural traits and waste management potential. *Springer Plus, 1*(1), 26.

Peña-Méndez, E. M., Havel, J., & Patočka, J., (2005). Humic substances–compounds of still unknown structure: Applications in agriculture, industry, environment, and biomedicine. *J. Appl. Biomed., 3*(1), 13–24.

Prakash, M., Gunasekaran, G., & Elumalai, K., (2008). Effect of earthworm powder on antioxidant enzymes in alcohol induced hepatotoxic rats. *Eur. Rev. Med. Pharmacol. Sci., 12*(4), 237–243.

Rajiv, K. S., Sunita, A., Krunal, C., Vinod, C., & Brijal, K. S., (2010). Vermiculture technology: Reviving the dreams of Sir Charles Darwin for scientific use of earthworms in sustainable development programs. *Technology and Investment, 1*(3), Article ID:2490.

Rodríguez-Seijo, A., Da Costa, J. P., Rocha-Santos, T., Duarte, A. C., & Pereira, R., (2018). Oxidative stress, energy metabolism and molecular responses of earthworms (*Eisenia fetida*) exposed to low-density polyethylene microplastics. *Environmental Science and Pollution Research, 25*(33), 33599–33610.

Saba, S., Zara, G., Bianco, A., Garau, M., Bononi, M., Deroma, M., Pais, A., & Budroni, M., (2019). Comparative analysis of vermicompost quality produced from brewers' spent grain and cow manure by the red earthworm *Eisenia fetida. Bioresource Technology, 293*, 122019.

Sánchez, Ó. J., Ospina, D. A., & Montoya, S., (2017). Compost supplementation with nutrients and microorganisms in composting process. *Waste Management, 69*, 136–153.

Sarma, B. K., Singh, P., Pandey, S. K., & Singh, H. B., (2010). Vermicompost as modulator of plant growth and disease suppression. *Dynamic Soil, Dynamic Plant, 4*(1), 58–66.

Shen, Y., (2010). Earthworms in traditional Chinese medicine: (*Oligochaeta: Lumbricidae, Megascolecidae*). *Zoology in the Middle East, 51*(2), 171–173.

Sherman, R., (2018). *The Worm Farmer's Handbook: Mid-to Large-Scale Vermicomposting for Farms, Businesses, Municipalities, Schools, and Institutions*. Chelsea Green Publishing.

Shinya, F. U. K. E., Shoji, K. O. N. S. U., Kazuo, I. N. A., (1981). Identification of feeding stimulants for red sea bream in the extract of marine worm *Perinereis brevicirrus. Bulletin of the Japanese Society of Scientific Fisheries, B47*(12), 1631–1635.

Simsek-Ersahin, Y., (2011). The use of vermicompost products to control plant diseases and pests. In: *Biology of Earthworms* (pp. 191–213). Springer, Berlin, Heidelberg.

Singh, S., (2020). Entomophagy: Insects as human food. *Agriculture & Food, 2*(5), 75–79.

Sinha, R. K., Heart, S., Agarwal, S., Asadi, R., & Carretero, E., (2002). Vermiculture and waste management: Study of action of earthworms elsiniafoetida, *Eudrilus euginae* and *Perionyx excavatus* on biodegradation of some community wastes in India and Australia. *Environmentalist, 22*(3), 261–268.

Sinha, R. K., Valani, D., Sinha, S., Singh, S., & Heart, S., (2009). Bioremediation of contaminated sites: A low-cost nature's biotechnology for environmental clean up by versatile microbes, plants & earthworms. *Solid Waste Management and Environmental Remediation*, 1–72.

Slizovskiy, I. B., Kelsey, J. W., & Hatzinger, P. B., (2011). Surfactant-facilitated remediation of metal-contaminated soils: Efficacy and toxicological consequences to earthworms. *Environmental Toxicology and Chemistry, 30*(1), 112–123.

Tacon, A. G., (1995). Feed ingredients for carnivorous fish species: Alternatives to fishmeal and other fishery resources. In: Reinertsen, H., & Haaland, H., (eds.), *Sustainable Fish Farming* (pp. 89–114).

Venter, J. M., & Reinecke, A. J., (1988). The life-cycle of the compost worm (Oligochaeta). *African Zoology, 23*(3), 161–165.

Yadav, S. K., Miah, M. F., Makin, A. A., & Khan, Z. K., (2014). Small-Scale compost production through vermiculture biotechnology. *International Journal of Research in Agriculture and Forestry, 1*(2), 7–12.

Yang, F., Zhang, S., Cheng, K., & Antonietti, M., (2019). A hydrothermal process to turn waste biomass into artificial fulvic and humic acids for soil remediation. *Science of the Total Environment, 686*, 1140–1151.

Yao, X., Zhang, F., Qiao, Z., Yu, H., Sun, S., Li, X., Zhang, J., & Jiang, X., (2020). Toxicity of thifluzamide in earthworm (Eisenia fetida). *Ecotoxicology and Environmental Safety, 188*, 109880.

Yuvaraj, M., & Mahendran, P. P., (2020). Vermicompost. *Biotica Research Today, 2*(7), 556–558.

Zhang, L., Sintim, H. Y., Bary, A. I., Hayes, D. G., Wadsworth, L. C., Anunciado, M. B., & Flury, M., (2018). Interaction of *Lumbricus terrestris* with macroscopic polyethylene and biodegradable plastic mulch. *Science of The Total Environment, 635*, 1600–1608.

CHAPTER 10

Pineapple Processing Waste Utilization for Sustainable Development in North-Eastern States of India

TH. ANAND SINGH, PRAKASH KUMAR SARANGI, and NG. JOYKUMAR SINGH

Directorate of Research, Central Agricultural University, Imphal, Manipur-795004, India, E-mail: anan4thangjam@gmail.com (Th. Anand Singh)

ABSTRACT

Agrowaste utilization in the fruits and vegetable processing industries is challenging and arduous exertion. The discarded portions need to be converted into profitable endeavors to augment the income of the farmers. Various processes like extraction of bioactive components, enzymes and pigments, fermentation, application of biotechnology for bioconversion, etc. were successfully employed to exploit the pineapple processing waste (PPW). In this chapter, the focus is on the extraction of bromelain enzyme, vermicomposting, feed and silage, pectin extraction, and fiber production from PPW. This work is an attempt to solve the pineapple waste menace in Northeast India by highlighting the ongoing work and technologies on the utilization of pineapple waste that is appropriate for the region.

10.1 INTRODUCTION

The pineapple (*Ananas comosus*) is one of the most important fruits in the world belonging to the *Bromeliaceae* family. Pineapple juice is also among

Biotechnology for Waste Biomass Utilization. Prakash Kumar Sarangi & Latika Bhatia (Eds.)

the most preferred juice worldwide after orange and apple juice (Cabrera et al., 2000). The area under pineapple plantation is about 9,20,349 ha with an estimated production of more than 18 million tons (FAO, 2007). Commercially, the most important product is the canned fruits which are consumed all over the world (Tran, 2006). The pineapple is known to have essential minerals and vitamins with some medicinal value. Pineapple also contains considerable calcium, potassium, fiber, and vitamin C (Hemlatha and Ambuselvi, 2013). This pan-tropical fruit crop is abundantly available in Manipur and many other states of Northeastern states in India. About more than 40% of the total pineapple production of the country was made from the NE region and 90 to 95% of the produces were organic. 'Kew' and 'Queen' are the common cultivars grown in this region. Manipur contributes about 7.37% of the total pineapple production of India with a productivity of 9.5 MT/Ha having a production of 124,146.00 MT. It is one of the most promising horticultural crops with huge potential for the export market, particularly in the region as the pineapple produced here is known for its extraordinary taste and juiciness. The demand for pineapple in the international market is expanding day by day. The high-quality pineapples are usually consumed fresh or as canned products. The perishable and seasonal nature of the fruit limits the economic returns of the pineapple farmers of the state. Rough handling of fruits and exposure to adverse environments during transportation and storage may lead to 55% of waste (Nunes et al., 2009). The natural advantage may be exploited more fruitfully by processing and value-addition of pineapple for products like juice, slices, osmo-dehydrated ring, and pineapple powder to enhance the economic returns of these farmers. Other process products include juices, concentrates jelly, and jams. Pineapple slices have also been preserved after freezing (Larrauri et al., 1997). However, low-quality fruits do not fetch market and are left on farms. Besides, during pineapple processing, a large amount of unusable waste material is generated (Tanaka et al., 1999). During pineapple processing, the crown and stem are cut off before peeling. The core is then removed for further processing. These wastes (peel, core, stem, crown, and leaves) generally account for 56% (w/w) of total pineapple weight (Singh). The increasing production of pineapple processed items results in massive waste generations mainly due to the elimination of components unsuitable for human consumption. It has been calculated that about 50,000–60,000 MT of pineapple wastes are produced annually causing immense environmental pollution. These wastes are usually prone to microbial spoilage thus limiting further exploitation. Further, the drying, storage, and shipment of these wastes are not cost-effective, and hence efficient, inexpensive, and eco-friendly exploitation is becoming more than a simple necessity. In addition, the residues also cause serious environmental problems while disposing of as the pineapple wastes have a high level of BOD and COD. Reports have shown that 40–80% of pineapple fruit is discarded as waste having high biological oxygen demand (BOD) and chemical oxygen demand (COD) values (Bankoffi and Han, 1990).

Owing to the above facts, various researchers have designed various approaches for the efficient utilization of pineapple processing waste materials. Extraction of Bromelain enzyme (Murachi, 1976), vermicomposting (Mainoo et al., 2009), animal feed and ensilaging for organic acids (Tran et al., 1998; Ueno et al., 2003), pectin extraction, fiber extraction from crown leaves of pineapple, bioethanol (ICIS Chemical Business, 2006), and cellulose enzyme extraction (Saravanan et al., 2013) was attempted with positive results. In this chapter, we will focus on the technologies relevant in the Northeastern States of India, particularly- bromelain enzyme, vermicomposting, feed and silage for cattle and pig, Pectin and fiber from crown leaves of a pineapple.

10.2. BROMELAIN

Bromelain is a group of cystiene proteolytic enzymes derived from pineapples that have similar but distinctly different amino acid sequences. Stem, peel, juice crown, and core of pineapple contain a considerable amount of bromelain. Many studies have shown that pineapple by-products could be exploited as cheap sources of bromelain enzyme (Murachi, 1967). Bromelain has found applications in the medicinal and cosmetic industries as anti-inflammatory, antithrombotic, fibrinolytic activities, skin cleaning agent, and anticancer agent. Furthermore, bromelain is finding wide applications in pharmaceutical and food uses (Hebbar et al., 2008).

Bromelain enzyme is widely employed in food industries, particularly in baking and meat industries where meat and dough were treated to get the desired effect in a quick time. The effect was validated by determining the physicochemical and quality characteristics such as pH, water holding capacity, moisture content, soluble peptide content and cooking yield, etc. (Singh et al., 2018). These physicochemical changes observed after treatment with BE 0.7% are listed in Table 10.1.

BE is also used in baking industries for solubilizing gluten in the flour which allows relaxation of dough and prevents shrinkage of dough and thus dough rises more evenly during baking (Kong et al., 2007). BE treated flour was reported to be hypoallergic and safe to those people who are allergic to wheat flour (Watanabe et al., 2000). BE find application as an anti-browning agent in fruits (Srinath et al., 2012) and its application as clarifying agent to prevent haze formation in beer (Benucci et al., 2011). Animal and fish waste were treated to obtain protein hydrolysate for animal feed industries (Tanuja et al., 2012). In cosmetic industries, proteolytic enzymes like papain and Bromelain are used to alleviate skin problems like acne, wrinkle, swelling by digesting the protein in the dead skin cells and allowing them to grow fresh skin cells (Ozlen and Chatsworth, 1997).

TABLE 10.1 Meat Tenderization by Bromelain Enzyme (BE) Treatment–Changes in Physico-Chemical Properties

Physico-chemical properties of Meat	Chicken		Beef	
	No treatment	0.7% of BE	No treatment	0.7% of BE
pH	6.67	6.11	6.93	6.29
Water holding capacity (%)	37.4	24.69	35.34	25.94
Moisture content %	76.8	65.9	75.44	68.09
Cooking yield %	85.12	75.9	79.44	68.09
Soluble Peptide content µmol/g	328	798	709	238

The values presented in Table 10.1 are the mean of three readings each.

10.3 VERMICOMPOSTING

The environmental hazards that are associated with improper disposal of pineapple processing waste due to high COD and BOD were highlighted earlier. The problem could be effectively dealt with by composting pineapple wastes into vermicompost (Mainoo et al., 2009). They have reported that vermicomposting rapidly decomposed about 99% of pineapple pulp wet mass while peel had a loss in weight by almost 87%. The pH of the waste changed from acidic to neutral to alkaline during composting.

The pineapple waste was collected and chopped before putting in plastic bins perforated at the bottom that acts as a reactor for pre-composting (Figure 10.1). The holes in the bottom allow the leachate to pass out while water is sprinkled on every fourth day to maintain humidity. After 25 days, the precompost pineapple waste was removed from the reactor and passed onto the vermicomposting reactor with a perforated hole along with an equal amount of other precompost load materials consisting of other vegetable and agricultural wastes. Before adding the precompost, the reactor bed was lined with a 3 cm thick soil layer where the precompost was added on it. The thoroughly mixed precompost materials were inoculated with exotic earthworm (*Eisenia fetida*) or local species were inoculated at a rate of 100 earthworms per 10 kg of precompost. After inoculation 5 cm thick waste layer was added and mixed to homogenize. The reactor is sprayed with water every alternate day while manual mixing was done every day to aerate properly. Fist test must be employed to ensure excess water (when the sample is squeezed in your hand no water should come out from the sample). After 45 days, the vermicompost was ready while the earthworms were harvested for sale.

FIGURE 10.1 Precomposting of PPW for vermicomposting.

10.4 FEED AND SILAGE

Utilization of agrowaste as animal feed alleviates the problems due to shortage of forage as well as potential environmental hazards due to improper disposal seems to serve a dual purpose. Many studies reported the utilization of PPW as feed for ruminants (Tran, 2006). However, it is also reported that by-product of the pineapple processing industry is not considered attractive as an animal feed because of high fiber content and soluble carbohydrates with low protein content (Correia et al., 2004). The quality of feed could be enhanced and made more attractive to the animals by ensilaging (Gowda et al., 2015). Pineapple waste from fields and cannery were given as feed in China and ensiled and dried PPW could replace 50% roughage dairy cattle feed (Sruamisri, 2007). Dietary fiber powder prepared from PPW has 70.6% total dietary fiber with enhanced sensory properties than commercial dietary fibers from apple and citrus fruits (Larrauri et al., 1997). 99% of ethnic populations of Northeast India are non-vegetarian and pork is the most common meat in the region. While pigs are voracious eaters and could survive by almost everything thrown at them, a proper feed with adequate nutrients will give healthy pigs, and hence, pork. Luckily, both cattle and pigs consume raw pineapple processing waste but ensilaging ensures more nutrients and micronutrients. Moreover, the presence of sharp and pointy leaves in raw pineapple waste harms the mouth of the animals (Tran, 2006). Ensilaging with selected microorganisms could be utilized to produce specific biochemicals like ferulic acids, acetic acid, etc. along with silage for the farm animals. PPW silage is nutritious and reported to produce more milk in dairy cattle due to its high fiber content (Gowda et al., 2015). The neutral detergent fiber and acid detergent fiber were 56.04% and 19.76%, respectively. The crude protein was also increased to 7.5% from 4.3% on dry matter (DM) basis. The relatively higher crude protein content (7.5%) and micronutrient content of the PPW silage vindicate the good quality of feed. The metabolizable energy value was calculated to be 10.79 MJ/Kg DM and organic digestibility was 72.01%. Further experiments also prove that higher return to the farmers by producing more milk and meat when the animals were fed with or supplemented with PPW silage. Not only does it increase digestibility, but it also makes the feed softer.

10.5 SILAGE PREPARATION FROM PPW

Silage was prepared by natural fermentation of the PPW in batches. PPW other than the crown is diced together and crowns were chopped separately as the fiber content is very high while moisture content is lowest in the crown of pineapple. For silage preparation, 3–4 parts of crown leaves were mixed thoroughly with one part of other waste materials. The mix is packed tightly in a 20L plastic container using sticks (Figure 10.2). Afterward, the container is covered tightly. Clay may be also used to seal the packed. The container was then kept at an ambient temperature for 22 days. The silage thus form is kept in the storage bin or feed to the animal directly. Rice wine residue (Figure 10.3), which is the staple feed of pigs in rural Northeast India, is supplemented with the PPW silage to obtain a complete diet.

FIGURE 10.2 Silage preparation from PPW, (a) pineapple crown; (b) peel, core, pomace waste; (c) packing in the plastic container for fermentation.

FIGURE 10.3 Rice wine residue and PPW silage (complete diet for pig).

10.6 FIBER FROM THE PINEAPPLE CROWN LEAVES

Fibers from pineapple fruit have been reported by several researchers (Lund and Smoot, 1982; Bartolome and Ruperez, 1995; Gorinstein et al., 1999). The pineapple crown leaves could be converted to coarse textile fiber which could be used for making bags and other packing materials (Tran,

2006). Superior quality and higher yield of 2.1 g fiber/100 g pineapple pulp waste have been reported by alkaline processing method than the mechanical and semi-chemical method employed (Sreenath et al., 1996). Besides, pineapple leaf fibers were exploited in making fiber-reinforced polymeric composites owing to high cellulosic content, abundance, and cost-effectiveness (Devi et al., 1997; Arib et al., 2006). Their investigation on the tensile, flexural, and impact behavior of pineapple leaf fiber-reinforced polyester composites as a function of fiber loading, fiber length, and fiber surface modification found that the mechanical properties of the composites were superior in quality.

10.7 CONCLUSION

Pineapple processing waste could be converted to various high-value products. The PPW has huge potential for bye product utilization and heading towards a buoyant future. Bromelain extraction, silage, and vermicomposting along with emerging technologies, such as green technology for biogas or bioethanol production from PPW could change the dwindling scope of pineapple production and processing in northeast India. Fiber-rich PPW may be exploited for dietary fiber in the dairy industry and course textile fiber from the crown may be an alternate to jute fiber. The application of novel scientific and technological methods could harness the environmental polluting by-products of the pineapple processing industry could be converted into products with a high economic value other than the major product.

KEYWORDS

- **bromelain**
- **feed and silage**
- **pineapple processing waste**
- **vermicomposting**
- **waste utilization**

REFERENCES

Arib, R. M. N., Sapuan, S. M., Ahmad, M. M. H. M., Paridah, M. T., & Zaman, H. M. D. K., (2006). Mechanical properties of pineapple leaf fiber reinforced polypropylene composites. *Mater. Des., 27*, 391–396.

Ban-Koffi, L., & Han, Y. W., (1990). Alcohol production from pineapple waste. *World J. Microbiol. Biotechnol., 6*, 281–284.

Bartolome, A. P., & Ruperez, P., (1995). Dietary fiber in pineapple fruit. *J. Clin. Nutr., 49*(S2), 61–S263.

Benucci, I., Liburdi, K., Garzillo, A. M. V., & Esti, M., (2011). Bromelain from pineapple stem in alcoholic–acidic buffers for wine application. *Food Chem., 124*(4), 1349–1353.

Cabrera, H. A. P., Menezes, H. C., Oliveira, J. V., & Batista, R. F. S., (2000). Evaluation of residual levels of benomyl, methyl parathion, diuron, and vamidothion in pineapple pulp and bagasse (Smooth cayenne). *J. Agric. Food Chem., 48*, 5750–5753.

Correia, R. T. P., McCue, P., Magalhaes, M. M. A., Macedo, G. R., & Shetty, K., (2004). Production of phenolic antioxidants by the solid-state bio-conversion of pineapple waste mixed with soy flour using *Rhizopus oligosporus*. *Process Biochem., 39*, 2167–2172.

Devi, L. U., Bhagawan, S. S., & Thomas, S., (1997). Mechanical properties of pineapple leaf fiber-reinforced polyester composites. *J. Appl. Polym. Sci., 64*, 1739–1748.

Devi, R. Y., (2012). Biotechnological application of proteolytic enzymes in post cocoon technology. *Int. J. Sci. Nat., 3*(2), 237–240.

Gorinstein, H., Zemser, M., Haruenkit, R., Chuthakorn, R., Grauer, F., Martin-Belloso, O., & Trakhtenberg, S., (1999). Comparative content of total polyphenols and dietary fiber in tropical fruits and persimmon. *J. Nutr. Biochem., 10*, 367–371.

Gowda, N. K. S., Vallesha, N. C., Awachat, V. B., Anandan, S., Pal, D. T., & Prasad, C. S., (2015). Silage from pineapple waste study on evaluation of silage from pineapple (*Ananas comosus*) fruit residue as livestock feed. *Trop. Anim. Health Prod., 47*, 557–561.

Hebbar, H. U., Sumana, B., & Raghavarao, K. S. M. S., (2008). Use of reverse micellar systems for the extraction and purification of bromelain from pineapple wastes. *Bioresour. Technol., 99*, 4896–4902.

Hemalatha, R., & Anbuselvi, S., (2013). Physicohemical constituents of pineapple pulp and waste *J. Chem. Pharm. Res., 5*(2), 240–242.

ICIS Chemical Business, (2006). Pineapple rings true for bioethanol. *Focus on Catalysts, 7*, 6.

Kong, X., Zhou, H., & Qian, H., (2007). Enzymatic hydrolysis of wheat gluten by proteases and properties of the resulting hydrolysates. *Food Chem., 102*(3), 759–763.

Kumar, D., Jain, V. K., Shanker, G., & Srivastava, A., (2003). Utilization of fruits waste for citric acid production by solid state fermentation. *Process Biochem., 38*, 1725–1729.

Larrauri, J. A., Ruperez, P., & Calixto, F. S., (1997). Pineapple shell as a source of dietary fiber with associated polyphenols. *J. Agric. Food Chem., 45*, 4028–4031.

Lund, E. D., & Smoot, J. M., (1982). Dietary fiber content of some tropical fruits and vegetables. *J. Agric. Food Chem., 30*, 1123–1127.

Luo, S., & Netravali, A. N., (1999). Mechanical and thermal properties of environment-friendly "green" composites made from pineapple leaf fibers and poly (hydroxybutyrate-co-valerate) resin. *Polym. Compos., 20*, 367–378.

Mainoo, N. O. K., Barrington, S., Whalen, J. K., & Sampedro, L., (2009). Pilot-scale vermicomposting of pineapple wastes with earthworms native to Accra, Ghana. *Bioresour. Technol., 100*, 5872–5875.

Murachi, T., (1976). Bromelain enzymes. In: Lorand, L., (ed.), *Methods in Enzymology* (Vol. XLV, pp. 475–485). New York: Academic Press.

Negesse, T., Makkar, H. P. S., & Becker, K., (2009). Nutritive value of some non-conventional feed resources of Ethiopia determined by chemical analyses and *in vitro* gas method. *Anim. Feed Sci. Tech., 154*, 204–217.

Nigam, J. N., (2000). Continuous ethanol production from pineapple cannery waste using immobilized yeast cells. *J. Biotechnol., 80*, 189–193.

Nunes, M. C. N., Emond, J. P., Rauth, M., Dea, S., & Chau, K. V., (2009). Environmental conditions encountered during typical consumer retail display affect fruit and vegetable quality and waste. *Postharvest Biol. Technol., 51*, 232–241.

Ozlen, S. N., & Chatsworth, C., (1995). *Cosmetic Composition Containing Alpha Hydroxyacids, Salicyclic Acid, and Enzyme Mixture of Bromelain and Papain.* U.S. Pat. 1995, 5,441,740.

Rani, D. S., & Nand, K., (2004). Ensilage of pineapple processing waste for methane generation. *Waste Manage., 24*, 523–528.

Saravanan, P., Muthuvelayudham, R., & Viruthagiri, T., (2013). Enhanced production of cellulase from pineapple waste by response surface methodology. *J. Eng., 1*–8.

Schieber, A., Stintzing, F. C., & Carle, R., (2001). By-products of plant food processing as a source of functional compounds-recent developments. *Trends Food Sci. Technol., 12*, 401–413.

Sreenath, H. K., Sudarshanakrishna, K. R., Prasad, N. N., & Santhanam, K., (1996). Characteristics of some fiber incorporated cake preparations and their dietary fiber content. *Starch-Starke, 48*, 72–76.

Srinath, R., Ramalingam, C., & Nasimun, I. N., (2012). Isolation and characterization of bromelain from pineapple (*Ananas comosus*) and comparing its anti-browning activity on apple juice with commercial anti-browning agents. *Elixir Food Sci., 45*, 7822–7826.

Sruamisri, S., (2007). Agricultural wastes as dairy feed in Chiang Mai. *Anim. Sci. J., 78*, 335–341.

Tanabe, S., Arai, S., & Watanabe, M., (1996). Modification of wheat flour with bromelain and baking hypoallergenic bread with added ingredients. *Biosci, Biotechnol, Biochem., 60*(8), 1269–1272.

Tanaka, K., Hilary, Z. D., & Ishizaki, A., (1999). Investigation of the utility of pineapple juice and pineapple waste material as low cost substrate for ethanol fermentation by *Zymomonas mobilis*. *J. Biosci. Bioeng., 87*, 642–646.

Tanuja, S., Viji, P., Zynudheen, A. A., & Joshy, C. G., (2012). Composition, functional properties and antioxidative activity of hydrolysates prepared from the frame meat of striped catfish (*Pangasianodon hypophthalmus*). *Egypt J. Biol., 14*, 27–35.

Tilay, A., Bule, M., Kishenkumar, J., & Annapure, U., (2008). Preparation of ferulic acid from agricultural wastes: Its improved extraction and purification. *J. Agric. Food Chem., 56*, 7664–7648.

Tran, A. V., (2006). Chemical analysis and pulping study of pineapple crown leaves. *Ind. Crops Prod., 24*, 66–74.

Tran, C. T., & Mitchell, D. A., (1995). Pineapple waste- a novel substrate for citric acid production by solid state fermentation. *Biotechnol. Lett., 17*, 1107–1110.

Tran, C. T., Sly, L. I., & Mitchell, D. A., (1998). Selection of a strain of Aspergillus for the production of citric acid from pineapple waste in solid-state fermentation. *World J. Microbiol. Biotechnol., 14*, 399–404.

Ueno, T., Ozawa, Y., Ishikawa, M., Nakanishi, K., & Kimura, T., (2003). Lactic acid production using two food processing wastes, canned pineapple syrup and grape invertase as substrate and enzyme. *Biotechnol. Lett., 25*, 573–577.

Wang, C. H., Lin, P. J., & Chang, J. S., (2006). Fermentative conversion of sucrose and pineapple waste into hydrogen gas in phosphate-buffered culture seeded with municipal sewage sludge. *Process Biochem., 41*, 1353–1358.

Watanabe, M., Watanabe, J., Sonoyama, K., & Tanabe, S., (2000). Novel method for producing hypoallergenic wheat flour by enzymatic fragmentation of the constituent allergens and its application to food processing. *Biosci. Biotechnol. Biochem., 64*(12), 2663–2667.

Weng, C. H., Lin, Y. T., & Tzeng, T. W., (2009). Removal of methylene blue from aqueous solution by adsorption onto pineapple leaf powder. *J. Hazard. Mater., 170*, 417–424.

CHAPTER 11

Biogasification of Agricultural Biomass: Towards a Zero-Waste Circular Agrieconomy

PUNEET K. SINGH,[1] DEBOSMITA CHAKRABORTY,[2] SLIPA KANUNGO,[2] SNEHASISH MISHRA,[1] and RITESH PATTNAIK[2]

[1]Bioenergy Lab and BDTC, School of Biotechnology, KIIT University, Bhubaneswar-751024, Odisha, India, E-mail: snehasish.mishra@gmail.com

[2]School of Biotechnology, KIIT University, Bhubaneswar-751024, Odisha, India, E-mail: riteshpattnaik@gmail.com

ABSTRACT

An alarming increase in the global human population with a constant energy demand vis-a-vis health supplements need have drawn attention towards the generation and extensive use of biomass residue-based renewable energy worldwide. As fossil fuels have a tremendous negative impact on environmental health, the use of biomass energy resources that encourage carbon sequestration has gained popularity primarily because of its potential in (zero-waste) circular economy. Indian economy is primarily agrarian and leads to the generation of a huge quantity of agricultural organic residue that are primarily lignocellulose. This otherwise-a-waste agricultural biomass can be a promising alternative for energy production, emphasizing more on minimizing agrowaste generation and greenhouse gas (GHG) emission in line with the circular bioeconomy. Anaerobic digestion (AD) techniques assist in the valorization of such wastes into energy. The energy thus generated is utilizable at the household or the community level and could also be transmitted through the national grid thereby adding to the economic growth

Biotechnology for Waste Biomass Utilization. Prakash Kumar Sarangi & Latika Bhatia (Eds.)

and employment opportunities. The digestate thus obtained as a by-product could be an organic alternative for soil conditioning, fertilizer/manure for the field crop, and integrated nutrient management. The present chapter outlines the potential of agricultural biomass residues for biogasification, the various challenges/mitigation strategies essential towards positive environmental impact, and the circular bioeconomy.

11.1 INTRODUCTION

With the ever-increasing population of this planet, the energy demand is additionally increasing. Energy is vital for the improvement of any nation because it has its requirement in various sectors like agricultural, industrial, development of transport and generation of power and many others such sectors (Silitonga et al., 2011; Tripathi et al., 2016). The utilization of petroleum derivatives might undergo a quick increment as the world's energy mainly relies on it (Coyle Eugene and Simmons, 2014; Naqvi et al., 2018). As a result, it will lead to an exhaustion of resources of fossil fuel that are not renewable, a worldwide temperature alteration, and energy security in the future. To incompletely substitute non-renewable resources with renewable ones and to lessen the reliance on fossil fuels and emission of Greenhouse gas, various technologies and initiatives are in progress (Naqvi et al., 2018). Gradually, various attempts were undertaken so that renewable energy sources such as hydro-power, wind energy, and solar energy can be put to use. Even though this option was a promising one, it had its restrictions on price, adaptability in the production of energy, storage and transportation, and land availability (Singh et al., 2019). It is thought that lignocellulosic biomass can act as a potential climate-friendly alternative to chemicals and fuels of carbon origin. Currently, scientists are giving incredible consideration to the usage of biomass to deliver value-added items Moreover, a negligible amount of inorganic compounds are contained in the biomass, with nitrogen ash and sulfur quantity being very less. In contrast to other conventional fuels, less amount of harmful gases (like sulfur dioxide, nitrogen oxides, and smoke) are delivered to the environment during the burning of biomass. As a result, the burning of biomass is comparatively favorable to the environment by reusing carbon dioxide *via* photosynthesis, its outflow can also be constrained (Zaman et al., 2017). Lignocellulosic biomass has several benefits as a huge organic renewable resource. To mention a few: (1) the biomass has cellulose and hemicellulose aplenty having biofuel bioconversion potential; (2) it is available in excess without any appreciated alternate use; and (3) the expenses of disposing of them and carbon emission could be decreased.

Cellulose, hemicellulose, and lignin are the three main constituents of lignocellulose. The expense in processing lignocellulosic biomass is overpriced because of its recalcitrance and the complexity of its hierarchical structure. Particularly, a lignol-based protective covering on the plant cell wall hinders microbial activity and also restricts the surface area for hydrolytic enzymes to act upon for fuel biotransformation. As a result, pretreating biomass before transforming lignocellulosic biomass into fuel is essential. Pretreatment is done to disarrange the lignocellulose layers and raise the absorbability of the biomass. There are three methods to pretreat, physical, chemical, and biological. Biological pretreatment receives more acceptability due to its several benefits. These include zero involvement of noxious chemicals, low energy requirements, decreased pollution, and increased yield. These days, a huge quantity of agrowaste produced is this lignocellulosic biomass, leading to more research attention in waste management and its value-addition (Singh et al., 2019). This paper focuses on the biogasification of agricultural biomass toward a zero-waste circular agrieconomy.

11.2 LIGNOCELLULOSE AS A BIOGASIFYING SUBSTRATE

Any organic material which is capable of storing its energy via the cycle of photosynthesis is referred to as biomass. Biomass comprises algae, crops, wood, waste, and so on. On the basis of the type, biomass-based biofuel could be categorized into four generations:

1. First-generation biomass: Agricultural crops that are food to humans.
2. Second-generation biofuels: Biofuels obtained from biomass of lignocellulosic nature comprising cellulose, hemicellulose, and lignin.
3. Third-generation biofuels: Biofuels derived from algal biomass (microalgae and macroalgae).
4. Fourth-generation biofuels: Biofuels derived from engineered plants and organisms designed to possess the desired qualities (Krishna et al., 2019).

Plant biomass that can be classified into four categories – hardwood, softwood, agricultural wastes, and grasses are referred to as lignocellulosic biomass (Hassan et al., 2018). Glucose (C6 sugar) and C5 sugars are the monomeric units for cellulose and hemicellulose respectively. In the case of lignin, the monomeric units are coumaryl alcohol, sinapyl alcohol, and coniferyl alcohol. These three types of alcohol are connected to one another in a complicated manner (Krishna et al., 2019). Strangely enough, there is a production of an extremely huge amount of residues from agricultural fields every year throughout the planet, but most of these residues are either disposed of or burned down. Any food item that is thrown away obtained from the food supply chain is referred to as food waste. This food waste might be either retrieved so that it can be used alternatively or is discarded away (Östergren et al., 2014; Hassan et al., 2018). Food processing and agriculture produce a large amount of lignocellulosic waste which can be used as feedstock for the second-generation of sustainable biorefineries. Polysaccharides (like cellulose and hemicellulose) and lignin are the major constituents of plant biomass. Being polymers of sugar polysaccharides can be a probable origin of fermentable sugars. On the other hand, various chemicals can be produced from lignin. In general, a greater portion of lignocellulosic biomass is contained in residues of cereal crops like wheat straw, rice, straw, sugarcane bagasse. As a result, they are the most favorable feedstock for biorefineries. On the other hand, the content of lignocellulosic biomass in the wastes of vegetables, fruits, and grasses is comparatively quite low. On the basis of dry weight lignocellulosic biomass is comprised of cellulose (50), hemicellulose (10–40% in woods and herbs), and lignin 10–40% in woods and herbs) (Sharma et al., 2015). Nevertheless, various factors like the season of harvesting, the condition of culturing the plant, and the plant's age may lead to variation in the proportion between cellulose, hemicellulose, and lignin in a single plant (Hassan et al., 2018).

11.3 NATURE OF LIGNOCELLULOSE BIOMASS

11.3.1 BIOCHEMISTRY

Cellulose, hemicellulose, and lignin are the three main constituents that make up a plant cell wall. A natural lignocellulosic substance also has these three organic compounds as its primary components. For the determination of the structure of the plant cell wall, there is a consistent arrangement of cellulose gathering into several bundles. On the other hand, hemicellulose and lignin fill up the fibers. The ratio of cellulose, hemicellulose, and lignin in a plant cell wall is 4:3:3. Depending on the source of lignocellulosic material (like hardwood, softwood, and herbs), this particular ratio varies (Chen, 2014). A schematic diagram of the main components of lignocellulosic biomass, its subunits, and the chemical bonding between them is shown in Figure 11.1.

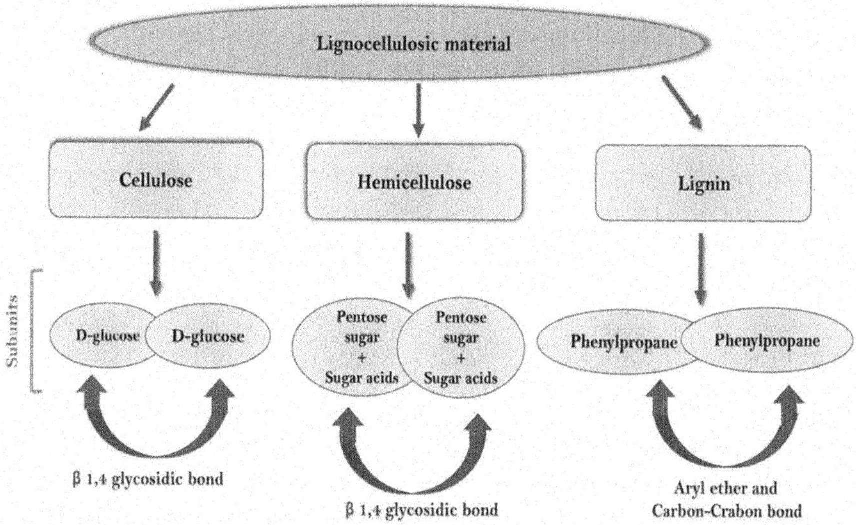

FIGURE 11.1 A schematic diagram of the main components of lignocellulosic biomass, its subunits, and the chemical bonding between them.

11.3.1.1 CELLULOSE

The main component that forms the structure of a plant cell wall is cellulose. It is made up of subunits of d-glucose, and these subunits are linked together via β-1,4 glycosidic bonds (Pérez et al., 2002; Sharma et al., 2017). Microfibrils of cellulose are formed by the polysaccharide chains of high length, which are not branched and are organized in a parallel form. The natural biopolymer having the highest abundance on earth is cellulose. The formation of a firm amorphous or crystalline structure is seen by the tight binding between the cellulose microfibrils connected by both intra- and inter-molecular hydrogen bonds (Sharma et al., 2017).

11.3.1.2 HEMICELLULOSE

Hemicellulose, by nature, is a complex carbohydrate. It is a branched polymer. It comprises a heterogeneous mix of pentose sugar like arabinose and xylose, hexose sugar like glucose and galactose, and sugar acids like glucuronic and galacturonic acids. The chemical bond that links these sugars is mainly the β-1,4-glycosidic bond. Occasionally the β-1,3-glycosidic bond also links these sugars (Joy et al., 2016; Sharma et al., 2017). The composition of hemicellulose differs according to the forms of wood. In the case of hardwood, glucomannans and xylans are present. Whereas in the case of softwood, xylans, xyloglucans, glucans, glucomannans, and arabinogalactans are present. The binding between cellulose microfibrils and hemicelluloses is formed due to pectin and lignin, which forms a network of cross-linking of a heterogeneous mix of pentose sugar and hexose sugar in the plant cell wall (Zhang et al., 2012; Sharma et al., 2017).

11.3.1.3 LIGNIN

By nature, lignin is complex, amorphous hetero-biopolymer, and is not soluble in water. It is made up of units of phenylpropane connected by aryl–ether and carbon-carbon chemical bonds. As a biopolymer, lignin, together with, cellulose is thought to have the highest abundance on earth (Pérez et al., 2002; Sharma et al., 2017). Transconiferyl alcohol, trans-sinapyl alcohol, trans-p-coumaryl alcohol are the three monolignols that make up the lignocellulose. These monolignols are linked among themselves by oxidative coupling. When polymers are formed from these monomers, the units of phenylpropane are referred to as guaiacyl, syringyl and p-hydroxyphenyl units (Cesarino et al., 2012; Sharma et al., 2017).

11.3.2 THE MICROBIOLOGY OF DEGRADATION

11.3.2.1 CELLULOSE DEGRADATION

Cellulose, which is a polysaccharide comprised of D-glucopyranose chains, is linked via β-1,4 glycosidic bonds. To degrade this bond three classes of enzymes are needed, which are-exoglucanases/cellobiohydrolases (CBH), β-glucosidase (BGL), and β-1,4-endoglucanases (EGL). The initial degradation is done by endoglucanases, which cleaves the internal position of the cellulose chain, primarily the amorphous portion. After this degradation is done by endoglucanases further units are released that needs degradation by BGLs and/or CBHs (Souza and Rodrigo, 2013). The cellulose-derived disaccharide units (cellobiose units) are cleaved from the furthermost part of the polysaccharide chains, by the enzyme cellobiohydrolases (Aron et al., 2005; Souza and Rodrigo, 2013). Ultimately, the hydrolysis of cellobiose to glucose is done by the enzyme β-glucosidase (BGL). After this, the monomer formed (carbon source) can be readily metabolized by fungi (Fontes and Gilbert, 2010; Souza and Rodrigo, 2013).

11.3.2.2 HEMICELLULOSE DEGRADATION

Due to its complex structure, hemicellulose needs a coordinated activity of three types of enzymes for its degradation. The first is the endo-enzymes that will cleave the main chain from an internal position. Second is the exo-enzymes whose action will lead to the release of monomeric sugars. The third is the accessory enzymes that will cleave the polymer side chains and connected oligosaccharides releasing several monosaccharides and disaccharides based on the type of hemicellulose. Xylan, the degradation of this type of hemicellulose is done by the activity of *β*-1,4-endoxylanase. This enzyme cleaves the backbone of xylan into smaller oligosaccharides. These oligosaccharides are further cleaved to form xylose by the *β*-1,4-xylosidase enzyme (Souza and Rodrigo, 2013). Xxyloglucan, the degradation of this type of hemicellulose is done by the action of two enzymes – endoglucanases (xyloglucanases) and *β*-glucosidases (Master et al., 2008; Desmet et al., 2007; Souza and Rodrigo, 2013). Mannans: This type of pf hemicellulose is degraded by the activity of two enzymes-*β*-endomannanases (*β*-mannanases) and *β*-mannosidases (de Vries et al., 2001; Souza and Rodrigo, 2013).

11.3.2.3 LIGNIN DEGRADATION

For the degradation of lignin polymer, the most suitable enzymes are – lignin peroxidase (LiP), laccase (Lac), manganese peroxidase (MnP), versatile peroxidase, and enzymes that produce hydrogen peroxide like aryl alcohol oxidase (AAO) and glyoxal oxidase (GLOX) (Souza and Rodrigo, 2013). A variety of oxidative reactions which require hydrogen peroxide are catalyzed by Lignin peroxidases (LiP) and manganese peroxidases (MnP) (Hattaka A., 2001, Souza and Rodrigo., 2013). Phenolic compounds are oxidized by the enzyme laccases to reduce molecular oxygen to water (Wong, 2009; Souza and Rodrigo, 2013). For purpose of lignin degradation, extracellular sources of hydrogen peroxide for supporting oxidative turnover of enzyme LiPs and MnPs that causes ligninolysis. Aryl alcohol oxidase (AAO) and glyoxal oxidase (GLOX) are the most suitable enzymes for the generation of extracellular hydrogen peroxide (Souza and Rodrigo, 2013).

11.4 CHALLENGES AND MITIGATION STRATEGIES IN BIOGASIFICATION

One of the most significant challenges in bio gasification is that the yield of biomass is less as it produces a low quantity of end product (methane). Consequently, one of the methodologies for expanding the yield of methane per unit area is to increase and enhance the yield of biomass (Sawatdeenarunat et al., 2014). The yield of biomass can also be enhanced by the determination of suitable species of biomass for a specific geographic area with enhanced effectiveness of photosynthesis, higher productivity for input usage (like irrigation and fertilizer), and immune to pests and diseases (Sims et al., 2006; Sawatdeenarunat et al., 2014). Additionally, another significant difficulty in biogasification is the transformation of lignocellulosic biomass into methane because of its complex structure. As a result, it is important to harvest biomass at a suitable stage of maturity. Due to this, two important outcomes can be achieved. One is the improved yield of biomass, and the second is the easy conversion of lignocellulosic biomass in the absence of severe pretreatment of the biomass. Also, another restriction in processing lignocellulosic biomass is the absence of a good digester for dealing with lignocellulosic feedstocks. The treatment plan for wastewater contributed to the understanding of the recent designs of anaerobic digester. The main purpose of developing this anaerobic digester was to handle low solids feedstocks. A lot of energy is required in operating this kind of digester for the digestion of lignocellulosic material, particularly while the biomass is mixed. As a result, the final energy yield is quite low. As a result, to improve the energy equilibrium of the digestion of lignocellulosic material to methane, a suitable digester plan fit for the proficient treatment of high solids feedstocks is required. Ultimately, there is yet a need to develop a complete framework for productive usage of both the residue of digestate (which is the solid residue left after anaerobic digestion) and also for effluents. The process of enzymatic hydrolysis can break down the residues obtained from digestate (generally having a high content of cellulose) into simple sugar. After this process of enzymatic breakdown, products of high worth are obtained. However, because of its high nitrogen content, there is a need to treat the effluent before its discharge into the environment. Currently, the integration of anaerobic digestion with the production of microalgae has caused algae to utilize the nutrients in the effluents. Biodiesel can be produced from the lipids obtained from agal production while the effluents produced from agal production can further be reused to operate an anaerobic digester (Sawatdeenarunat et al., 2014).

11.5 PRETREATMENT

Various pretreatment strategies have been implemented to enhance the biomass digestibility for anaerobic digestion. To enhance the cellulose degradation, a key approach has been introduced by increasing the surface area of feed-stock where more microbes or enzymes are attached to the biomass surface (Yang et al., 2015). To outweigh the cost of particle size reduction, the size reduction in consortium with the consumption of energy needs to be thoroughly investigated to enhance the anaerobic digestion (Yang et al., 2015). Another bio-mechanism such as biomass de-crystallization is used to promote biogas digestion. To improve the digestibility and reduce biomass crystallinity, different pretreatment techniques have been used during the process of AD. A schematic representation of pretreatment of lignocellulosic biomass is provided in Figure 11.2 (Feng and Chen, 2008). Based on the efficiency of pretreatment in different biomass, pretreatment methods are classified into physical, biological, chemical, and physico-chemical methods. Tables 11.1–11.3 provide the different types of physical, chemical, and biological pre-treatment respectively, and their effects on biogas production of different biomass.

11.5.1 PHYSICAL PRETREATMENT

The surface area of the biomass can be increased by reducing the size of the particle with the help of physical methods of treatment. During AD, the size reduction can increase its susceptibility to the enzyme and microbial attacks and the accessibility of the biomass can be improved thus promoting biomass digestion. Importantly, there is no generation of any toxic compounds during physical treatments, which inhibit the process of AD (Zheng et al., 2014).

11.5.2 CHEMICAL PRETREATMENT

Based on the chemicals being used, chemical pretreatments are primarily classified into alkaline, organic, acid, and organo-solvent treatments. Biomass treatment involves major acidic agents which include hydrochloric, formic, sulfuric, and nitric acid. Agents such as ammonia and sodium hydroxide are commonly used during alkaline pretreatment. These treatments mainly aim to remove hemicellulose or lignin from the biomass.

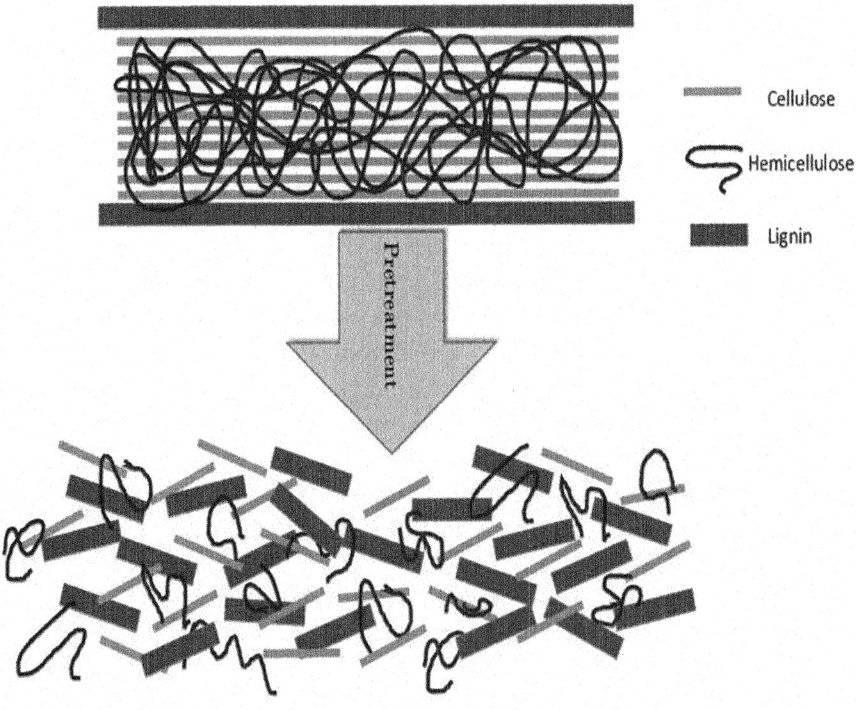

FIGURE 11.2 Pretreatment of lignocellulosic biomass.

11.5.2.1 ALKALINE PRETREATMENT

Alkaline such as NaOH pretreatment improved the digestibility of wheat straw by increasing the surface area, more porosity, and reduction in the degree of polymerization of lignocellulosic biomass thus enabling delignification thereby resulting in the methane and biogas yield by 112% and 88%, respectively (Chandra et al., 2012). It was further observed that grass silage when pre-treated with 5% NaOH, there was an increase in biogas yield by 38% (Xie, 2012). In another study, pretreatment of wheat straw when treated with 4% ammonia solution resulted in a 52% increase in biogas production as compared to a control (Yang et al., 2014). It was observed that when

TABLE 11.1 Physical Pretreatment and Their Effects on Biogas Production From Biomass

Pretreatments	Conditions	Biomass	AD	Methane/biogas	Reference	Advantages	Disadvantages
Irradiation	NaOH and microwave, 720 W; Alkaline and Microwave; Microwave, 400–1600 W	Paddy straw; Rice straw Oat; Wheat and Barley straw; Switch grass	Batch, 45 days, 37°C; Batch, 66 days, 35°C; Batch, 60 days, 37°C; Batch, 42 days, 35°C	55% increase in biogas production and 25% increase in biogas production; Zero net increase in biogas; No change in volume	Kaur and Phutela (2016); Qian et al. (2019); Sapci (2013); Jackowiak et al. (2011)	Increases surface area, Destruction of cellulose structure	High energy consumption
Steam	Steam explosion, 5, 10, 15 min, 160–220°C; Steam explosion, 1–4 min; Steam explosion, 5, 10, 15, 20 min, 160–220°C; Steam explosion, up to 15 min, 160–205°C, and 5–15 bar Steam explosion, 10 min, 210°C	Late harvested hay; Rice straw; Reed biomass; Rice straw; Birch wood	Batch, 37.5°C; Batch, 21 days, 38°C; Batch, 63 days, 37.5°C; Batch, 60 days, 37.5°C; Batch, 50 days, 62°C	16% increase in methane production and 51% increase in biogas production; 89% increase in methane production and increase in biogas production up to 147%; 118% increase in biogas production	Bauer et al. (2014); Zhou et al. (2016); Lizasoain et al. (2016); Aski et al. (2019); Mulat et al. (2018)	Destruction of cellulose crystallinity, hydrolysis of lignin and hemicellulose	Recalcitrant compounds formation, high energy demand
Hydrothermal	Hydrothermal, 60 min, 120–180°C; Hydrothermal, 15 min 125–200°C; Hydrothermal, 150 min, 90–130°C; Hydrothermal, 30–50 min, 120–180°C	Wheat straw; Napier grass; Rice straw; Safflower straw	Batch, 45 days, 35°C; Batch, 42 days, 37°C; Batch, 30 days, 35°C; Batch, 45 days, 37°C	53% increase in biogas production and 35% increase in methane production; 22% increase in methane production and 70% increase in biogas production	Rajput and Visvanathan (2018); Phuttaro et al. (2019); Luo et al. (2019); Hashemi et al. (2019)	Destruction of cellulose structure, hydrolysis of lignin and hemicellulose	High energy input, High water consumption

TABLE 11.2 Physical Pretreatment Measures for Different Biomass and Their Effect on Biogas Production

Pretreatments	Conditions	Biomass	AD	Methane/biogas	Reference	Advantages	Disadvantages
Extrusion	Twin-screw extruder; Single screw extruder; Variable hydro-module, 0.45–2.7 MPa, 4–12 min Co-rotating twin screw extruder	Straw, manure, grass, and deep litter; Maize straw silage; Hay Biomass; Deep litter, Wheat straw	Batch,90–110 days, 35°C; Batch, 27 days, 38°C; Batch, 20 days, 38°C; Batch, 90 days, 35°C	After 90 days 9–28% increase in methane 35% increase in methane production; 33% higher biogas production; after 90 days 1–16% increase in methane	Hjorth et al. (2011); Pilarski et al. (2016); Maroušek (2012); Wahid et al. (2015)	Disruption of cellulose crystallinity and hydrogen bonds, fast heat transfer, increase surface area, short reaction time	Scalability issues, High electricity demand
Grinding/ Milling	Plate Method; Mechanical knife mill 300 μm and 1200 μm particle size; Brush and a steel mesh conditionerGrinding: 10 and 25 mm, 0–100°C	Ensiled meadow grass Mexican sunflower, Siam weed, Elephant grass, Wheat straw, Meadow Grass, Rice Straw	Batch, 20 days, 54°C; Batch, 30 days, 37°C; Batch, 28 days, 37°C; Batch,27 days, 54°C; Batch, 25 days, 35°C	25% increase in methane yield 22% increase in methane production; 49.3% increase in methane yield; 27% increase in biogas production; 17.5% increase in biogas production	Tsapekos et al. (2015); Dahunsi (2019); Dell'Omo and La Froscia (2018); Tsapekos et al. (2018)	Reduction of particle size and crystallinity, ease substrate management, increased surface area	High electricity demand

TABLE 11.3 Chemical Pretreatments and Their Effect on Biogas Production

Pretreatments	Conditions	Biomass	AD	Methane/biogas	References	Advantages	Disadvantages
Acid	Di. H_2SO_4 (2,4,6%) 30°C, 2 days; Dil. H_2SO_4, HCl, CH_3COOH and H_2O_2 (1,2,3,4%), 25°C, 7 days; Dil. H_2SO_4 (1%), 121°C, 10–120 min; Dil. H_2SO_4 (5%), 121°C, 60 min.	*Salvinia molesta;* Corn stover; Wheat straw; Water hyacinth	Batch, 30 days, 30°C; Batch, 35 days, 37°C; Batch, 30 days, 37°C; Batch, 90 days, 28–30°C	5% increase in biogas production; 115% increase with 3% H_2O_2; 16% increase in methane production; 131% increase in biogas production	Syaichurrozi et al. (2019); Song et al. (2014); Taherdanak et al. (2016); Sarto et al. (2019)	Alternation of cellulose structure, Hydrolysis of hemicellulose	Formation of inhibitors, High cost of acids
Oxidative	Ozone (6–81 mg O_3/g CH), pH 3–11; Ozone, Fenton and ozone combined with Fe(II) and H_2O_2; H_2O_2 (2.68%), 6–18 days; H_2O_2 (1%), 50°C, 6 h; H_2O_2 (4%), 55°C, 24 h	Hemicellulosic hydrolysate of coffee husk; Agricultural wastes; Rice straw; Greenhouse crop waste; Sunflower stalk	Batch, 40 days, 35°C; Batch, Semi-batch,20 days, 35°C; Batch, 41 days, 37°C; Batch, 62 days, 35°C; Batch, 35 days, 35°C	27% increase of methane in two-stage anaerobic digestion; 30% increase in methane production; 88% increase in methane production; 77% increase in methane production; 33% increase in methane production	Santos et al. (2018); Almomani et al. (2019); Song et al. (2013); Perendeci et al. (2018); Monlau et al. (2011))	Removal of lignin and hemicellulose	Formation of inhibitors, Chemical cost
Alkali	NaOH (5%), 100°C, 4–48 h; NaOH (4%), 37°C, 5 days; Ammonia (2,4,6%), 35°C, 7 days; Ammonia (0–30%), 20–80°C, 6–48 h; $Ca(OH)_2$ (7.5%), 10°C, 20 h; NaOH (1.6%), 30°C, 24 h	Grass silage; Wheat straw; Maize straw, sprout stem, and grass; Wheat straw	Batch, 90 days, 35°C; Batch, 60 days, 37°C; Batch, 60 days, 35°C; Batch, 28 days, 30°C; Batch, 30 days, 37°C; Batch, 40 days, 37°C	38% increase in methane production; 87.5% increase in biogas production; 52% increase in methane production; 40% increase in biogas production; 37% increase in methane production; 15% increase in methane production	Xie (2012); Chandra et al. (2012); Yang et al. (2014); Li et al. (2015a,b); Khor et al. (2015); Mancini et al. (2018)	Alternation of cellulose structure, hydrolysis of lignin	Formation of inhibitors, Cost of alkali

TABLE 11.3 *(Continued)*

Pretreatments	Conditions	Biomass	AD	Methane/biogas	References	Advantages	Disadvantages
Organic Solvent pretreatment	NMMO, 120°C, 2 h; NMMO, 130°C, 1–15 h; NMMO, 90°C, 7 h; BMIM.Cl/DMSO, 120°C, 120 min NMMO, 90 and 120°C, 1,3,5h;75% ethanol and 1% H_2SO_4; 50% or 70% ethanol; NMMO, 120°C, 3 h	Waste textiles; Softwood spruce, rice, and triticale Straw; Wheat straw; Water hyacinth Oil palm empty fruit bunch; Rice straw, pine & elm wood; Sweet Sorghum Stalks;	Batch, 6 days, 55°C; Batch, 20 days, 55°C; Batch, 45 days, 55°C; Batch, 30 days, 35°C; Batch, 50 days, 55°C; Batch, 55 days, 39°C; Batch, 50 days, 37°C; Batch, 40 days, 37°C; Batch, 40 days, 37°C	16% increase in methane yield; 400–1200% increase in methane yield; 47% increase in methane production; 98% increase in biogas production; Up to 167% increase in methane production; 35 – 58% increase in biogas production; 270% increase in biogas production; 11% increase in methane production; 15% increase in methane yield	Jeihanipour et al. (2010); Teghammar et al. (2012); Akhand and Blancas (2012); Gao et al. (2013); Purwandari et al. (2013); Mirmohamadsadeghi et al. (2014); Ostovareh et al. (2015);	Solubilization of cellulose or hemicellulose/ lignin	Solvent removal steps, Solvent cost

rice straw was pre-treated with KOH and NaOH, there was an increase in 50% of the production of biogas as compared to that of control (Dong et al., 2009). Despite, pretreatment with NaOH being beneficial for the production of biogas enhancement, effective measures should be taken in the formation of Na^+ ions to avoid AD inhibition.

11.5.2.2 ACID PRETREATMENT

When the biomass is pre-treated with acid, covalent bonds, hydrogen, and Vander walls forces get disrupted as a result cell lysis and disintegration of biomass takes place leading to the hemicellulose elimination from the lignocellulosic biomass. Dilute sulfuric acid when pretreated with wheat straw at high temperature, there was an increase in 16% of methane yield (Taherdanak et al., 2016). During the AD process, when rice straw and *Salvinia molesta* were pre-treated with an acid, it resulted in an increase of 5% in the production of biogas (Syaichurrozi et al., 2019). A comparative study was made between three different alkaline reagents and four different acid reagents from corn straw with respect to methane yield (Song et al., 2014) with 8% $Ca(OH)_2$ and 3% H_2O_2 providing the maximum yields of 105% and 115%, respectively. Higher methane yield was produced when pretreatment techniques like acid, alkaline treatment, hot water were used as compared to others (Ikeda et al., 2019).

11.5.2.3 ORGANIC SOLVENT PRETREATMENT

Separation of high purity cellulose from lignocellulosic biomass can be done with the use of organic solvents without causing degradation of the cellulose. When elmwood, pinewood, and rice straw were pre-treated with 1% H_2SO_4 and 75% ethanol for 0.5–1 hr at 150–180°C, there was an increase in the production of biogas in 73, 85, and 35% after the process of AD (Mirmohamadsadeghi et al., 2014). When sweet sorghum stalks were pre-treated using 50% ethanol for 30min at 160°C, there was an increase in 270% of biogas production than control (Ostovareh et al., 2015). Cellulose can get dissolved with the help of Ionic Liquids (ILs) which can be recovered completely after the process of pretreatment leaving no negative effects on the process of AD. Among the different ILs available, the most commonly used ILs for biomass pretreatment is N-methyl morpholine-N-oxide (NMMO) (Kabir et al., 2014) which when pre-treated with wheat straw has been shown in the increase in production of methane to 47% than control conditions during AD process (Akhand et al., 2012).

11.5.2.4 OXIDATIVE PRETREATMENT

The structure of hemicellulose and lignin get destructed by oxidative Pretreatment in lignocellulosic biomass thereby promoting the side chains dislocation, electrophilic substitutions, breakage of aromatic nuclei, and alkyl aryl ether bonds cleavage (Paudel et al., 2017). When lignocellulosic biomass is acted upon during oxidative pretreatment, hydroxyl radicals (OH^-) are formed by the conversion of hydrogen peroxide. After the process of pretreatment, the portions of cellulose and hemicellulose get lost (Almomani et al., 2019). When agricultural wastes were pretreated with ozone, Fe(II), fenton, and ozone combined with H_2O_2, and was observed that there is a 25% increase in the digestion process and 30% increase in production of methane when $O_3|H_2O_2$ was used for oxidative pretreatment. The performance of AD was improved by performing the oxidative treatment using ozone on coffee husks resulted in a production of methane of 85mL/g (Santos et al., 2018).

11.5.3 PHYSICOCHEMICAL PRETREATMENTS

A combined method before the AD process is utilized for the degradation of lignin polymers or hemicellulose within lignocellulosic biomass at a temperature reported from 150°C to 180°C temperature (Dahadha et al., 2017). where the surface area is increased for efficient microbial action towards biomass by heat resulting in the breakage of hydrogen bonds between the complex polymers (Rodriguez et al., 2017).

11.5.3.1 HYDROTHERMAL PRETREATMENT

Biomass is treated with the help of water at 200°C under high-pressure conditions, during the hydrothermal process (Mansini et al., 2018). leading to the formation of furan derivatives and undesirable phenolic compounds, thereby resulting in a 35% increase in the production of methane for incubation of 15min at 175°C temperature (Phuttaro et al., 2019). A study was made from wheat straw at an optimum temperature of 180°C, which resulted in the production of methane to 53% as compared to control (Rajput et al., 2018). A comparative study was made between safflower at 100°C and wheat straw at 100°C and 120°C which resulted in a 98% and 20% rise in production of methane as compared to control (Hashemi et al., 2019).

11.5.3.2 STEAM-EXPLOSION PRETREATMENT

Under high pressure at 160–250°C, the structure of lignocellulosic biomass is altered by the steam explosion pretreatment (Paudel et al., 2017). A case study was made where reed biomass was pre-treated with steam for 15min at 200°C and resulted in 89% of methane production as compared to the control (Lizasoain et al., 2016). Birchwood biomass was pre-treated by a combined effect of bioaugmentation and steam explosion, which increased the biogas yield by 140% than the control whereas the single effect of steam explosion on the biomass of birch wood increased the yield of methane by 118% indicating that the methane production is higher in case of combined effect as compared to single effect (Mulat et al., 2018). When rice straw was steam exploded, methane production was observed up to a recovery rate of 147% (Aski et al., 2019).

11.5.3.3 EXTRUSION PRE-TREATMENT

In extrusion pre-treatment, the biomass containing lignocellulose undergoes a series of operations such as mixing, heating, and sudden drop in pressure during the release of the material from the extruder causing the structural breakdown and release of intracellular water from the feedstock, thereby improving the digestion of biomass (Zheng et al., 2014). The operating parameters during extrusion pretreatment include moisture content, reaction time, and pressure of the biomass (Marousek et al., 2012). During the process of AD, in extrusion pretreatment of deep litter and wheat straw, methane production was increased by 1–16% after 90 days (Wahid et al., 2015). Extrusion pretreatment of maize silage increased the production of methane by 35% (Pilarski et al., 2016). Excessive supply of heat during the production of methane results in toxic compound formation such as furfural (Zheng et al., 2014).

11.5.4 BIOLOGICAL PRETREATMENT

During the process of AD, the yield of biogas is enhanced with the help of microbial pretreatment (Mutschlechner et al., 2015, Table 11.4). Lignocellulosic substrates involve biological pretreatments which include the direct mode of by-products and metabolism of microbes (Zheng et al., 2014). As compared to the chemical or physical process, biological pretreatment

TABLE 11.4 Effect of Biological Pretreatment on the Biomass and Their Subsequent Outcome on the Biogas Production

Pretreatments	Microorganism (s)	Biomass	AD	Methane /biogas	Reference	Advantages	Disadvantages
Enzyme	Mn Peroxidase + Versatile Peroxidase; Endoglucanase+Exo glucanase+Xylanase; Laccase; Cellulase; Cellulase+Cellobiase; Endoglucanase+Xylanase +Pectinase; Endoglucana se+Xylanase+Pectinase	Corn stover; Sorghum forage; Corn stover; Switch grass; Spent hops; Sugar beet pulp silage	Batch, 30days, 37°C; Batch, 30days, 35°C; Batch, 30 days, 37°C; Batch, 18 days, 37°C; Batch, 30 days, 50°C;	17% higher methane production;15% higher methane production; 25% higher methane production; 36.9% higher biogas production; Methane yield 274.28 ml g^{-1} (VS); 13% higher biogas production	Schroyen et al. (2014); Rollini et al. (2014); Schroyen et al. (2014); Wang et al. (2018); El-Mashad (2015); Zieminski et al. (2012); Zieminiski and Kowalska (2015)	Alternation of cellulose structure, delignification, fast process, partial hydrolysis of hemicellulose, low energy demand	High cost of enzymes, continuous addition may require
Bacterial	*Bacillus subtilis* *Bacillus* sp. *Citrobacter werkmanii* VKVVG4	Corn straw; Rice straw; Water hyacinth	Batch (SS), 50 days, 37°C; Batch (SS), 50 days, 37°C; Batch (SS), 80 days, Mesophilic	17.35% higher methane production 76% higher biogas production 3.07 times higher biogas production	Xu et al. (2018) Shah et al. (2019) Barua et al. (2018)	Low energy consumption, no inhibitory compound formation	Slow process
Microbial consortium	Rumen fluid; Microbial consortium TC-5; Microbial consortium	Rice straw; Saw dust; Wheat straw; Wheat straw	Batch, 30 days, 35°C; Batch, 28 days, Mesophilic; Batch, 35 days, 45°C; Batch, 20 days, 37°C	82.6% higher methane production and 25.6% higher biogas production; 36.6% higher methane production and 80.34% higher methane production	Zhang et al. (2016); Alia et al. (2017); Kong et al. (2018); Zhong et al. (2016)	Low energy consumption, no inhibitory compound formation, Hydrolysis of lignin and hemicellulose	Carbon losses, slow process, the necessity of a large sterile area

TABLE 11.4 *(Continued)*

Pretreatments	Microorganism (s)	Biomass	AD	Methane /biogas	Reference	Advantages	Disadvantages
Fungal	*Trametes versicolor; Polyporusbrumalis; Flammulina velutipes; Pleurotus eryngii; Trichoderma reesei; Pleurotus ostreatus; Ceriporiopsis subvermispora; Ceriporiopsis subvermispora*	Corn silage; Wheat straw; *Agropyrone longatum;* Corn stover Rice straw; Yard trimmings;	Semi continuous, 21 days, 37°C; Batch, 57 days, 36°C; Batch, 24 days, 37°C; Batch, 40 days, Mesophilic; Batch (SS), 45 days, 37°C; Batch (SS), 45 days, 37°C; Batch (SS), 45 days, 37°C	Methane generation rate 0.236 m^3 CH4 $KgVS^{-1}$; 52% higher methane and 120% biogas production; 19% higher biogas and 78.3% higher methane production; 120% higher methane production; 106% higher methane production	Tisma et al. (2018); Rouches et al. (2018); Lalak et al. (2016); Wyman et al. (2018); Mustafa et al. (2016); Mustafa et al. (2016); Zhao et al. (2014); Ge et al. (2015)	Low energy consumption, no inhibitory compound formation	Slow process

is considered to be an eco-friendly process due to the use of mild process conditions and the need for less consumption of energy (Shrestha et al., 2017). During the process of AD, strategies involved in biological pretreatment are more compatible as compared to non-biological pretreatment as no formation of toxic by-products occurred in biological pretreatment but in the case of non-biological pretreatment, a processing step is not required for AD after the generation of the hydrolysate from biological pretreatment (Wagner et al., 2018). The primary objective involved in biological pretreatment is delignification which can expose the fractions of hemicellulose and cellulose capable of efficient digestion (Shrestha et al., 2017).

11.5.4.1 ENZYME PRETREATMENT

Under anaerobic conditions, the degradation of lignocellulosic biomass can be enhanced by the exogenous application of oxidative or hydrolytic classes, enzymes (Koupaie et al., 2019). During the process of digestion, the demand for enzyme pretreatment has been increased due to minimal sugar loss and short-reaction times (Romero-Güiza et al., 2016). The efficiency of pretreatment of enzymes is determined by the specificity of the enzyme towards the substrate, incubation time, the anaerobic digestion system, tolerance of the enzymes to various inhibitors, the activity of the enzyme, stability of the enzymes at various pH and temperatures (Kiran et al., 2015; Thomas et al.,2016). Depending upon the activity towards lignocellulosic biomass, the enzymes are categorized into lignin-removing enzymes/ hemicellulose degrading/ cellulose. Production of biogas can be imposed with the help of enzyme pretreatment from recalcitrant lignocellulosic biomass (Koupaie et al., 2019). During the AD process, the digestibility of a complex substrate can be increased by the breakdown of complex polymers with the help of hemicellulose or cellulase enzymes. A pretreatment study on corn-stover was made which resulted in the enhancement in the biogas yield by up to 36.9% and improvement in the rate of hydrolysis with the help of cellulase (Wang et al., 2018). The oxidative degradation of lignin which is H_2O_2-dependent is catalyzed by the hem-containing enzyme known as LiP. Oxidation of Mn^{+2} to Mn^{+3} is due to the Mn peroxidase enzyme which helps in the breakage of the bond of lignin as a result, LiP oxidizes the region of non-phenolic rings, and the region of phenolic regions are oxidized by MnP (Houtman et al., 2018). while both lignin polymer compounds are oxidized by VP (Zavarzina et al., 2018).

11.5.4.2 MICROBIAL PRE-TREATMENT

The microbial pre-treatment strategy is widely used as a fungal treatment for lignocellulosic biomass (Rouches et al., 2016). During anaerobic digestion, when wheat straw was microbial pre-treated with *Polyporus brumalis* BRFM985, it was observed that biogas yield was improved by 52% and the process of AD was influenced by the incubation time and addition of metal (Rouches et al., 2018). Corn silage was pretreated with the use of the fungus *Trametes versicolor* and was observed that there was an enhancement in the yield of biogas up to the rate of 0.236m^3 CH_4/KgVS (Tisma et al., 2018). The efficiency of lignocellulosic biomass pre-treatment was screened by using fungi such as *Trametes versicolor, Pleurotus eryngii, and Pleurotus ostreatus* and was studied in the case of *Pleurotus eryngii*, production of ligninolytic enzyme was higher with a 19% increase in the yield of biogas (Wyman et al., 2018). It has been reported that several aerobic species of bacteria show a high potential of degradation towards biomass containing lignocellulose and a faster rate of growth as compared to fungi (Xu et al., 2018). Different bacterial species show some unique features on mechanisms of degradation of lignin as compared to those of fungal species in terms of Cβ- Cβ bond cleavage and Cα-oxidation in the lignin polymer. Fungi have a higher yield of lignolytic enzyme efficiency as compared to bacteria, hence for better lignocellulosic biomass delignification, more research is needed.

In rice straw, lignin content was reduced by the introduction of co-culture of Bacillus sp., and was observed that methane production was increased by 76% via AD (Shah et al., 2019). During the process of AD, stabilization of water hyacinth was improved by a bacterial strain, *Citrobacter werkmanii* VKVVG4, which is isolated from the gut of silverfish and resulted in the yield of biogas by three-fold rise as compared to control conditions (Barua et al., 2018). In another study, methane yield was increased by 72.6% from a microbial consortium, isolated from rotten saw dust as compared to the control treatment (Alia et al., 2017).

11.6 AGRICULTURAL BIOMASS-BASED CIRCULAR BIOECONOMY

A circular bioeconomy is the production of a value-added output from any process in a sustainable way. The process must be resource-efficient by proper utilization of available resources. When it comes to the production of materials and energy supply, biomass is of high importance in circular bioeconomy. Out of all the different forms of bioenergy, bio gasification is perceived as the most cost-proficient one, which requires a low skill set and is also not environmentally and chemically harmful (Chandra et al., 2012; Mishra et al., 2018). As more than 30–40% of demand for fuel is satisfied by the production of biogas, an altogether huge number of studies all over the world utilizing various advancements are accounted for. Smoke and particulates produced due to the usage of traditional fuels have caused various health-related problems and biomethanation. As a result, the use of biofuels remarkably helps in diminishing air contamination and respiratory infections. Moreover, a high input of energy is required for most of the pretreatments like the physical, chemical, and physicochemical ones. Due to this, several byproducts are produced which are harmful to the environment and also have an unfavorable effect on the production of methane (Paudel et al. 2017; Mishra et al., 2018). To cater to such problems, different industrial enzymes that can degrade cellulose and lignin can be used for the degradation of lignocellulosic material for biomethanation (Lopez et al., 2007; Mishra et al., 2018). As the unit price of industrial enzymes is high, their usage for biological pretreatments is not very profitable. Nevertheless, using raw enzymes to degrade lignocellulose indirectly via a biological pathway utilizing a consortium of microbes, can likely be considered as suitable candidates for the production of biomass and its pretreatment (Mishra et al., 2018).

11.7 POTENTIAL AGRICULTURAL BIOMASS FOR BIOGASIFICATION

The chief biofuel items obtained from the distinctive biochemical courses of several agricultural biomasses include biogas, bioethanol, and biodiesel. Unlike the first-generation biofuel, the main focus of the second-generation biofuel is on agricultural residues and non-food biomass. The agricultural residues incorporate waste from vegetable and fruit, crops and crop by-products, manure, wood and wood waste, and residues from various other processes of agriculture (Paudel et al., 2017).

11.7.1 AGRICULTURAL RESIDUES

Residue from cereal is a plentifully available resource renewable in nature, consisting of residues both while processing and produced on-site. Materials left behind in the crop field constitute the on-site residues while residues post-harvesting of crop includes leaves, straw, seedpods, and stems. Materials that are left behind post-processing the cereal crop like seeds, husk, roots, and bagasse constitute the processing residue. A large quantity of agricultural residue comes from cereal crop harvest (Paudel et al., 2017). If biomass availability is taken into consideration, the primary agricultural residues would include cereal residues like wheat straw, rice straw, sugarcane bagasse, and corn stover (Kim and Dale., 2004; Paudel et al., 2017).

11.7.2 FRUIT AND VEGETABLE WASTES

All over the world, there is a production of a huge volume of by-products and wastes obtained from vegetables and fruits. These by-products and wastes are generated at the processing industries of certain crops like grapes, bananas, apples, tomatoes, potatoes, and citrus. Even though these unconventional resources have all the qualities of a suitable bioenergy source, a large part of these by-products are thrown away, used in a landfill, on just reused as fertilizer (Barbosa-Cánovas et al., 2003; Paudel et al., 2017). Combustion or thermochemical transformation is not a method suitable for the treatment of food waste because the moisture content is high in it. As a result, a preferable treatment method in the case of food waste is anaerobic digestion (Zhang et al., 2007; Paudel et al., 2017).

11.7.3 GRASSES AND GRASS SILAGE

Another substitute for bioenergy feedstock, which is gaining popularity, is biomass from grassland since its utilization as a feed for animals, in general, is diminishing (Tsapekos et al., 2015; Paudel et al., 2017). Based on the type and purpose of harvesting, grassland can be categorized into two types: temporary grassland and permanent grassland (Meyer et al., 2017; Paudel et al., 2017). In the case of temporary grasslands, the total production of biomass from grass is evaluated to be 300 Mt, and the amount of feedstock energy produced from the residues is evaluated to be 10% of the toral production (Paudel et al., 2017). Similarly, for permanent grasslands, the total production of biomass from grass is evaluated to be 11,124 Mt, and the amount of feedstock energy produced from the residues is evaluated to be 30% of the total production (Meyer et al., 2017, Paudel et al., 2017).

11.7.4 ANIMAL MANURE

Various factors determine the production of animal manure and its attributes. These factors include animal type, climate, feed distribution. The health condition and age of the animals, etc. Out of the many techniques of recycling, bioenergy produced from animal manure turned out to be most popular with several advantages. These advantages include the reduction of greenhouse gases (Nitrous oxide, methane, carbon dioxide), the prevention of environmental pollution, and acquiring valuable energy in the form of residue (Paudel et al., 2017).

11.8 DIGESTATE MANAGEMENT AND APPLICATIONS

11.8.1 METHODS OF DIGESTATE APPLICATION

For digestate application generally, the same equipment is utilized that is used to apply raw slurry and the isolated liquid on the fields. Similarly, the equipment utilized to spread manure obtained from farmyards can likewise be utilized for spreading isolated solids. To guarantee its ideal use in the form of fertilizer, the application of digestate to fields must be done during the crop growing season. And also, the application of digestate must be done with such a piece of equipment that will guarantee even application over the entire field and also ensure the precise rate of application. The process of ammonia volatilization will also get reduced if this methodology is followed. Despite the fact that splash plate application is utilized in numerous places even now, this technique is not suggested for applying digestate to fields. This is because splash plate application conveys a high danger of volatilization of ammonia which is an unwanted ecological impact. Due to this, the technique of splash application isn't just ill-suited, but on the other hand, is restricted in certain nations. In comparison with crude slurry, digestate spreads less smell, permeates all the more rapidly into the soil, and has a much lesser threat of inconvenience caused due to odor during and post spreading. Notwithstanding, because the content of ammonia is higher in digestate as compared to crude slurry the capability for the volatilization of ammonia during and post digestate application is more prominent. Minimization of the surface zone presented to air and guaranteeing contact with the soil are henceforth the most appropriate strategies for the application of digestate. These strategies are namely the trailing hoses, trailing-shoes, and injection. Advantages like lower pollution, lower loss of nutrients, increased usage of nutrients in the digestate counterbalances the greater expenses of these strategies in comparison to the splash plate method of spreading (Lukehurst et al., 2010).

11.8.2 MANAGEMENT OF NUTRIENTS IN BIOGAS DIGESTATE

When the uptake of water is almost nothing during autumn and winter, the application of digestate, in general, prompts the draining of nutrients into both groundwater and surface water. Therefore, it is crucial to store digestate till the suitable season of application arrives. Generally, the requirement of potassium, phosphorus, and nitrogen by crops is satisfied by digestate application (Singh et al., 2019). Before the process of digestate application, the demand for nutrients by crops must be taken into consideration (Cavalli et al., 2016; Singh et al., 2019). Some certainly crucial points that must be taken into consideration before applying digestate are the application timing, period of storage, and the rate at which nitrogen is applied. Any sort of negative influence upon the environment can be reduced by following the above-mentioned points. A danger of fertilizing the soil in excess persists, which may lead to pollution in both the surface water and groundwater, thereby degrading the quality of drinking water. Partial N absorption is caused due to an unsuitably high quantity of fertilizer. To get rid of such issues, the management of nutrients is a crucial measure. The maximum level at which a particular nutrient must be present in digestate is decided by the feedstock (input material) (Singh et al., 2019). To avoid all sorts of environmental contamination caused by pathogens, toxic substances, and heavy metals; the feedstock source for anaerobic digestion must be chosen very cautiously (Neshat et al., 2017; Singh et al., 2019). Regular sampling of digestate is necessary to analyze pathogens, toxic elements, heavy metals, and organic pollutants (Singh et al., 2019).

11.9 ENVIRONMENTAL IMPACT OF BIOGAS TECHNOLOGY

Being a sustainable and legitimate option in contrast to petroleum derivatives is considered as one of the most crucial advantages that biogas technology provides to the environment (Cecchi et al., 2015; Paolini et al., 2018). Because the energy potential of biogas is high, it can decrease the emission of Greenhouse gas along with improving the energy security of energy (Tricase et al., 2009; Angelidaki., 2009; Paolini et al., 2018). Being a renewable source of energy, it permits the exploitation of by-products obtained from the zootechnical, municipal, and agricultural sectors. Also, the influence it has on the quality of air is lower in comparison to burning-based systems for this biomass (Fruergaard et al., 2010; Giovanis et al., 2015; Domingo et al., 2015; Paolini et al., 2018). Besides, although there is hardly any application of remains from ignition in agriculture (Freire et al., 2015; Vakalis et al., 2017; Paolini et al., 2018), the by-product obtained from AD (digestate) is considered a suitable substance for farming employments (Tambone et al., 2015; Paolini et al., 2018). One more significant benefit of the biogas innovation is its easy quantification of biogas, which permits the exploitation of the energy capacity of decentralized biomass sources (Delzeit et al., 2013; Sarkilahti et al., 2017; Paolini et al., 2018). Ultimately, biogas can be reformed to biomethane, appropriately utilized as a fuel for the vehicle, or infused into national natural gas grids (Yang et al., 2014; Bekkering et al., 2015; Paolini et al., 2018). Reducing the emissions of Greenhouse gas is another significant benefit that biogas provides. Nevertheless, the unwanted emissions of nitrous oxide and methane must be given proper attention. The budget of emission of nitrous oxide and methane barely has any relation with the direct delivery that occurs through the combustion of biogas or biomethane, while the storage of biomass and managing the digestate are important steps. Similarly, proper attention must be given in the case of ammonia as well, to decrease its effect on the formation of a secondary aerosol. For this purpose, the systematic storage of both biomass and digestate is suggested. According to various case studies, the level of nitrous oxide emissions was worth some worry among all other pollutants emitted from the burning of biogas. Volatile organic compounds, on the contrary, do not appear to account for an important concern. When the effect of spreading digestate on the soil quality is considered, additional investigations are required to completely survey the long-term effect. When the medium-short term is considered, digestate is by all accounts ideal contrasted with untreated biomass. Improvement of the quality of air and reduction of greenhouse gas is generally caused due to the up-gradation of biogas to biomethane. Nevertheless, the sustainability of the entire process methane might get affected due to the loss of methane in the off-gas (Paolini et al., 2018).

11.10 CHALLENGES AND FUTURE PROSPECTS OF BIOGAS-BASED CIRCULAR BIOECONOMY

Farming practices are related to a significant ecological burden influencing every section of the environment. It leads to degradation of the quality of soil and water by consuming the scanty resources (like water, energy, land, etc.) both directly and indirectly. It also causes a threat to the environment by producing a large and diversified stream of waste which are not used proficiently. The most developed and extensive technology for the transformation of agricultural by-products is anaerobic digestion, being capable of producing almost 17,264 GWh energy (Gontard et al., 2018). Anaerobic Digestion still has various technological constraints and shortcomings. These include the low yield of the transformed lignocellulosic material (Ahring et al., 2015). Low financial incentive and also problems associated with the supply of feedstock and the storage and handling of digestate (Gontard et al., 2018). Developing bio-products from agricultural byproducts has huge opportunities. However, from a technological perspective, the process of transformation of biomass requires an upgrade and more extensive information about their capability and also economic, societal, and environmental sustainability. Regardless of extreme developing worries about environmental impacts like contamination of plastic and global warming, the most internationally utilized energy, chemicals, and multiuse plastic material are provided in the global economy by the petrochemical industry. More or less all chemicals and building materials for plastics can be produced utilizing materials that are raw and renewable, from a biotechnological and chemical point of view. Nevertheless, neither are these processes attainable in a commercial perspective nor are they well efficient (Gontard et al., 2018). Moreover, inadequate purity and excessive price are frequently displayed by the products. What is still awaited is an appropriate approach of evaluation approach which gives legitimate understanding and also provides guidance on the regional and seasonal effect and the complex nature of managing agricultural residues. As a result, the development of multi-criteria, multi-actor environmental, multi-scale, economic, and cross-disciplinary strategies for evaluating performances is important. It is crucial to implement these methodologies at the very beginning of the research and development process, which means to support the eco-design at a bench scale. Proper consideration must be given to the indirect effects that the technological improvements of the future have on society, to judge the obvious possibilities of huge interests in bioconversion technologies. The perfect resource capable of producing chemicals of high value is residues from the agricultural sector. Despite the complexity and heterogeneity in their structure, these agricultural residues are handled by suitable transformation into items that are competitive in the market. Hence there is a need to develop an upstream or downstream Anaerobic digestion process that is economical, ecologically efficient, and innovative (Gontard et al., 2018).

11.11 CONCLUSION

An alarming increase in the global human population with a constant energy demand has drawn attention towards the generation and extensive use of biomass residue-based renewable energy worldwide. Any organic material which is capable of storing its energy via the cycle of photosynthesis is referred to as biomass. Second-generation biofuels: Biofuel obtained from biomass of lignocellulosic nature comprising of cellulose, hemicellulose, and lignin. Plant biomass that can be classified into four categories, hardwood, softwood, agricultural wastes, and grasses are referred to as lignocellulosic biomass. A natural lignocellulosic substance has cellulose, hemicellulose, and lignin as its primary components. Nevertheless, there are several challenges in the process of bio gasification. Various studies have been done and are still going on about how to overcome these challenges. One of the important techniques to overcome such challenges includes the treatment of lignocellulosic biomass to enhance the biomass digestibility for anaerobic digestion. Based on the efficiency of pretreatment in different biomass, pretreatment methods are classified into physical, biological, chemical, and physio-chemical methods. Because it can produce value-added output sustainably, biomass is of high importance in the circular bioeconomy. Out of all the different forms of bioenergy, bio gasification is perceived as the most cost-proficient one, which requires a low skill set and is also not environmentally and chemically harmful. Unlike the first-generation biofuel, the main focus of the second-generation biofuel is on agricultural residues and non-food biomass. The agricultural residues incorporate waste from vegetable and fruit, crops and crop byproducts, manure, wood and wood waste, and residues from various other processes of agriculture. Because the energy potential of biogas is high, it can decrease the emission of Greenhouse gas along with improving the energy security of energy. Despite being so vigorous, Anaerobic Digestion still comes up with various technological constraints and shortcomings. These mainly relate to low transformed lignocellulosic yields, low financial incentive, and problems associated with the supply of feedstock and the storage and handling of digestate. As a result, the development of multi-criteria, multi-actor environmental, multi-scale, economic, and cross-disciplinary strategies for evaluating performances is important. It is crucial to implement these methodologies at the very beginning of the research and development process, which means to support the eco-design at the bench scale. Proper consideration must be given to the indirect effects that the technological improvements of the future have on society, to judge the obvious possibilities of huge interests in bioconversion technologies.

KEYWORDS

- **biomass**
- **carbon sequestration**
- **circular economy**
- **lignocellulose**
- **renewable energy**
- **zero waste**

REFERENCES

Ahring, B. K., Biswas, R., Ahamed, A., Teller, P. J., & Uellendahl, H., (2015). Making lignin accessible for anaerobic digestion by wet-explosion pretreatment. *Bioresour. Technol., 175*, 182–188.

Akhand, M., & Méndez, B. A., (2012). *Optimization of NMMO Pre-treatment of Straw for Enhanced Biogas Production*. Master Thesis.

Al Mamun, M. R., & Torii, S., (2017). Comparison effect of feedstock to inoculum ratios on biogas yields of cafeteria, vegetable, fruit wastes with cattle manure using co-digestion process. *International J. Green Ener., 14*(7), 665–673.

Ali, S. S., Abomohra, A. E. F., & Sun, J., (2017). Effective bio-pretreatment of sawdust waste with a novel microbial consortium for enhanced biomethanation. *Bioresour. Technol., 238*, 425–432.

Almomani, F., Bhosale, R. R., Khraisheh, M. A. M., & Shawaqfah, M., (2019). Enhancement of biogas production from agricultural wastes via pre-treatment with advanced oxidation processes. *Fuel, 253*, 964–974.

Amiri, H., & Karimi, K., (2015). Improvement of acetone, butanol, and ethanol production from woody biomass using organosolv pretreatment. *Biopro. Biosys. Eng., 38*(10), 1959–1972.

Angelidaki, I., Alves, M., Bolzonella, D., Borzacconi, L., Campos, J. L., Guwy, A. J., & Van, L. J. B., (2009). Defining the biomethane potential (BMP) of solid organic wastes and energy crops: A proposed protocol for batch assays. *Wat. Sci. Tech., 59*(5), 927–934.

Aro, N., Pakula, T., & Penttilä, M., (2005). Transcriptional regulation of plant cell wall degradation by filamentous fungi. *FEMS Micro. Rev., 29*(4), 719–739.

Aski, A. L., Borghei, A., Zenouzi, A., Ashrafi, N., & Taherzadeh, M. J., (2019). Effect of steam explosion on the structural modification of rice straw for enhanced biodegradation and biogas production. *BioResources, 14*(1), 464–485.

Barua, V. B., Goud, V. V., & Kalamdhad, A. S., (2008). Microbial pretreatment of water hyacinth for enhanced hydrolysis followed by biogas production. *Renew. Ener., 126*, 21–29.

Cavalli, D., Cabassi, G., Borrelli, L., Geromel, G., Bechini, L., Degano, L., & Gallina, P. M., (2016). Nitrogen fertilizer replacement value of undigested liquid cattle manure and digestates. *Eur. J. Agro., 73*, 34–41.

Cecchi, F., & Cavinato, C., (2015). Anaerobic digestion of bio-waste: A mini-review focusing on territorial and environmental aspects. *Waste Manag. Res., 33*(5), 429–438.

Cesarino, I., Araújo, P., Domingues, Jr. A. P., & Mazzafera, P., (2012). An overview of lignin metabolism and its effect on biomass recalcitrance. *Bra. J. Bot., 35*(4), 303–311.

Chandra, R., Takeuchi, H., Hasegawa, T., & Kumar, R., (2012). Improving biodegradability and biogas production of wheat straw substrates using sodium hydroxide and hydrothermal pretreatments. *Energy, 43*(1), 273–282.

Chen, H., (2014). Chemical composition and structure of natural lignocellulose. In: *Biotechnology of Lignocellulose* (pp. 25–71). Springer; Dordrecht.

De Souza, W. R., (2013). Microbial degradation of lignocellulosic biomass. *Sustainable Degradation of Lignocellulosic Biomass-Techniques, Applications and Commercialization, 15*, 207–247.

De Vries, R. P., & Visser, J. A. A. P., (2001). Aspergillus enzymes involved in degradation of plant cell wall polysaccharides. *Microbiol. Mol. Bio. Rev., 65*(4), 497–522.

Delzeit, R., & Kellner, U., (2013). The impact of plant size and location on profitability of biogas plants in Germany under consideration of processing digestates. *Biomass Bioener., 52*, 43–53.

Desmet, T., Cantaert, T., Gualfetti, P., Nerinckx, W., Gross, L., Mitchinson, C., & Piens, K., (2007). An investigation of the substrate specificity of the xyloglucanase Cel74A from hypocrea jecorina. *FEBS J., 274*(2), 356–363.

Domingo, J. L., Rovira, J., Vilavert, L., Nadal, M., Figueras, M. J., & Schuhmacher, M., (2015). Health risks for the population living in the vicinity of an integrated waste management facility: Screening environmental pollutants. *Sci. Tot. Environ., 518*, 363–370.

Dongyan, Y., Yunzhi, P. A. N. G., Hairong, Y., Shulin, C. H. E. N., Jingwei, M. A., Liang, Y. U., & Xiujin, L. I., (2014). Enhancing biogas production from anaerobically digested wheat straw through ammonia pretreatment. *Chinese J. Chem. Eng., 22*(5), 576–582.

Escobedo-Avellaneda, Z., Guerrero-Beltrán, J. Á., Tapia, M. S., Barbosa-Cánovas, G. V., & Welti-Chanes, J., (2018). Minimal processing of fruits. In: *Fruit Preservation* (pp. 67–92). Springer, New York, NY.

Feng, L., & Chen, Z. L., (2018). Research progress on dissolution and functional modification of cellulose in ionic liquids. *J. Mol. Liq., 142*(1–3), 1–5.

Fontes, C. M., & Gilbert, H. J., (2010). Cellulosomes: Highly efficient nanomachines designed to deconstruct plant cell wall complex carbohydrates. *Ann. Rev. Biochem., 79*, 655–681.

Fruergaard, T., Hyks, J., & Astrup, T., (2010). Life-cycle assessment of selected management options for air pollution control residues from waste incineration. *Sci. Tot. Environ., 408*(20), 4672–4680.

Giovanis, E., (2015). Relationship between recycling rate and air pollution: Waste management in the state of Massachusetts. *Waste Manag., 40*, 192–203.

Gontard, N., Sonesson, U., Birkved, M., Majone, M., Bolzonella, D., Celli, A., & Schaer, B., (2018). A research challenge vision regarding management of agricultural waste in a circular bio-based economy. *Cri. Rev. Environ. Sci. Tech., 48*(6), 614–654.

Harmsen, P. F., Hackmann, M. M., & Bos, H. L., (2014). Green building blocks for bio-based plastics. *Biof. Biopro. Bioref., 8*(3), 306–324.

Hashemi, S. S., Karimi, K., & Mirmohamadsadeghi, S., (2019). Hydrothermal pretreatment of safflower straw to enhance biogas production. *Energy, 172*, 545–554.

Hassan, S. S., Williams, G. A., & Jaiswal, A. K., (2018). Emerging technologies for the pretreatment of lignocellulosic biomass. *Bioresour. Technol., 262*, 310–318.

Houtman, C. J., Maligaspe, E., Hunt, C. G., Fernández-Fueyo, E., Martínez, A. T., & Hammel, K. E., (2018). Fungal lignin peroxidase does not produce the veratryl alcohol cation radical as a diffusible ligninolytic oxidant. *J. Biol. Chem., 293*(13), 4702–4712.

Kabir, M. M., Niklasson, C., Taherzadeh, M. J., & Horvàth, I. S., (2014). Biogas production from lignocelluloses by N-methylmorpholine-N-oxide (NMMO) pretreatment: Effects of recovery and reuse of NMMO. *Bioresour. Technol., 161*, 446–450.

Kim, S., & Dale, B. E., (2004). Global potential bioethanol production from wasted crops and crop residues. *Biomass Bioener., 26*(4), 361–375.

Kiran, E. U., Trzcinski, A. P., & Liu, Y., (2015). Enhancing the hydrolysis and methane production potential of mixed food waste by an effective enzymatic pretreatment. *Bioresour. Technol., 183*, 47–52.

Koupaie, E. H., Dahadha, S., Lakeh, A. B., Azizi, A., & Elbeshbishy, E., (2019). Enzymatic pretreatment of lignocellulosic biomass for enhanced biomethane production: A review. *J. Environ. Manag., 233*, 774–784.

Krishna, B. B., Biswas, B., Kumar, J., Singh, R., & Bhaskar, T., (2016). Role of reaction temperature on pyrolysis of cotton residue. *Waste Biomass Valo., 7*(1), 71–78.

Lamb, J. J., (2020). Analysis and pre-treatment of feedstocks. *Anaerobic Digestion: From Biomass to Biogas*.

Lamb, J. J., (2020). Digestate and biogas cleaning and processing. *Anaerobic Digestion: From Biomass to Biogas*.

Lizasoain, J., Rincón, M., Theuretzbacher, F., Enguídanos, R., Nielsen, P. J., Potthast, A., & Bauer, A., (2016). Biogas production from reed biomass: Effect of pretreatment using different steam explosion conditions. *Biomass Bioener., 95*, 84–91.

Lopez, M. J., Vargas-García, M. D. C., Suárez-Estrella, F., Nichols, N. N., Dien, B. S., & Moreno, J., (2007). Lignocellulose-degrading enzymes produced by the ascomycete *Coniochaeta ligniaria* and related species: Application for a lignocellulosic substrate treatment. *Enz. Microb. Technol., 40*(4), 794–800.

Mancini, G., Papirio, S., Lens, P. N., & Esposito, G., (2018). Increased biogas production from wheat straw by chemical pretreatments. *Renew. Energ., 119*, 608–614.

Maroušek, J., (2012). Finding the optimal parameters for the steam explosion process of hay. *Revista Técnica de la Facultad de Ingeniería. Universidad del Zulia., 35*(2), 170–178.

Martínez, Á. T., Speranza, M., Ruiz-Dueñas, F. J., Ferreira, P., Camarero, S., Guillén, F., Martínez, M. J., et al., (2005). Biodegradation of lignocellulosics: Microbial, chemical, and enzymatic aspects of the fungal attack of lignin. *Int. Microbiol., 8*(3), 195–204.

Master, E. R., Zheng, Y., Storms, R., Tsang, A., & Powlowski, J., (2008). A xyloglucan-specific family 12 glycosyl hydrolase from *Aspergillus Niger*: Recombinant expression, purification and characterization. *Biochem. J., 411*(1), 161–170.

Mirmohamadsadeghi, S., Karimi, K., Zamani, A., Amiri, H., & Horváth, I. S., (2014). Enhanced solid-state biogas production from lignocellulosic biomass by organosolv pretreatment. *BioMed Res. Int., 2014*.

Mishra, S., Singh, P. K., Dash, S., & Pattnaik, R., (2018). Microbial pretreatment of lignocellulosic biomass for enhanced biomethanation and waste management. *3 Biotech, 8*(11), 458.

Montgomery, L., (2014). Knowledge unlatched: A global library consortium model for funding open access scholarly books: Full report on the proof of concept pilot 2014. *Cult. Sci., 7*(2), 1–66.

Mulat, D. G., Huerta, S. G., Kalyani, D., & Horn, S. J. (2018). Enhancing methane production from lignocellulosic biomass by combined steam-explosion pretreatment and bioaugmentation with cellulolytic bacterium *Caldicellulosiruptor bescii. Biotechnol. Biofuels, 11*(1), 19.

Mutschlechner, M., Illmer, P., & Wagner, A. O., (2015). Biological pre-treatment: Enhancing biogas production using the highly cellulolytic fungus *Trichoderma viride. Waste Manag., 43*, 98–107.

Naqvi, S. R., Jamshaid, S., Naqvi, M., Farooq, W., Niazi, M. B. K., Aman, Z., & Afzal, W., (2018). Potential of biomass for bioenergy in Pakistan based on present case and future perspectives. *Renew. Sustain. Energ. Rev., 81*, 1247–1258.

Neshat, S. A., Mohammadi, M., Najafpour, G. D., & Lahijani, P., (2017). Anaerobic co-digestion of animal manures and lignocellulosic residues as a potent approach for sustainable biogas production. *Renew. Sustain. Energ. Rev., 79*, 308–322.

Nikoo, M. B., & Mahinpey, N., (2008). Simulation of biomass gasification in fluidized bed reactor using aspen plus. *Biomass Bioenerg., 32*(12), 1245–1254.

Östergren, K., Gustavsson, J., Bos-Brouwers, H., Timmermans, T., Hansen, O. J., Møller, H., Anderson, G., O'Connor, C., et al. (2014). *FUSIONS Definitional Framework for Food Waste.* Full Report 2014. Project report fusions.

Paolini, V., Petracchini, F., Segreto, M., Tomassetti, L., Naja, N., & Cecinato, A., (2018). Environmental impact of biogas: A short review of current knowledge. *J. Environ. Sci. Health, Part A., 53*(10), 899–906.

Paudel, S. R., Banjara, S. P., Choi, O. K., Park, K. Y., Kim, Y. M., & Lee, J. W., (2017). Pretreatment of agricultural biomass for anaerobic digestion: Current state and challenges. *Bioresour. Technol., 245*, 1194–1205.

Phuttaro, C., Sawatdeenarunat, C., Surendra, K. C., Boonsawang, P., Chaiprapat, S., & Khanal, S. K., (2019). Anaerobic digestion of hydrothermally-pretreated lignocellulosic biomass: Influence of pretreatment temperatures, inhibitors and soluble organics on methane yield. *Bioresour. Technol., 284*, 128–138.

Pilarski, K., Pilarska, A. A., Witaszek, K., Dworecki, Z., Żelaziński, T., Ekielski, A., & Michniewicz, J., (2016). The impact of extrusion on the biogas and biomethane yield of plant substrates. *Journal of Ecological Engineering, 17*(4), 264–272.

Rajput, A. A., & Visvanathan, C., (2018). Effect of thermal pretreatment on chemical composition, physical structure and biogas production kinetics of wheat straw. *J. Environ. Manag., 221*, 45–52.

Rodriguez, C., Alaswad, A., Benyounis, K. Y., & Olabi, A. G., (2017). Pretreatment techniques used in biogas production from grass. *Renew. Sustain. Energ. Rev., 68*, 1193–1204.

Romero-Güiza, M. S., Vila, J., Mata-Alvarez, J., Chimenos, J. M., & Astals, S., (2016). The role of additives on anaerobic digestion: A review. *Renew. Sustain. Energ. Rev., 58*, 1486–1499.

Rouches, E., Herpoël-Gimbert, I., Steyer, J. P., & Carrere, H., (2018). Improvement of anaerobic degradation by white-rot fungi pretreatment of lignocellulosic biomass: A review. *Renew. Sustain. Energ. Rev., 59*, 179–198.

Rouches, E., Zhou, S., Sergent, M., Raouche, S., & Carrere, H., (2018). Influence of white-rot fungus *Polyporus brumalis* BRFM 985 culture conditions on the pretreatment efficiency for anaerobic digestion of wheat straw. *Biomass Bioener., 110*, 75–79.

Särkilahti, M., Kinnunen, V., Kettunen, R., Jokinen, A., & Rintala, J., (2017). Replacing centralized waste and sanitation infrastructure with local treatment and nutrient recycling: Expert opinions in the context of urban planning. *Technol. Forecast. Soc. Change, 118*, 195–204.

Sawatdeenarunat, C., Surendra, K. C., Takara, D., Oechsner, H., & Khanal, S. K., (2015). Anaerobic digestion of lignocellulosic biomass: Challenges and opportunities. *Bioresour. Technol., 178*, 178–186.

Shah, T. A., Lee, C. C., Orts, W. J., & Tabassum, R., (2019). Biological pretreatment of rice straw by ligninolytic Bacillus sp. strains for enhancing biogas production. *Environ. Prog. Sust. Ener., 38*(3), e13036.

Sharma, A., Pareek, V., & Zhang, D., (2015). Biomass pyrolysis—A review of modelling, process parameters and catalytic studies. *Renew. Sustain. Energ. Rev., 50*, 1081–1096.

Sharma, H. K., Xu, C., & Qin, W., (2019). Biological pretreatment of lignocellulosic biomass for biofuels and bioproducts: An overview. *Waste and Biomass Valor., 10*(2), 235–251.

Shrestha, S., Fonoll, X., Khanal, S. K., & Raskin, L., (2017). Biological strategies for enhanced hydrolysis of lignocellulosic biomass during anaerobic digestion: Current status and future perspectives. *Bioresour. Technol., 245*, 1245–1257.

Silitonga, A. S., Atabani, A. E., Mahlia, T. M. I., Masjuki, H. H., Badruddin, I. A., & Mekhilef, S., (2011). A review on prospect of *Jatropha curcas* for biodiesel in Indonesia. *Renew. Sust. Ener. Rev., 15*(8), 3733–3756.

Sims, R. E., Hastings, A., Schlamadinger, B., Taylor, G., & Smith, P., (2006). Energy crops: Current status and future prospects. *Glob. Change Bio., 12*(11), 2054–2076.

Song, Z., Liu, X., Yan, Z., Yuan, Y., & Liao, Y., (2014). Comparison of seven chemical pretreatments of corn straw for improving methane yield by anaerobic digestion. *PloS One, 9*(4), e93801.

Syaichurrozi, I., Villta, P. K., Nabilah, N., & Rusdi, R., (2019). Effect of sulfuric acid pretreatment on biogas production from *Salvinia molesta. J. Environ. Chem. Eng., 7*(1), 102857.

Taherdanak, M., Zilouei, H., & Karimi, K., (2016). The influence of dilute sulfuric acid pretreatment on biogas production from wheat plant. *Int. J. Green Ener., 13*(11), 1129–1134.

Tambone, F., Terruzzi, L., Scaglia, B., & Adani, F., (2015). Composting of the solid fraction of digestate derived from pig slurry: Biological processes and compost properties. *Waste Manag., 35*, 55–61.

Thomas, L., Parameswaran, B., & Pandey, A., (2016). Hydrolysis of pretreated rice straw by an enzyme cocktail comprising acidic xylanase from Aspergillus sp. for bioethanol production. *Renew. Ener., 98*, 9–15.

Tišma, M., Planinić, M., Bucić-Kojić, A., Panjičko, M., Zupančič, G. D., & Zelić, B., (2018). Corn silage fungal-based solid-state pretreatment for enhanced biogas production in anaerobic co-digestion with cow manure. *Bioresour. Technol., 253*, 220–226.

Tripathi, M., Sahu, J. N., & Ganesan, P., (2016). Effect of process parameters on production of biochar from biomass waste through pyrolysis: A review. *Renew. Sustain. Energ. Rev., 55*, 467–481.

Tsapekos, P., Kougias, P. G., & Angelidaki, I., (2015). Anaerobic mono-and co-digestion of mechanically pretreated meadow grass for biogas production. *Ener. Fuels, 29*(7), 4005–4010.

Vakalis, S., Boschiero, M., Moustakas, K., Sotiropoulos, A., Malamis, D., Zerbe, S., & Baratieri, M., (2017). Assessing the suitability of biomass ashes from combustion in boilers as soil fertilizers: Statistical entropy analysis and introduction of the potassium utilization potential factor. *Waste and Biomass Valor., 8*(5), 1569–1576.

Van, E. R. H., De Souza, V. J. N., & Brasil Jr. A. C. P., (2012). The Brazilian experience of rural electrification in the Amazon with decentralized generation–the need to change the paradigm from electrification to development. *Renew. Sustain. Energ. Rev., 16*(3), 1450–1461.

Vickers, N. J., (2017). Animal communication: When i'm calling you, will you answer too?. *Cur. Bio., 27*(14), R713–R715.

Wagner, A. O., Lackner, N., Mutschlechner, M., Prem, E. M., Markt, R., & Illmer, P., (2018). Biological pretreatment strategies for second-generation lignocellulosic resources to enhance biogas production. *Energies., 11*(7), 1797.

Wahid, R., Hjorth, M., Kristensen, S., & Møller, H. B., (2015). Extrusion as pretreatment for boosting methane production: Effect of screw configurations. *Ener. Fuels, 29*(7), 4030–4037.

Wang, S., Li, F., Wu, D., Zhang, P., Wang, H., Tao, X., & Nabi, M., (2018). Enzyme pretreatment enhancing biogas yield from corn stover: Feasibility, optimization, and mechanism analysis. *J. Agri. Food Chem., 66*(38), 10026–10032.

Wong, D. W., (2009). Structure and action mechanism of ligninolytic enzymes. *App. Biochem. Biotechnol., 157*(2), 174–209.

Wyman, V., Henríquez, J., Palma, C., & Carvajal, A., (2018). Lignocellulosic waste valorization strategy through enzyme and biogas production. *Bioresour. Technolo., 247*, 402–411.

Xu, W., Fu, S., Yang, Z., Lu, J., & Guo, R., (2018). Improved methane production from corn straw by microaerobic pretreatment with a pure bacteria system. *Bioresour. Technol., 259*, 18–23.

Yang, D., Zheng, Y., & Zhang, R., (2009). *Alkali Pretreatment of Rice Straw for Increasing the Biodegradability* (p. 1). American Society of Agricultural and Biological Engineers. Reno, Nevada.

Yang, L., Ge, X., Wan, C., Yu, F., & Li, Y., (2014). Progress and perspectives in converting biogas to transportation fuels. *Renew. Sust. Ener. Rev., 40*, 1133–1152.

Yang, L., Xu, F., Ge, X., & Li, Y., (2015). Challenges and strategies for solid-state anaerobic digestion of lignocellulosic biomass. *Renew. Sust. Ener. Rev., 44*, 824–834.

Zavarzina, A. G., Lisov, A. V., & Leontievsky, A. A., (2018). The role of ligninolytic enzymes laccase and a versatile peroxidase of the white-rot fungus *Lentinus tigrinus* in biotransformation of soil humic matter: Comparative *in vivo* study. *J. Geophy. Res. Biogeosci., 123*(9), 2727–2742.

Zheng, Y., Zhao, J., Xu, F., & Li, Y., (2014). Pretreatment of lignocellulosic biomass for enhance biogas production. *Prog. Ener. Combust. Sci., 42*, 35–53.

CHAPTER 12

Socioeconomic Impacts and Rural Development of Biogas Technology

DEBASMITA PANDA,[1] BHABJIT PATTNAIK,[2] and MOUSUMI MEGHAMALA NAYAK[3]

[1]*Department of Botany, Banki Autonomous College, Banki, Cuttack, India, E-mail: debasmita.pinky87@gmail.com*

[2]*Department of Botany, Christ College, Cuttack–753008, India*

[3]*Procyto Labs Pvt Ltd, KIIT Technology Business Incubator, Bhubaneswar–751024, India*

ABSTRACT

Biogas is an important source for supplying clean and sustainable energy. For becoming self-dependent and acquiring highly organic alternatives for cooking gas, the biogas plant is a very needed unit for households having feed material. It can solve various household problems like air pollution within the house and can reduce the cost of using LPG cylinders as well. The production process includes the conversion of organic materials into gases (fuel) and nitrogen and organic fertilizer, i.e., sludge by the means of a microbiological reaction under anaerobic conditions. The fuel is composed of different constituents like methane (63%), CO_2 (30%), nitrogen (4%) followed by hydrogen sulfide, traces of hydrogen, oxygen, and carbon monoxide (1%). Methane alone holds about 90% of the vitality of the substrate. Cooking and lighting are vital applications of biogas. It is without smell and ignites with a blue fire without smoke. Cellulose and hemicellulose are the two major components of biomass because they both can be bio-converted into valuable products. Biomethane is one of the bioenergy gases which can be produced

Biotechnology for Waste Biomass Utilization. Prakash Kumar Sarangi & Latika Bhatia (Eds.)

by the valorization of wastes using different bioconversion technologies. In this chapter, we are focusing on the bioconversion technologies for producing biogas from agro wastes focusing on its implantation in rural areas.

12.1 INTRODUCTION

In many developing countries biogas plants are playing a key role to fulfill the exigency of fuel. Among various countries, India is continuously leading for the last 30 years in the production of biogas. The research was first to be established at the IARI, New Delhi, in 1939, followed by some other institutes such as the Gobar Gas Research Station (Ajitmal, Etawah), khadi and village industries commission (KVIC), Bombay, national environmental engineering research institute (Nagpur), and the Planning Research and action institutes (Lucknow). Biogas plant construction was first developed by KVIC with a production of 1,00,000 units by 1978 according to the government of India coordinated project. An institute named as non-Conventional Energy Development Agency (NEDA) of Uttar Pradesh has manufactured a total of 100-night soil-based biogas plants all over the state.

Biogas is a combined source of energy made up of natural cow dung (Gobar), which is eco-friendly to nature. It can be majorly used for producing heat and to produce electricity. Biogas plants are mostly designed for the better improvement of the environment by providing means for safe solid waste, human and animal feces in the rural and urban areas. It has been evaluated that the cattle dung processed in India, which is approximately 980 million tons (by wet weight basis) per year, could be sufficient for manufacturing around 41,000 million cubics of biogas per annum (i.e., directly proportional to 196 MW of electric power). It can be a leading solution for problems like global warming and restrain greenhouse gas emissions. It is known for environment-friendly technology.

12.2 BIOGAS: A VIABLE ENERGY SOURCE

Energy sources are categorized into two types, one is renewable and the other one is non-renewable. Renewable energy comprises wind, hydro, solar, geothermal, oceanic, naturally produced solar energy, and biomass the inexhaustible resources that have the intrinsic ability of renewable resources to reappear or replenish by own self through recycling, reproduction, regeneration, or replacement. Despite the sun's definite life period, solar energy is considered a renewable resource because of two specific reasons: first is the faster supply of solar energy as compared to the rate it is used and second is the emitting of solar energy that could evolve up to numerous human generations before the sun virtually ceased.

Biogas is a type of renewable energy source, as solar energy is directly used for the formation of it. Herbivores consume both dry plant biomass which is generally the residuals of oilseeds, cereals, pulses, crop straw, that are usually obtained once the crop produced is being harvested, while the grasses from forest areas and permanent pastures in the mountain regions, usually form the feed of these animals.

Biogas can be a product from animal waste by merely exposing it to the anaerobic digestion process. Within the biogas plant, primary carbohydrates, lipids, and proteins that are organic, are unitedly fermented to obtain biogas, comprised of carbon dioxide and methane, which is collectively known as Bio methanation. This is a multistep complex process that generally occurs in the environment by the combination of three important physiological categories of microbes, primary fermenting bacteria (hydrolytic-acidogenic), anaerobic oxidizing bacteria (acetogenic), and methanogenic archaea (Hattori, 2008; Thauer et al., 2008). These microorganisms decompose the complex organic matter (lipids, carbohydrates, proteins, etc.) into simple monomers (amino acids, glycerol, etc.) that are later transformed into short-chain fatty acids, alcohols (ethanol and methanol), and carbon dioxide. These short-chain fatty acids are hydrogen, acetate, and carbon dioxide and at the end converted to methane and carbon dioxide by the methanogenic archaea (Figure 12.1) (Angelidaki, et al., 2011).

It is apparent that conventional energy sources make an effect on the environment by causing pollution. The major advantages of acquiring renewable energy comprise its unending nature and extensible, low gestation and less capital-intensive nature, easily available resources, and effective usage of power for the development of rural areas.

12.3 BIOMETHANATION

In India, cattle dung is used as a principal substrate to operate biogas plants. In India, the combined population of domestic animals is about 974.01 million heads, comprising 485 million livestock and 489 million poultry. These animals produce about 1386.23 million tons of animal excreta and drop annually. Assuming that 75% of the total animal waste is compiled, and further presuming 25% of the waste as not worth using for biogas generation due to pattern of holding the animal and other uses, per year about 693.12 million tons of animal excreta is available, which is used for six biogas technology: in the purpose of sustainable development bio-methanation. About 290 × 108 m^3 of gas having 188.66 × 108 m^3 of methane can be generated. At present, for the production value of a daily supply of about 2.2 million m^3 of natural gas with a capacity of 16 m^3, more than 3.7 million biogas plants have been installed and estimated.

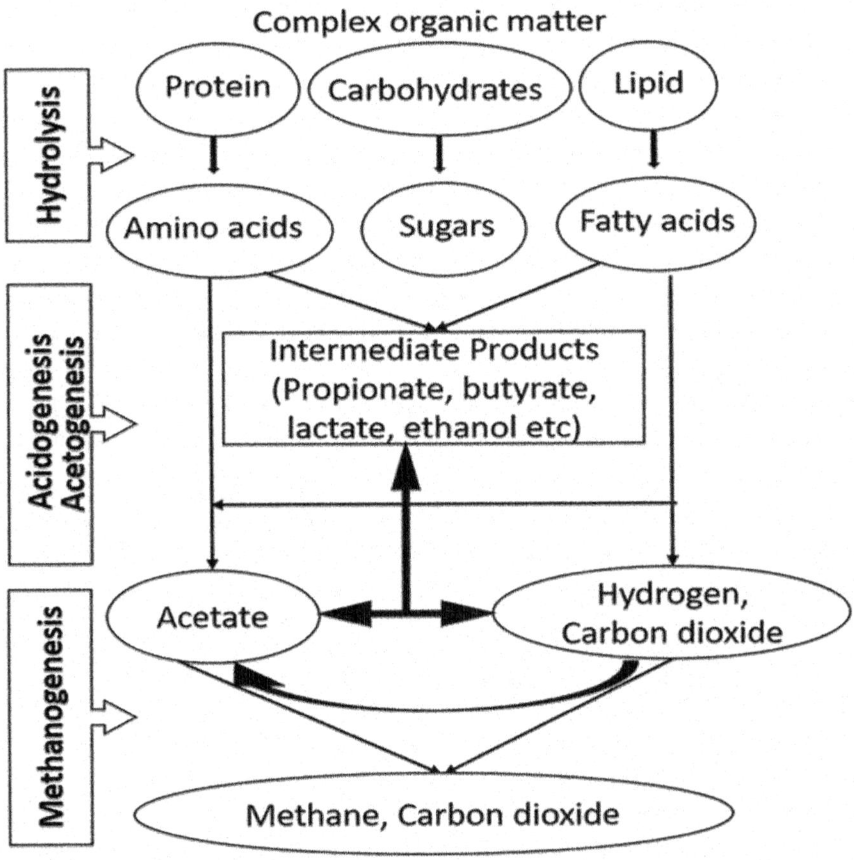

FIGURE 12.1 Biomethanation processes and mechanisms.

Technically the cattle dung, organic waste available in the country are used to generate biogas. Over 273.67 million tons of crop residues are available each year in our country. About 45 million tons wastage of fruit and vegetable congregation each year. A huge part of this biomass is kept unusable and creates a problem for waste disposal which leads to environmental pollution. Many advanced biogas designs were developed for treating wastes, such as by-products related to food industries, wastes (agricultural, garden, market, municipal, etc.). If all this biomass is taken into consideration as feedstock, the feasibility for biogas generation in the country could be virtually endless.

Besides gas yield, organic manure production using biogas plants can experience a substantial quantity of nutrient demand in the country. If the whole amount of the cattle dung produced in the country is accessible for bio-methanation, the resulting digested slurry can generate 1 million tons of nitrogen and phosphate as well, and 0.5 million tons of potash, which can also be used as fertilizer (Mital, 1996). To address the pressing social biogas technology has that enormous calibers, environmental, and economic problems in rural areas. Usage of biogas can save time and women labor is engaged for many productive activities. From an economic point of view, the digested slurry can act as an additive for chemical fertilizers, improve soil, and boost agricultural production. From an environmental perspective, technology can mitigate the problems of indoor air pollution, also reduce soil pollution. Therefore, the biogas system provides a broader scope in the rural areas of India.

12.4 HISTORICAL DEVELOPMENT OF BIOMETHANATION

The enigmatic appearance of featherhead lights and flames from below the surface of swamps was noted by Plinius (Van Brakel, 1980). Van Helmont stated the consequence of inflammable gas from depleting organic matter in 1630. He mentioned among 15 various kinds of gases, an inflammable gas that evolves during putrefaction and is also a part of intestinal gases. Shirley, in 1667 has quoted about the discovery of marsh gas (methane). In 1776, Alessandro Volta wrapped up the fact that there was a direct connection between the amount of rotting organic matter and the volume of flammable gas produced.

In 1801, it has been proved by Cruikshank that methane does not carry oxygen. In 1808, Sir Humphry Davy stated that the gases produced during the process of anaerobic digestion of cattle manure contain methane gas. He collected around 0.3 liters of methane and double the amount of carbon dioxide from cattle manure placed inside a retort under a vacuum. Dalton, Henry, and Davy, through their study, concluded that the chemical composition of methane and coal gas is confirmed to be very similar to Volta's marsh gas. Bunsen (1856) and Hoppe-Seyler (1886) made major contributions to the first microbiological knowledge base on anaerobic digestion. By the time Söhngen wrote his thesis in 1906, it was taken into account that organic materials were hydrolyzed by enzymes and broken down into alcohols and also as fatty acids, whereas methane was a production of these products (Van Brakel, 1980).

After looking into the sanitation concerns of individuals and municipalities, the adoption of anaerobic technology started in the late 19th century and early 20th century. Anaerobic microbes (Clostridium) were initially explained by Pasteur during a course on butyric fermentation. Gayon, fermented manure at around 35°C and acquired as much as 100 liters of methane per cubic meter of manure and was also a student of Pasteur. During demonstrating this finding at the Academy of Science in Paris in 1884, Pasteur then finalized that the process of fermentation could be an origin of heating and lighting. And at the same time (1875), the Dutch farmer Outer Slays became the first to use methane for the purpose of illumination.

From 1860 onwards, a septic tank was found in sewage purification. Although it was a known fact that methane was formed in the tanks, it was not collected for use. By the 1900s, anaerobic digesters were used all across the world, majorly in anaerobic ponds. In 1904, Travis introduced a two-stage process, in which the submerged material from the wastewater was charged into a separate hydrolyzing chamber.

The report of connectivity of methanogenesis with microbial activity was not very known until the end of the 19th century. In 1868, Bechamp found an organism that was responsible for the production of methane from ethanol. In 1876, Herder stated that acetate found in sewage sludge was converted stoichiometrically into equal amounts of carbon dioxide and methane (Zehnder, 1978). In 1906, Söhngen was up to enrich two distinct acetate-utilizing bacteria, and he noticed that format and hydrogen, with the help of carbon dioxide, can be precursors for methane. Buswell and others in the 1930s developed microbiology as a science for doing extensive research in identifying anaerobic bacteria and their conditions under which they promote methane production.

Buswell explained the stoichiometry of the reaction, the fate of nitrogen in anaerobic digestion, the use of the process for industrial waste, and the production of energy from farm wastes (Buswell and Hatfield 1930, 1936). To perform basic biochemical studies, Barker significantly contributed his knowledge of methane bacteria, and enriched cultures (Barker, 1956). Schnellen in 1947 found two methane bacteria, named, *Methano sarcina barker* and *Methano bacterium formicicum*.

12.5 HISTORY OF BIOMETHANATION IN INDIA

The theories of getting methane from the organic waste using anaerobic digestion have already been explored in India since 1897 when a methane digester was introduced in Mumbai at Matunga Leper Asylum. Methane was collected during the period 1895–1920 from the digesters having septic type tanks and was used to apply in a gas engine for discharging the sewage, as well as for cooking and lighting purposes (Fowler, 1934). By the early 1920s, findings on the practicality of producing methane using plant biomass started in the Indian Institute of Science, Bangalore (UN, 1984). They ended up with the fact that biogas plants might turn up as a better alternative for the Western Coast of India, which has no coal (Fowler and Joshi, 1920).

When the installation of a sewage gas plant at Dadar, Mumbai, in 1938 inspired Desai and Biswas, in 1939 they started looking on small-scale biogas plants relied on cattle dung at the Indian Agricultural Research Institute (IARI), New Delhi (MNES, 1993). Starting with bottle experiments, Desai made the first successful plant functioning on cow dung in 1941 at the IARI. The exclusive property of the plant was a well-type digester and a mild steel gasholder that was stabilized by a counterweight connected through ropes and supporting poles.

A presidential address was delivered in front of the agricultural section of the Indian Science Congress in 1945 by Dr. N. V. Joshi, who was formerly associated as a scientist at IARI and professor at the Poona Agricultural College (Joshi, 1945). Joshi patented a biogas design in 1946. However, the former plants developed in the country were exceptionally expensive in terms of gas output (KVIC, 1993). More or less the early models were quite responsive to bursting; Hence, the technology was quite impracticable for large-scale applications.

Jashbhai J. Patel developed the Gramlakshmi Plant in 1951, based on similar works done earlier by scientists like N. V. Joshi, S. V. Desai, S. C. Biswas, Y. N. Kotwal, and C. N. Acharya. With additional modifications by S. C. Das Gupta, Swami Vishwakarmanda, and C. N. Acharya, a much-modified version of Gramlakshmi turned up in 1954. This improved design had a two-chambered digester; the gasholder was planned in such a way so that it maintains the uniformity of the pressure of the order of 34 inches of the water column. The counterweight used in the gasholder was substituted by a central common pole, enhancing the probability of rotary and vertical movement. This design was embraced by the Khadi and Village Industries Commission (KVIC) for its countrywide diffusion during the 1960s and

1970s. In 1965/66, the only responsibility for controlling the government funds for making biogas plants all over the country was given to KVIC.

During 1952 and 1954, S. C. Das Gupta of the Khadi Pratisthan at Sodepur near Kolkata tried effortlessly to develop cost-efficient plants bottomed on materials like bamboo-thatch plastered along with earth and indigenous materials. The same efforts were accomplished by S. Vishwakarmananda at Ramakrishna Mission at Belur Math near Mumbai. In 1962, at the National Sugar Research Institute, Kanpur, Uttar Pradesh, a pilot plant was established to study the anaerobic digestion of bagasse (Bartha, 1965; Gupta, 1965). The Central Fuel Research Institute at Bihar designed a pilot plant in 1964 as a result of an experiment with the digestion of rice straw and (Goswami and Choudhary, 1967).

On the lines of the drumless, pig dung-based biogas plants in China, the Gobar Gas Research Station at Ajitmal, Etawah, Uttar Pradesh, experimented, developed, and released Janata type biogas plants in 1977. The research station at Ajitmal was constructed in 1960 under the guidance of the Planning Research and Action Development (PRAD), Uttar Pradesh. Since the Janata plant had no moving gasholder, it was also called the fixed dome type biogas plant (Jain, 1993).

Further evolution of the fixed dome plant by Action for Food Production Organization (AFPRO), New Delhi, led to another cost-efficient plant, namely, the Deenbandhu biogas plant, in 1986. Later modification in the above-mentioned plants led to another array of plants. The plan for setting up institutional, night-soil-based, and community biogas plants was commenced in 1982/83. It caters to the requirements of the community as a whole and also institutions.

12.6 DIFFUSION OF BIOGAS PLANTS IN INDIA

In India, the biogas plant's development and penetration have initiated about half a century ago. This process has become integrated with the start of the National Project on Biogas Development (NPBD) in 1981 and the Department of Non-conventional Energy Sources (DNES) in 1982.

The NPBD was initiated for mass diffusion of biogas plants along with a multiagency, multi-model strategy. The program that synchronizes the application and research and development of biogas digesters, is centrally regulated by the MNRE (Ministry of New and Renewable Energy). MNRE approves designs and allocates budgets for training, renders financial aid to buyer's subsidies and service charges to state governments and the KVIC, makes available turnkey construction fees and incentives to promoters and organizes for rectifying of plants with structural problems. The National Biogas and Manure Management Program (NBMMP) was replaced by NPBD in 2002/03. The main objectives of the NBMMP are to dispense fuel for cooking purposes and supply rural households with organic manure through family-type biogas plants and benefits with the improvement of sanitation in extremely rural areas by making links between sanitary toilets with biogas plants.

The NBMMP provides diverse forms of financial help, including central subsidy on the capital value of plants. Subsidies are allowed on plants up to 10 m^3 capacities and generally for the models which are recognized by the government. The depth of subsidy relies on the socio-economic status of the user, the size of the plant, and geographical region, as shown in Table 12.1. The amount of turnkey job fee per plant is Rs. 800 and Rs. 700 for hilly regions and other regions, respectively. For connecting the dung-based biogas plant with a sanitary plant, the government was giving an additional central subsidy of Rs. 500. The plants that have already been functioning for at least five years or more than that and have started showing some structural defects, were assigned a maximum of 50% of the central subsidy for their repairing and maintenance. Financial aid for organizing different kinds of training courses is given to training users, staff, and construction-cum maintenance and turnkey workers. Approximately 500 plants of the IARI model were installed by the year 1955 seeking government support. The KVIC and PRAD, Lucknow, tried promoting IARI design.

Well, it did not become popular due to alignment issues while tilting of the gasholder sideways for which uneven and interrupted gas supply occurs. In 1961/62, the KVIC added in its program, the Gobar Gas Scheme, and adopted the Gramlakshmi-III biogas plant for propagation. KVIC operated as a determined promoter of biogas in India. The KVIC dispensed high incentives for the installation of this plant by giving grants and loans. Progress was quite slow but steady. In 1962/63, 315 plants were installed, and by 1970, in total, about 3000 plants were installed, mostly in Gujarat. By 1975, about 17000 plants had been installed in India. By the early 1980s, about 80,000 biogas systems were constructed by the KVIC. Till 1986, a total of 642,900 digesters came into existence, and in 1985/86 alone, the number of digesters manufactured was 185,800. Total expenditure of Rs 6.7 million was made by the government in the year 1985/86 (75% subsidies, 25% training), though less amount was allocated for the next year. The Central Government Department of Science and Technology had spent Rs 5.6 million on its project on biogas which was of three-year and also coordinated through all over India.

12.7 FACTORS HINDERING THE DISTRIBUTION OF BIOGAS PLANTS

It is persuasively noticed that the biogas technology comprising huge potential to meet two basic rural requirements that are energy and manure, its wide-scale penetration is not without problems. Biogas technology diffusion in India accomplished several setbacks because of the non-operation of a large proportion of the plants installed. Operational simplicity is through a remarkable feature of the biogas plant technology, issues like a host of chemical, microbiological, engineering, and socio-economic have to be tackled and to be ensured the large-scale adaptation of those plants. Appropriate limitations are considered as the unavailability of sufficient feedstock, followed by defects in construction and microbiological failure in plants. Despite all, the cost of biogas installation maintained earlier prohibitively high and needs average economic status of the potential users. The requirement of having a requisite number of animals, land resources for plant installation, and the ability to mortgage for easy credit are the legitimate causes for concern.

Until the beginning of the 1980s underdeveloped technology of plants and a diffusion technique which was only barely succeeded and which eventually did not acknowledge the value of user training and follow up services until much later acted as stumbling blocks. Lacking servicing facilities, spare parts, and repair and maintenance facilities, inadequate access to technical information, often posed many a problem. Occasionally, the plants show some technical problems or faulty indications in their construction, which creates their non-functioning behavior. The travel budgets for the technical personnel are comparatively meager because of which plant operators are rarely informed to act over the technical issues. At higher altitudes in mountains, low temperature reduces the productivity of biogas.

There are chances where plants may show some technical issues during the entire year because of the minimum temperatures during winter forsaken methanogenesis (Singh, 1985; Sudhakar and Gusain, 1991).

Biogas technology is not welcomed by certain communities because they have reservations against the handling of animal dung for creating problems that are socially acceptable. In the case of using toilets in biogas systems, cultural practices may also clog the technology (Singh, 1988). Even though it is technically quite feasible for biogas plants to join hose hold toilets, people show some disinterest in the general idea of disposal and interrupt it on the grounds of the impure night soil. The extensive utilization of biogas has been extended up to a level that is hampered indirectly with the low cost of fossil-based energy, restricted over waste disposal regulation, and signified knowledge and well-known aerobic processes.

The problems mentioned above may be resolved through proper extension and support services, and effective selection processes for the technology. Another basic requirement for the successful promotion of the technology is monitoring and evaluation. There is a requirement to move out from open grazing to stall feeding of cattle. Requirements of coordination among the various agencies and benefits monitoring can lead to promoting and sustaining the technology. In addition to that, an enhancement of promotional campaign based on the advertisements in mass media, exhibitions, films, and posters should be used to create interest and educate people on this concept and the benefits of biogas plants.

12.8 TYPES OF BIOMASS AND THEIR POTENTIAL

Biomass is considered a living organic matter. Biomass can be any type of organic matter which are denoted as a source/feedstock. After the processing or preparation of this biomass, the fuel form which is obtained is called biofuel, the outcoming energy of biogas or bio- solid is called bioenergy, which is a measure of the energy potential of the biomass which is used. A large scope of biomass is accessible for the potential sources for CH_4 production.

12.8.1 TERRESTRIAL BIOMASS

12.8.1.1 BIOGAS FROM WOODS AND WEEDS

The process of anaerobic digestion for producing biogas using woody biomass has been considered unfeasible without any prior treatment (Nallathambi, 1997) due to its anaerobic biodegradability, which depends on the following factors: relative lignin; cellulose, and hemicellulose content; lower moisture content; the proportion of structural and non-structural carbohydrates; degree of connection between lignin and carbohydrates; cellulose crystallinity; particle size; wood-to-bark ratio; and toxic elements (Turick et al., 1991). The use of weedy plants is considered potential biomass and also a recent concept for the reason that it has the capability to trap a sufficient amount of solar energy (Nallathambi, 1997).

Results show some of the weeds studied as a source of CH4, these weeds include Parthenium hysterophorus, Lantana camara, and Ageratum. According to Gunaseelan (1994), the batch codigestion of cow manure (CM) and Parthenium has been shown to enhance the production of biogas using Parthenium. Anaerobic digestion of Parthenium in CSTR while pre-treated Parthenium increased the CH4 yield by 95%. Lantana camara, it's a weed plant that grows extravagantly on the slope of Himalaya, India, is treated with sodium hydroxide (NaOH) and mixed with cow manure to nourish batch digesters for 37 days at a temperature range of 2831°C produced 62% higher CH_4 yield compared to CM alone (Dar and Tandon, 1987; Gunaseelan, 1994). It is already reported that Ageratum alone (mono digestion) yielded 0.24 m^3/kg VS added of CH4 yield in batch digesters at a temperature of 30°C (Kalia and Kanwar, 1990).

12.8.1.2 BIOGAS FROM LEAVES AND GRASS

According to Chynoweth et al. (1993), methane manufactured from leafy biomass are normally higher in comparison to that produced from the stems. As reported by Sharma et al. (1988), *Ipomoea jistulosa* leaves yielded more CH_4 compared to that of the stem. According to Gunaseelan (1988), when Gliricidia leaves, i.e., green-leaf manuring found in India, undergoes anaerobic digestion it produces methane of 0.18 m^3/kg. Though, some leaves with the existence of some toxic compound produced low CH_4 due to limited inhibition of the digestion task; for instance, Calotropis is one such leaf (Mahamat et al., 1989). Research conducted by Shyam and Sharma (1994) showed that as compared to digestion of CM, batch digestion can be done of high solids with the utilization of mango leaves and higher biogas yield.

According to Some literature report, it is taken into account that grasses like Napier grass, energy cane (ball milled), Alemangrass-6A, turf grass, maize crop, paddy straw, millet straw, wheat straw, sorghum, oats crop, and corn stover presented methane production as high as 0.3 m^3/kg (Chynoweth et al., 1993). According to a report by Turick et al., (1991), the grass with the highest yield of CH_4 is sweet sorghum. the age of the grass plays a significant role in the grass as it produces younger grass compared to a higher amount of methane than the previous ones, probably just because the younger one comprises tissues, which are less lignified (Shiralipour and Smith, 1984).

12.9 FUTURE STUDIES

The need for energy is steadily escalating all across the globe along with the rapid growth in the world population, increased energy usage, and economic growth. Recent reports are showing that there is an increase in the utilization of fossil fuels by mankind and very soon the outcome will be the depletion of fossil fuel resources. Hence the foremost thing to do now is to look after the recent research studies made and to search for alternative energy sources that are renewable and sustainable for future generations. The most important application of renewable energy produced during anaerobic digestion of biomass is the erosion of any waste materials or toxic compounds. Although, recently there was increased interest in the manufacturing of biogas using carbohydrate-rich energy crops by means of anaerobic digestion. Because cassava is exceptionally grown in Africa, substantial experimental studies into distinct nomenclature which can give a high yield of biogas from cassava can be done. In the future, reports can be prepared on how the host communities get affected by the production at biomass plants, selection of technologies for small-scale biogas production, perceptions of farmers on biomass crops. In addition, one can also look into the employment-related issues and judgment on subsidies on small-scale biogas production in rural communities. Doing inquiries into these groundwork breaches will tone up the management for biogas production and sustainability in rural communities.

12.10 CONCLUSION

Energy is one of the essentials for human beings. Due to population increase and rapid industrial growth, there is a huge shortage of energy, another important reason for this shortage is the fast deterioration of fossil fuels and excessive requirement for energy. The sources of energy available are divided into two major categories, renewable and non-renewable. Renewable sources of energy are being reformed endlessly in nature. The concept of biogas energy is proposed to deal with the environmental issues linked with non-renewable energy sources. India is always on the cutting edge in its exertion towards improving these kinds of technologies to perfection and in finding effective solutions to various problems coming from the present-day energy crisis. On the other page, the current state of biogas production does not provide a complete package of the solution to the energy nutrient environmental pollution crisis in rural areas, but it provides a limited answer. Anaerobic digestion in biogas plants is not a very new technology for mankind, most countries already knew it by mid of the 20th century. Since 1960, India has been implementing biogas systems, but it was with the establishment of the NPBD in 1981 that the drive to step up their propaganda was commenced. While the country has been setting up its benchmark by achieving an installation capacity of over 3.8 million biogas plants so far, a major constraint still there is the high rates of non-functioning plants that are available in the environment, which may risk their further propagation. despite the vast potential, the importance of biogas as a fuel remains immoderate.

KEYWORDS

- **bioconversion**
- **biogas technology**
- **biomethane**
- **rural development**
- **socioeconomic impacts**

REFERENCES

Angelidaki, I., Karakashev, D., Batstone, D. J., Plugge, C. M., Stams, A. J., (2011). Biomethanation and its potential. *Methods Enzymol., 494*, 327–351.

Chynoweth, D., Turick, C., Owens, J., Jerger, D., & Peck, M., (1993). Biochemical methane potential of biomass and waste feedstocks. *Biomass Bioenergy, 5*, 95–111.

Dar, G. H., & Tandon, S., (1987). Biogas production from pretreated wheat straw, lantana residue, apple, and peach leaf litter with cattle dung. *Biological Wastes, 21*, 75–83.

Gunaseelan, V. N., (1988). Anaerobic digestion of *Gliricidia* leaves for biogas and organic manure. *Biomass, 17*, 1–11.

Gunaseelan, V. N., (1994). Methane production from *Parthenium hysterophorus* L., a terrestrial weed, in semi-continuous fermenters. *Biomass Bioenergy, 6*, 391–398.

Hattori, A., (2008). Syntrophic acetate-oxidizing microbes in methanogenic environments. *Microbes Environ., 23*, 118–127.

Hills, D., & Roberts, D., (1982). Conversion of tomato, peach and honeydew solid waste into methane gas. *Trans. ASAE., 25*, 820–826.

Iyilade, Y., (2010). *An Investigation into the Mechanical Biological Pretreatment of Garden Waste Using Forced Aeration and its Impact on Carbon Emissions Reduction Potential.* Durban. Ph.D. Thesis.

Kalia, A. K., & Kanwar, S. S., (1990). Anaerobic fermentation of *Ageratum* for biogas production. *Biological Wastes, 32*, 155–158.

Knol, W., Van, D. M. M. M., & De Waart, J., (1978). Biogas production by anaerobic digestion of fruit and vegetable waste. A preliminary study. *J. Sci. Food Agric., 29*, 822–830.

Mahamat, A., Gourdon, R., Leger, P., & Vermande, P., (1989). Methane recovery by anaerobic digestion of cellulosic materials available in Sahel. *Biological Wastes, 30*, 181–197.

Mata-Alvarez, J., Cecchi, F., Pavan, P., & Llabres, P., (1990). The performances of digesters treating the organic fraction of municipal solid wastes differently sorted. *Biological Wastes, 33*, 181–199.

Nallathambi, G. V., (1997). Anaerobic digestion of biomass for methane production: A review. *Biomass Bioenergy, 13*, 83–114.

Scherer, P., Neumann, L., Demirel, B., Schmidt, O., & Unbehauen, M., (2009). Long term fermentation studies about the nutritional requirements for biogasification of fodder beet silage as mono-substrate. *Biomass Bioenergy, 33*, 873–881.

Sharma, S. K., Mishra, I., Sharma, M., & Saini, J., (1988). Effect of particle size on biogas generation from biomass residues. *Biomass, 17*, 251–263.

Shiralipour, A., & Smith, P. H., (1984). Conversion of biomass into methane gas. *Biomass, 6*, 85–92.

Shyam, M., & Sharma, P., (1994). Solid-state anaerobic digestion of cattle dung and agro-residues in small-capacity field digesters. *Bioresour. Technol., 48*, 203–207.

Thauer, R. K., Kaster, A. N., Seedorf, H., Buckel, W., & Hedderich, R., (2008). Methanogenic archae: Ecologically relevant differences in energy conservation. *Nat. Rev. Microbiol., 6*, 579–591.

Turick, C. E., Peck, M. W., Chynoweth, D. P., Jerger, D. E., White, E. H., Zsuffa, L., & Kenney, W. A., (1991). Methane fermentation of woody biomass. *Bioresour. Technol., 37*, 141–147.

Index

D

E

F

G

T

Z